Transport Phenomena in Capillary-Porous Structures and Heat Pipes

Transport Phenomena in Capillary-Porous Structures and Heat Pipes

Henry Smirnov

CRC Press
Taylor & Francis Group
Boca Raton London New York

CRC Press is an imprint of the
Taylor & Francis Group, an **informa** business

CRC Press
Taylor & Francis Group
6000 Broken Sound Parkway NW, Suite 300
Boca Raton, FL 33487-2742

First issued in paperback 2019

© 2010 by Taylor and Francis Group, LLC
CRC Press is an imprint of Taylor & Francis Group, an Informa business

No claim to original U.S. Government works

ISBN-13: 978-1-4200-6203-8 (hbk)
ISBN-13: 978-0-367-38535-4 (pbk)

Library of Congress Cataloging-in-Publication Data

Smirnov, H. F. (Henry F.)
  Heat transfer at vaporization in capillaries and capillary-porous structures / author, Henry Smirnov.
    p. cm.
  "A CRC title."
  Includes bibliographical references and index.
  ISBN 978-1-4200-6203-8 (hardcover : alk. paper)
    1. Ebullition--Experiments. 2. Heat--Transmission--Experiments. 3. Capillarity--Experiments. 4. Mesoporous materials--Heating--Experiments. I. Title.

  QC320.36.S65 2009
  621.402'2--dc22   2009014839

Visit the Taylor & Francis Web site at
http://www.taylorandfrancis.com

and the CRC Press Web site at
http://www.crcpress.com

Transport Phenomena in Capillary-Porous Structures and Heat Pipes

Henry Smirnov

CRC Press
Taylor & Francis Group
Boca Raton London New York

CRC Press is an imprint of the
Taylor & Francis Group, an **informa** business

CRC Press
Taylor & Francis Group
6000 Broken Sound Parkway NW, Suite 300
Boca Raton, FL 33487-2742

First issued in paperback 2019

ISBN-13: 978-1-4200-6203-8 (hbk)
ISBN-13: 978-0-367-38535-4 (pbk)

Library of Congress Cataloging-in-Publication Data

Smirnov, H. F. (Henry F.)
    Heat transfer at vaporization in capillaries and capillary-porous structures / author, Henry Smirnov.
        p. cm.
    "A CRC title."
    Includes bibliographical references and index.
    ISBN 978-1-4200-6203-8 (hardcover : alk. paper)
    1. Ebullition--Experiments. 2. Heat--Transmission--Experiments. 3. Capillarity--Experiments. 4. Mesoporous materials--Heating--Experiments. I. Title.

QC320.36.S65 2009
621.402'2--dc22   2009014839

Visit the Taylor & Francis Web site at
http://www.taylorandfrancis.com

and the CRC Press Web site at
http://www.crcpress.com

Dedication

This book is dedicated to my mother, Rachel Khaletskaya, one of Stalin's many victims. She was imprisoned in the Gulag for more than nine years, and it was not expected that she would ever be able to see her son again, not to mention her future granddaughters. Fortunately, she survived, and was liberated in 1945. Until she passed away in 1981, she remained an energetic, optimistic woman, in spite of the many troubles she was dealt.

Contents

Foreword xi
Preface... xiii
Introduction.. xv
Nomenclature xix

Chapter 1 Hydrodynamics and Heat Transfer at Single-Phase
Flow in Capillary and Slit Channels 1

    1.1   Hydrodynamics and Heat Transfer of Steady
Single-Phase Flow in Capillaries and Slits 1
    1.2   Hydrodynamics and Heat Transfer of
Nonsteady Single-Phase Laminar Flow 9
    1.3   Hydrodynamics and Heat Transfer of Single-Phase
Turbulent Flow in Capillaries and Slits.................................... 12

Chapter 2 Hydrodynamics and Heat Transfer at
Single-Phase Flow through Porous Media 17

    2.1   Physical Fundamentals, Models, and Equations
of Momentum and Heat Transfer at Single-Flow
through Porous Media 17
    2.2   Effective Thermal Conductivity
of Porous Structures 20
    2.3   Internal Heat Transfer in Porous Media 28
    2.4   Heat Transfer on the Porous
Structure External Surface 33

Chapter 3 Thermohydrodynamics at Vaporization
inside Capillary-Porous Structures 41

    3.1   Vaporization Conditions in Capillaries
and Capillary-Porous Structures 41
    3.2   Hydrodynamics at Vaporization in
Capillary-Porous Structures 45
    3.3   Fundamental Principle of "Irreversibility Minimum"
in Two-Phase Filtration Modeling.. 55
    3.4   Investigations of Vaporization in Porous Structures
with Internal Heat Generation 63
    3.5   Experimental Investigations of Internal Characteristics
and Mechanisms of Thermohydrodynamic Phenomena......... 71

Chapter 4 Thermohydrodynamics at Vaporization
in Slit and Capillary Channels 93

4.1   Experimental Studies on Heat Transfer at
Boiling in Slits and Capillary Channels 93
4.2   Hydrodynamic Phenomena and Vaporization
in Slit and Capillary Channels 115
4.3   Experimental Studies of Vaporization Mechanism
in Plain Slits and Annular Channels 121
4.4   Physical Imaginations and Theoretical
Models for Heat Transfer at Vaporization
in Vertical Slits and Capillaries... 147
4.5   Physical Imaginations and Theoretical Models
for Heat Transfer at Vaporization in Horizontal
and Short Slits and Capillaries 153

Chapter 5 Heat and Mass Transfer at Vaporization on
Surfaces with Capillary-Porous Coverings 163

5.1   Experimental Investigations of Vaporization Heat
Transfer on Surfaces with Porous Coatings 163
5.1.1   Experimental Investigations of
Boiling Heat Transfer on Coated
Surfaces in Subcritical Thermal Regimes................ 163
5.1.2   Experimental Investigations of
Boiling Heat Transfer on Surfaces
Covered by Screen Wicks 168
5.1.3   Experimental Investigations of Vaporization
Heat Transfer on Fiber–Metal Surfaces 182
5.1.4   Experimental Investigations of Vaporization
Heat Transfer on Surfaces Covered by
Corrugated Structures 189
5.1.5   Experimental Investigations of Heat Transfer
inside Evaporators of Loop Heat Pipes 194
5.1.6   Experimental Investigations of Boiling
Heat Transfer on Surfaces Covered by
Sintered and Gas-Sprayed Coatings......................... 201
5.2   Experiments on Heat Transfer at Vaporization on
Surfaces with Sintered Coatings: The Malyshenko
Phenomenological Theory of Boiling 214
5.3   Physical Imaginations and Models of Vaporization
on Surfaces with Porous Coatings.. 226
5.4   General Model of Vaporization Processes
on the Coated Surfaces (the Third Stage).............................. 260

5.5   Heat Transfer at Vaporization on Surfaces
      Covered by Movable Capillary Structures 271
5.6   Models of Heat Transfer inside Evaporators of
      LHP and Capillary Pumped Loops 291

Chapter 6 Heat Transfer Crises at Vaporization Inside Slits, Capillaries,
               and on Surfaces Covered by Capillary-Porous Structures 299

6.1   Physical Explanations and Semiempirical
      Models of Boiling Heat Transfer Crisis 299
6.2   Modified Hydrodynamic Theory of
      Boiling Crises in Restricted Spaces 303
6.3   Experiments on Boiling Heat Transfer Crisis at
      Forced Liquid Flow in Slits and Capillary Tubes 305
      6.3.1   Boiling Crisis at Forced Flow in Capillaries 306
      6.3.2   Boiling Heat Transfer Crisis at Low-Velocity
              Flow inside Annular Capillary Channels 318
      6.3.3   Experimental Investigation of the CHF at Forced
              Flow in Capillaries: Modified Hydrodynamic
              Theory of Boiling Heat Transfer Crisis 319
      6.3.4   CHF in Narrow Annular Channels
              (Experimental Data and Semiempirical Models) 326
6.4   Experimental Research on the CHF at Pool Boiling in
      Slits, Capillaries, and Corrugated Capillary Channels 330
6.5   Experimental Research on Heat Transfer Crisis
      at Boiling on Surfaces with Porous Coverings 338
6.6   Maximum Heat Fluxes Inside Heat Pipes 347

References 355

Index 371

Foreword

This book is devoted to vaporization heat transfer and hydrodynamic processes occurring in capillary channels and porous structures. Particular attention is given to the physical mechanisms of these phenomena. Extensive experimental research activities on unique film and photo materials of boiling inside slits, capillaries, and capillary-porous structures were reviewed, and the most significant and notable among them are presented in this book. Corresponding analytical models were based on generalization of representative experimental data obtained by the author and his research group and by other investigators. Information about experimental methods and procedures, including direct experimental modeling of boiling at different conditions, is also included. In addition, numerous correlations helpful for engineering practice are presented. This material provides useful information resources for professionals in other engineering fields, including aerospace engineering, power and thermal engineering, air conditioning, refrigeration and cryogenics, chemical engineering, and food technology.

This book is also of specific interest to engineering graduate students, along with researchers, designers, and testing engineers of various areas of expertise. It can also be used as a reference book, and the recommendations presented here can be easily transformed into calculation algorithms and then implemented into computer software.

Preface

Fluid flow and heat and mass transfer at vaporization inside capillaries and capillary-porous structures are becoming increasingly important in many modern technologies. In comparison with vaporization at pool boiling or flow boiling in channels that are widespread in power engineering, refrigeration and air conditioning, food technology, and chemical engineering, the main distinctions of processes occurring in capillary structures are determined by restricted spaces for nucleation, development, and motion of vapor phase elements. As such, the average vapor bubble departure diameter is equal to or greater than the characteristic size of the structure in which boiling takes place.

The restricted conditions of vapor phase removal maintain a significant enhancement of heat transfer (from a few times to even an order of magnitude higher in particular cases). Consequently, vaporization in various forms of restricted conditions yields improvement of weight, geometric, and operating parameters of heat removal devices. On the other hand, application of restricted conditions in many cases has a negative influence on values of so-called critical heat fluxes.

However, in some conditions of vaporization inside capillary structures, the critical heat fluxes increase due to the capillary forces affect sustaining liquid phase microcirculation in the near-wall zone.

Specific interest into further research on the above-mentioned problems is associated with development of new forms of effective heat removal, thermal control, and thermo-stabilized devices that maintain the thermal regimes of electronic equipment. These also provide the basis for operating heat pipes, vapor chambers, heat exchangers with porous walls, and porous heat transfer surfaces used for heat protection.

The vapor-generating surfaces of evaporators in refrigeration and cryogenic systems, boilers, and other vaporization systems coated with porous coverings represent an attractive tool for heat and mass transfer enhancement or for increasing the productivity of these systems.

Therefore, the amount of corresponding experimental research increases continuously. Hence, improvement of physical explanations and theoretical models of vaporization at boiling on surfaces with porous coatings appears especially important.

Different assumptions about the physical nature of these processes and the theoretical models exist. However, a lack of critical comparative analysis of benefits and drawbacks of presented approaches restricts further progress in the field.

In this book, the author has made the first effort to overcome this problem by analyzing existing theoretical models in detail.

There are now many publications devoted to experimental investigations of corresponding heat transfer processes, whereas data on critical heat fluxes at vaporization in capillary structures are still rather limited. Most authors analyze their own data because of barriers connected to significant differences in manufacturing technologies of porous coatings for heat transfer surfaces with screen, fiber metal, sintered, gas-sprayed, corrugated, and many other types of capillary-porous coverings.

This book attempts to provide the common approach based on the reliable analysis of extensive experimental data on vaporization in capillary channels and inside porous structures.

Despite the fact that many books have been devoted to the discussion and examination of the different problems of heat pipes theory and practice, including the well-known monographs of Faghri [362], Silverstein [363], Dunn and Reay [348], and Peterson [364], as well as the works of such famous Russian experts in the field as Semena [19], Kovalev and Soloviev [36], and Sheckreladze [25], the analysis of heat and mass transfer problems at vaporization in capillaries and capillary-porous structures performed in these books was limited by the specific research results and heat pipe operating regimes. Generalization for known physical phenomena and regularities related to vaporization in capillaries and capillary-porous structures was not presented. To the same extent, this applies to the known monographs of Polyaev et al. [30], which dealt with the problem of vaporization at internal heat generation in massive metallic porous coverings.

The majority of experimental and theoretical studies on heat and mass transfer at vaporization in capillaries and capillary-porous structures presented in this book were completed by the author and his scientific team, which worked under his supervision in the Departments of Heat and Mass Transfer and Thermal Control Systems of the Odessa State Academy of Refrigeration, Ukraine, from 1970 to 1997. In addition, both known and original experimental and theoretical data published worldwide are reviewed in the book.

The personal input of Dr. Boris Afanasiev, Dr. Alexander Koba, and Dr. Yuri Kozhelupenko in obtaining the experimental data presented in the book is gratefully acknowledged. Their experimental visualizations yielded both the creation and justification of many of the models developed by the author.

Special thanks go to Prof. Yuri Maidanik, Prof. Vladimir Kiseev, and Prof. Valery Baranenko for providing extremely attractive original research data.

A word of appreciation is also due to the reviewer of the first Russian version of this book, Prof. Stanislav Malyshenko, for his helpful recommendations and critical comments.

The author is very grateful to Dr. Boris Kosoy for significant improvement of the English version of the book and many helpful suggestions.

Finally, the author would like to thank Prof. Sadik Kakac for his support and guidance during the entire process of preparing the English version of this book.

Introduction

The principles of two-phase vapor–liquid flow motion, and heat and mass transfer are considerably more complex than the transport phenomena associated with single-phase flow. Nevertheless, the development of physical explanations for two-phase hydrodynamic and heat transfer requires fundamental knowledge of thermal and hydrodynamic phenomena in the single-phase media.

This statement also relates to flow in capillary channels and capillary-porous structures. Therefore, the original content presented in Chapters 3, 4, 5, and 6 of this book is preceded by the first two chapters, which discuss known information about hydrodynamics and heat and mass transfer in channels at different flow regimes, boundary conditions, and geometry.

Since two-phase flow is initially unsteady, the influence of unsteady conditions on average regularities of hydrodynamics and heat and mass transfer is analyzed in Chapter 1, using information presented in the known reliable comprehensive studies devoted to this problem. The important conclusion to be drawn from the present analysis is that models of two-phase heat and mass transfer processes are quasi-steady in many cases.

Chapter 2 provides a basis for further development of physical explanations about two-phase thermohydrodynamics at vaporization in porous structures. The first part of this chapter reviews publications devoted to single- and two-phase filtration in porous media, internal heat transfer, and effective thermal conductivity of dry and wetted skeletons of porous structures.

Section 2.4 ("Heat Transfer on the Porous Structure External Surface") holds a special place in Chapter 2. For the first time, attention is given to the peculiarities of heat transfer at evaporation (condensation) on porous structure external surfaces. The corresponding conditions are classified from the viewpoint of local heat transfer mechanisms, and shows that significant quantitative and qualitative differences appear at vaporization on porous wetted structures of different geometry. Hence, the development of reliable heat and mass transfer models requires accounting for the mentioned peculiarities. Consequently, models developed by other authors were critically reviewed in this chapter.

Classification of regimes, thermohydrodynamic phenomena accompanying vaporization in porous structures, as well as conditions and forms of vaporization in capillary channels are presented in Chapter 3. In contrast to known statements of the filtration theory that as a rule neglects the capillary effects, the significance for the majority of thermohydrodynamic phenomena at vaporization inside porous structures is explicitly depicted. Experimental data reveal the formation inside capillary-porous structures of different two-phase "pictures," such as altering the vapor and liquid channels, existence of the continuous vapor layer of varied thickness in the near-wall zone, and the complete or partial filling of capillary-porous structure with vapor. Unfortunately, the geometry of two-phase structure inside both porous media and capillary channels could not be determined with certainty from the corresponding set of equations.

In Section 3.3 ("Fundamental Principle of 'Irreversibility Minimum' in Two-Phase Filtration Modeling"), the application of the fundamental principle of irreversible thermodynamics is justified for the description of two-phase structures inside porous media. According to this principle, the minimum deviation from equilibrium reveals the behavior of the nonequilibrium system in the stable stationary states that corresponds to the minimum entropy generation principle. The approach outlined in this chapter yields both real pictures of the two-phase structures inside porous media, as well as the values of two-phase structure geometric parameters, including the thickness of the near-wall vapor layer and the number and sizes of operating liquid and vapor channels inside porous media. This approach reveals a general principle for development of two-phase heat and mass transfer models in the case when their completeness is unachievable by using only the conservation equations with corresponding boundary conditions.

The generalized analysis of the most representative experimental investigations of boiling in capillaries and slits, including vertical and horizontal slit channels submerged in a liquid pool and cylindrical and annular channels at forced and combined convection conditions, is performed in Chapter 4. Special attention was devoted to the problem of two-phase flow visualization, including various phenomena that accompany the formation, and the development and obliteration of microlayer evaporation under vapor slugs appearing at boiling in slits and capillaries.

It is shown that specific types of the microlayer evaporation depend on geometry, liquid thermo-physical properties, and heat flux values. Under some conditions, the boiling of liquids with low thermal conductivities (such as refrigerants) causes secondary boiling nucleation centers to appear in microfilm. These centers destroy the microfilm and form a system of growing dry spots. This phenomenon causes a significant decrease in heat transfer intensity even in cases of very small heat fluxes (~ 1 W/m^2). In other cases, those formed under the vapor slugs evaporating the thin microlayer sustain high intensity of heat transfer up to the moment heat transfer crisis begins.

Semiempirical heat transfer models of boiling in slits and capillaries based on the data of visual observations and experimental investigations of the microlayer evaporation are considered in Sections 4.4 and 4.5.

Because there is so much experimental data on heat transfer at vaporization on surfaces with porous coverings, the most representative and systematic results were selected and generalized in Chapter 5. The goal was to provide a thorough analysis and discussion of heat transfer characteristics at vaporization on the surfaces with coverings manufactured by recognized and prospective technologies (i.e., dispersed bed and screen structures, sintered, fiber-metal, gas spraying coverings, etc.).

Analysis of the extensive experimental data, including visual observations, demonstrated that despite considerable differences in the structural and thermophysical parameters of porous coatings caused by technological peculiarities, a common approach to the modeling of two-phase heat transfer could be created. This approach is based on the known statements that connect between the structure and the wall; regularities of pore distribution and two-phase flow in single channels have a major influence on heat and mass transfer at vaporization on coated surfaces.

Correct accounting for temperature depressions appearing at vaporization in capillary channels of very small sizes (≤ 100 μm) plays a significant role in treatment of corresponding problems. Different physical explanations and theoretical approaches were also suggested for modeling of vaporization on surfaces covered by porous coatings. Unfortunately, only a very limited number of semiempirical models were checked experimentally. The author's physical explanations and theoretical concepts of boiling on surfaces with porous coverings are also presented in Chapter 5 in chronological order, starting with the model of the first level available since 1976 and to the third level model published in 1997.

Chapter 6 presents results of the first attempt at experimental data generalization and critical analysis of theoretical approaches on crisis phenomena at boiling transfer in capillaries, slits, and capillary-porous structures. The author believes that the Kutateladze hydrodynamic theory of boiling crises remains correct, despite numerous critical comments that have appeared in past few years. There are several reasons for this. First is because it is capable for such specific conditions as forced flow in cylindrical capillaries and annular channels, boiling in horizontal and vertical slits submerged in a liquid pool, and pool boiling on surfaces with thin-layer porous coverings. Moreover, it is shown that generalization of experimental data on critical heat fluxes at forced flow in capillaries and annular channels coincides well with calculations by the Kutateladze–Leontiev correlation. Using the energy form of the Kutateladze hydrodynamic theory of boiling crisis, the author has developed a theoretical concept of its modification for boiling in restricted spaces (i.e., slits, capillaries, corrugated, and thin-layered capillary-porous coatings).

The other types of crisis phenomena physical mechanisms occurred both at small flow velocities in annular channels and during the boiling of low thermal conductivity liquids (refrigerants) in slits that is associated with dry spots formation and development. Hence, a different approach to generalization of experimental data is required in these cases. The author's corresponding suggestions and analysis are presented in Chapter 6.

Regularities of heat transfer crises at boiling on coating surfaces are quite different in the following manner: interlayer crisis, hydrodynamic limit related to reaching porous structure dry-out in the evaporation regime, transition to film boiling on the external surface of submerged porous structure, and various combinations of mentioned crisis phenomena.

If the thickness of the porous coating significantly exceeds the elementary cell size (more than 10 average pore radii), the critical phenomena regularities and appearance of different stable stationary states of the two-phase layer are governed by the model of the third level. The experimental data on heat transfer crisis at boiling on thin-layer capillary-porous coverings with a thickness of less than 2 or 3 sizes of the elementary cell are generalized by the hydrodynamic theory of boiling crisis. Hence, different types of crisis (local crisis, integral crisis in heat pipes, etc.) are achievable depending on designs and manufacturing technologies of porous coating, and conditions of heat supply. Corresponding physical explanations and generalized correlations are considered in the final part of the book.

While writing this book, the author became aware of both the diversity and the complexity caused by the number of new prospective directions in the research on two-phase heat and mass transfer in capillary structures that appeared and have been developed extensively in recent years. Among them are investigations of porous coatings' influence on the so-called "second type crisis," which occurs in vapor–liquid flows at high vapor qualities.

The explanation and modeling of the positive influence of porous coatings yielding considerable heat transfer and enhancing and removing zones of poor heat transfer present ample opportunities for design improvements and decrease of expenditures related to the manufacturing of vapor generators in power engineering, evaporators in refrigeration, and cryogenics.

In addition, the application of porous coverings for improvement of thermal control of high heat-generating surfaces such as laser mirrors and high-frequency electronic devices is extremely promising. Thus, application of porous coverings in combination with intensive forced flow of subcooled liquid appears effective in these cases.

Nomenclature

a	Thermal diffusivity, m^2/sec; porous structure cell size, m
b	Width, m
c_p	Specific heat capacity, J/kg K
c	Concentration, $1/m^3$, mol/m^3
C_j	Constant
d, d_p	Diameter, m
D	Diffusion coefficient, m^2/sec; diameter, m
E	Density of vapor flow kinetic energy, J/m^3; elasticity module, Pa
$f(a), f(D)$	Pore size distribution
F	Surface area, m^2
g	Gravity acceleration, m/sec^2
g, j	Specific mass flow rate, kg/(m^2 sec)
G	Mass flow rate, kg/sec
H, h	Height, m
i	Specific enthalpy J/kg
k	Overall heat transfer coefficient, W/(m^2 K); numerical coefficient
$K, \bar{K}$	Permeability, m^2; relative permeability
l_n	Molecule free path, m
l, L	Length, characteristic size, m
m	Fin parameter, m^{-1}
M	Molecular mass
n	Circulation factor, concentration
n_F	Vaporization centers density, $1/m^2$
N	Mechanical pressure on contact surface, Pa
N_X	Particle contacts number in a single representative cell
p	Pressure, Pa, bar
ΔP	Pressure drop, Pa
q	Heat flux, W/m^2
q_v	Heat generation volumetric density, W/m^3
Q	Thermal power, W
R	Thermal resistance, m^2 K/W; vapor bubble local radius, m; gas constant J/(kg K)
r	Latent heat, J/kg; radius, m
s, S	Surface value, m^2; gap, tip, m
S_v	Specific surface, m^2/m^3
t, T	Temperature, °C, K
ΔT^*	Nucleation temperature drop, °C
t, B	Grooved structure tip, m
u	Velocity, m/sec
W	Saturation
w	Velocity, m/sec

x	Mass flow rate vapor volumetric concentration; vapor quality
x, y, z	Coordinates, m
$\overset{*}{X}; \overset{*}{R}; \dfrac{dx}{d\tau}; \dfrac{dR}{d\tau}$	Phase border motion velocities, m/sec
$\overset{**}{X}; \overset{**}{R}$	Phase border motion accelerations, m/sec^2
Z	Linear coordinate normal to the wall, m
α	Heat transfer coefficient, W/(m^2 K)
β	Volumetric vapor flow rate; volumetric expansion temperature coefficient
δ, δ_{CS}	Liquid microlayer thickness; capillary structure covering thickness, m
ε	Porosity
η, μ	Dynamic viscosity; Poisson coefficient, Pa s
Θ, θ	Dimensionless temperature
υ	Local value of excess temperature, K
ν	Kinematic viscosity, m^2/sec
λ	Thermal conductivity, W/(m K)
λ_E, λ_{SK}	Effective thermal conductivity, porous skeleton thermal conductivity, W/(m K)
U, Π	Perimeter, m
Π	Density of two-phase boundary layer potential energy, J/m^3
ρ	Density, kg/m^3
σ_0	Stefan–Boltzman constant, W/(m^2 K^4)
σ	Surface tension, N/m
τ	Time, sec; friction tension, N/m
φ	Vapor volume content; vapor quality
ω	Angular velocity, rad/sec; pulsation frequency, 1/sec

1 Hydrodynamics and Heat Transfer at Single-Phase Flow in Capillary and Slit Channels

1.1 HYDRODYNAMICS AND HEAT TRANSFER OF STEADY SINGLE-PHASE FLOW IN CAPILLARIES AND SLITS

Capillaries and slits are channels with a cross section characteristic diameter equal to or less than the departure bubble diameter at evaporation.

The following are the basic peculiarities of hydrodynamics and heat transfer at single-phase flow within such channels.

1. Flow frequently appears at moderate or low Reynolds numbers, Re, that is, at transient or laminar flow regimes. In fact, except for the low pressure region (i.e., $10^{-3} \leq p \leq 10^{-2}$, MPa), departure bubble diameters for most boiling liquids are approximately 0.1–2 mm. Feasible liquid-phase velocities range from 0.1 to 1 m/sec, and those for vapor phase range from 1 to 10 m/sec, that is, the maximum Re value for liquid flows is in the range of 100–10,000.
2. At small channel diameters, continuous media approximation cannot be justified in the near-wall region when a channel diameter d is compared with the mean free path length l_{pt}, that is, the Knudsen number is $Kn = l_{pt}/d > 0.001$.
3. In the vapor phase, when the molecule mass is in the range of 20–50, the length of the mean free path is $l_{pt} \approx 10^{-7}/\bar{p}_m$ ($\bar{p} = p/10^5$). Hence, inequality $Kn > 0.001$ at $\bar{p} = 1$ corresponds to the condition $d < 0.0001$ m. Consequently, the pressure and temperature drops at the interface should be accounted for in vapor flow inside micropores.
4. Because of the small liquid temperature changes through the channel cross section at evaporation in capillaries and capillary-porous structures, the liquid and vapor flows can be considered isothermal.

It is optional that single-phase flow regularities for capillaries and slits are the same as for conventional channel dimensions at laminar and transient flow regimes.

Details on laminar flow heat transfer and hydraulic resistance have been reported by Petukhov [1] and Isatchenko et al. [16]. Different two-phase flow regimes occur at evaporation in capillaries, slits, and capillary-porous structures, but the most feasible are slug, annular or dispersed-annular, and dispersed regimes. Depending on the heating mode, liquid filling, etc., at evaporation in capillary-porous structures the flow appears as two-phase inside vapor channels and single-phase within liquid channels. Single-phase quasi-steady flows are feasible inside numerous extended structures at exhaustive internal heat input, when the entire evaporation occurs within a porous structure volume.

Thus, for proper understanding of the heat transfer and hydrodynamics of two-phase flow in capillaries, it is essential to consider the following problems in single-phase flow in channels:

- Steady-state isothermal laminar flow in long and short channels at different boundary conditions
- Nonsteady, including pulsating, laminar flow in long and short channels
- Steady-state laminar flow in curved channels
- Transient and turbulent flow inside capillaries

Let us briefly depict known physical fundamentals and final correlations that characterize heat transfer and hydraulic resistance at laminar flow. Problems on heat transfer and hydraulic resistance at laminar flow can be related to those limited cases for which exact or approximate analytical or numerical solutions are available. The initial system of differential equations contains conservation of mass, momentum, and energy equations associated with their appropriate boundary conditions.

1. Mass conservation equation

$$\frac{\partial \rho}{\partial \tau} + div(\rho w) = 0 \tag{1.1}$$

Compression effects can be neglected for actual flow conditions, and Equation (1.1) simplifies to

$$divw = 0 \tag{1.2}$$

2. The momentum conservation equations of viscous Newton liquid with respect to the assumption of constancy of physical properties are

$$\rho \frac{dw_x}{d\tau} = \rho F_{Vx} - \frac{\partial p}{\partial x} + \mu \nabla^2 w_x;$$

$$\rho \frac{dw_y}{d\tau} = \rho F_{Vy} - \frac{\partial p}{\partial y} + \mu \nabla^2 w_y; \tag{1.3}$$

$$\rho \frac{dw_z}{d\tau} = \rho F_{Vz} - \frac{\partial p}{\partial z} + \mu \nabla^2 w_z$$

3. Energy conservation equation

$$\rho \frac{di}{d\tau} = div(\lambda \, gradT) + q_V + \frac{dp}{d\tau} + \mu \Phi \qquad (1.4)$$

Here, $\mu \Phi$ is a dissipation function.

In general, including capillary flow conditions, the terms $dp/d\tau$ and $\mu \Phi$ are negligibly small in comparison with terms $div(\lambda \, gradT)$ and q_V. Hence, the constancy of physical properties allows the following simplification for Equation (1.4)

Energy conservation equation

$$\frac{dT}{d\tau} = a\nabla^2 T + \frac{q_V}{\rho c_p} \qquad (1.5)$$

This set of equations (Equations 1.1–1.5), supplemented by appropriate boundary conditions, is complete, thereby allowing a subsequent thermal or hydrodynamic solution to problems.

Frequently, the influence of mass forces, including gravity, can be neglected for capillary channel flow, that is, $\rho F_{v_j} \ll \partial p / \partial_i$.

For laminar flow in small-diameter cylindrical channels at symmetry boundary conditions, when X axis coincides with channel axis, it is possible to assume that $W_x \gg W_y$ and $W_x \gg W_z$, and the simplified set of equations is

$$\frac{\partial w_x}{\partial x} = 0;$$

$$\frac{\partial w_x}{\partial \tau} = -\frac{1}{\rho} \frac{\partial p}{\partial x} + v\left(\frac{\partial^2 w_x}{\partial y^2} + \frac{\partial^2 w_x}{\partial z^2}\right);$$

$$\frac{\partial T}{\partial \tau} + w_x \frac{\partial T}{\partial x} = a\nabla^2 T + \frac{q_V(x, y, z)}{\rho c_p} \qquad (1.6)$$

For steady-state flow ($\partial w_x / \partial \tau = 0; \partial T / \partial \tau = 0$) in the absence of internal heat sources, the set of equations turns into a more simple form

$$v\left(\frac{\partial^2 w_x}{\partial y^2} + \frac{\partial^2 w_x}{\partial z^2}\right) = \frac{1}{\rho} \frac{dp}{dx};$$

$$w_x \frac{\partial T}{\partial x} = a\nabla^2 T \qquad (1.7)$$

The constancy of the dp/dx term and the option to change dp/dx to $\Delta p/l$ (Δp indicates pressure drop along length l) appeared from Equation (1.7). Assuming the temperature independence of physical properties, Equation (1.6) yields an independent solution for the hydrodynamic problem with boundary conditions based on velocity distribution at the channel inlet.

There are two different main cases:

1. Steady stabilized flow, which occurs when the velocity profile is constant along the channel length. The length of the section from the channel inlet when the flow becomes stabilized is hydrodynamic stabilization section, l_{In}. A channel is classified as long when its length $L \gg l_{In}$.
2. When $L \sim l_{In}$, then the velocity profile varies along the tube length, and such a flow is a nonstabilized one. Physical explanations about the nature of nonstabilized flow, corresponding to laminar flow in short channels, are as follows: Any arbitrary inlet flow velocity profile, for example, uniform, deforms under the action of the surface tension until it takes a defined shape. At the uniform velocity profile, the braking action of friction, first of all, influences the wall-adjoining liquid layer and reshapes the velocity profile. Such a layer is called hydrodynamic boundary layer (HBL). When liquid flows along the channel, HBL thickness δ increases until the whole channel cross section area will be filled with the HBL. The consequent cross section coordinate defines a margin of "initial section of nonstabilized flow." Correspondingly, a flow along the initial section can be divided into two main zones: the HBL and the flow core. Velocity in the first zone varies from zero at the wall to w_0 at the HBL external boundary, whereas velocity in the second zone is constant. Both methods of direct integration of momentum equation at initial section of nonstabilized flow and approximations based on analytical solution of integral equations are known. According to the above-mentioned technique, in cylindrical channels by implementing some dimensionless coordinates, $\bar{\delta} = \delta/d$; $X = 1/Re \times x/d$; $Re = w_0 d/v$ (w_0 is a constant velocity at the inlet cross section), it is possible to obtain dependencies of $\bar{\delta}, \bar{w}$ (average velocity) and dimensionless pressure drop $\dfrac{p_0 - p}{\rho w_0^2/2}$ on dimensionless channel length X as $\varphi(\bar{\delta}) = X$; $\dfrac{p_0 - p}{\rho w_0^2/2} = \psi(\bar{\delta})$.

A direct solution of the momentum conservation equation at the uniform inlet velocity profile is

$$\frac{p_0 - p}{\rho w_0^2 / 2} = 64X + 0{,}67 - 8 \sum_{n=1}^{1} \frac{1}{\beta_n^2} \exp(-4\beta_n^2 X) \tag{1.8}$$

Equation (1.8) takes the following form for viscosity coefficient

$$\xi Re = 64 + 32 \sum_{n=1}^{\infty} \exp(-4\beta_n^2 X) \tag{1.9}$$

Equation (1.9) shows that at $X \to \infty$, the left-side term of the equation approaches $\xi Re \to 64$. Actually, $\xi Re = 64$ when $X \geq 0.5$. Average values of $\overline{\xi Re}$ can be determined from the additivity of value $(p - p_0)$ from the formula $\overline{\xi Re} = \dfrac{l_{In}}{d} + \dfrac{\xi(X - l_{In})}{d}$. $\overline{\xi Re}$ is constant at $X \geq 0.5$. x_b is a relative channel length at which the effect of inlet conditions is evident:

$$x_b/d \cong 0.05 Re$$

According to Petukhov [1], for nonannular channels (e.g., a rectangular one) the following formula is appropriate

$$\frac{p_0 - p}{\rho w_0^2/2} = 96X + 0,4 - 4\sum_{n=1}^{\infty} \frac{1}{\gamma_n^2} \exp(-16\gamma_n^2 X)$$

that is, $\xi Re \rightarrow 96$, at $X \rightarrow \infty$.

The value of the constant in the equation $\xi Re = const$ depends slightly on the shape; the maximum variation between its values for annular and plane-parallel channel is equal to 1.5, but for other cases it does not exceed 10–20% [1].

A solution to the hydrodynamic problem with constant physical properties yields analytical solution to the heat transfer problem with first-type ($T = const$) or second-type ($q = const$) boundary conditions with the following assumptions:

1. Flow regime is steady state.
2. Medium is noncompressed, with constant physical properties.
3. Inlet liquid temperature is constant and equal to T_0; internal heat sources are absent. Dissipation effects can be neglected.
4. Flow is hydrodynamically stabilized.
5. Heat conduction in axial direction can be neglected in comparison with convection heat transfer in the same direction. Such a problem is known as Gretz–Nusselt problem. Its mathematical formulation is

$$w_x\left(\frac{\partial t}{\partial x}\right) = a\frac{\partial^2 T}{\partial r^2} + \frac{1}{r}\frac{\partial T}{\partial r} ; \qquad w_x = 2\overline{w}\left(1 - \frac{r^2}{r_0^2}\right)$$

and in dimensionless form it becomes

$$\frac{\partial^2 \theta}{\partial R^2} + \frac{1}{R}\frac{\partial \theta}{\partial R} = (1 - R^2)\frac{\partial \theta}{\partial X}$$

Here,

$$\overline{X} = \frac{2}{Pe}\frac{x}{d}; \quad \overline{R} = \frac{r}{r_0}; \quad \theta = \frac{T - T_c}{T_0 - T_c} = \frac{\vartheta}{\vartheta_0}$$

Boundary conditions are

$$X = 0 \quad \text{and} \quad 0 \leq R \leq 1: \ \theta = 1;$$

$$X \geq 0 \quad \text{and} \quad R = 0: \ \frac{\partial \upsilon}{\partial R} = 0;$$

$$X \geq 0 \quad \text{and} \quad R = 1: \ \theta = 0$$

For the solution of the problem, the Nusselt number is

$$Nu = \frac{\sum_{n=0}^{\infty} B_n \exp\left(-2\varepsilon_n^2 \frac{2}{Pe}\frac{x}{d}\right)}{2\sum_{n=0}^{\infty} \frac{B_n}{\varepsilon_n^2} \exp\left(-2\varepsilon_n \frac{1}{Pe}\frac{x}{d}\right)} \qquad (1.10)$$

Numbers ε_n produce an ascending sequence, that is, an increase in the value of X lowers the value of Nu. Such a decrease leads to the limit equal to $Nu_\infty = 3.66$. The zone where Nu is changing along the length is called the thermal initial section. The length of such a section (l_{In}) can be defined at the condition $\overline{X} = const$, where the constant value depends on the referred accuracy of approximation to automodel Nu number. The assumption that $Nu_x = l_{\text{In}} = 10.1 Nu_\infty$ for the problem considered here yields $l_{\text{In}}/d = 0.055 Pe$.

The exact solution for (1.10) in case of short channels $(x/d < 0.0001 Pe)$ can be approximated as

$$Nu = const \left(\frac{1}{Pe} \frac{x}{d} \right)^{-1/3} - 1,7$$

The average heat transfer coefficient (arithmetically averaged across the channel length) at $T_W = const$ is

$$\overline{Nu} = \frac{3,66 + 0,067 Pe \dfrac{d}{l}}{1 + 0,04 \left(Pe \dfrac{d}{l} \right)^{2/3}} \tag{1.11}$$

If the average heat transfer coefficient is related to the average temperature drop, then $\overline{Nu}$ is

$$\overline{Nu} = \frac{1}{2} Pe \frac{I}{d} \frac{1 - \overline{\theta}_{X=I}}{1 + \overline{\theta}_{X=I}} \quad \text{or at} \quad \overline{\theta} \to 0 \quad \overline{Nu} = \frac{1}{2} Pe \frac{l}{d}$$

However, it is necessary to note that at $\overline{\theta} \to$ the temperature drop cannot be arithmetically averaged. Similar correlations can be obtained for the plane channel. Only some variations occur because of the change in the minimum Nu_∞ value. Hence, using the slit-gap width as characteristic size yields $Nu_\infty = 3.77$.

Approximating models of heat transfer over the thermal initial section are similar to those of the initial section of nonstabilized flow, and suppose the existence of two zones: a thermal boundary layer (TBL) and flow core. The velocity profile is approximately linear for the greater part of the first zone, and temperature in the second zone is constant.

Based on this model, an analytical solution yields interpolation equations of initial section of nonstabilized flow:

$$Nu = 1,03 \left(\frac{1}{Pe} \frac{x}{d} \right)^{-1/3}$$

$$\overline{Nu} = 1,55 \left(\frac{1}{Pe} \frac{l}{d} \right)^{-1/3}$$

Assuming second-type boundary conditions $(q = const)$ and the above-mentioned statements, formulas for average and local heat transfer are similar to some numerical changes in the values of constants. Thus, the limit Nu_∞ value for annular channel is equal to 4.36, and for plane channel $Nu_\infty = 4.12$.

In almost all regions of the nonstabilized heat transfer, it is possible to use the following formula:

$$Nu = 1,3\left(\frac{1}{Pe}\frac{x}{d}\right)^{-1/3}\left(1+2\frac{1}{Pe}\frac{x}{d}\right) \tag{1.12}$$

The length of thermal initial sections can be calculated as $l_{In}/d = 0.07Pe$ and $l_{In}/d = 0.079Pe$ for annular channel and plane channel, respectively.

True heat transfer conditions for annular and plane slit channels are different from $T_W = const$ or $q = const$. Therefore, analysis of heat transfer features of steady laminar flow inside annular and plane channels at arbitrary temperature and heat flux changes along the channel length presented by Petukhov [1] deserves special attention. Hence, it was concluded that flow with heat transfer stabilization is feasible at some distance from the channel inlet, depending on the temperature change law, $T_W = f(x)$. Thus, at linear temperature change, it shows for annular tube that $Nu_\infty = 4.36$, similar to the case of $q_W = const$.

According to Petukhov [1], under channel length-varying boundary condition $(T_W = f(x))$, heat transfer stabilization occurs when $\lim\limits_{x \to \infty}\left(\frac{1}{f(x)} \times \frac{df(x)}{dx}\right) \le \infty$, that is, at any monotonous change of $f(x)$, which is valid for the majority of actual cases.

The Nu number depends on the channel shape and value of $\left(\frac{1}{f(x)} \times \frac{df(x)}{dx}\right)$. A change in heat flux value along the channel length causes an increase in the nonmonotonous character of heat transfer variance and extension of thermal initial section. However, the trend toward heat transfer stabilization persists at infinite increase of $X = (2/Pe)(x/d)$. Limit Nu_∞ is strongly dependent on parameter $K_0 = \lim\limits_{x \to \infty}\left\{\frac{1}{\psi(x)} \times \frac{d\psi(x)}{dx}\right\}$, where $\psi(x)$ characterizes the law of heat flux change. Modification of Nu along the annular tube is shown in Figure 1.1.

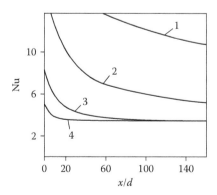

FIGURE 1.1 Modification of Nu along the annular tube at $Re = 2000$. ($1 - 4 - Pr = 100$; 10; 1; 0.1.)

Under third-type boundary conditions, it is helpful to introduce an overall Nusselt number, Nu_0 [1]

$$\frac{1}{Nu_0} = \frac{1}{Nu} + \frac{1}{Bi} \tag{1.13}$$

Here, $Nu_0 = kd/\lambda$, where k is a heat transfer coefficient; $Bi = k_1 d/\lambda$; $\frac{1}{k_1} = \frac{\delta}{\lambda_w} + \frac{1}{\alpha_0}$ (for plane channel); and $\frac{1}{k_1} = \frac{\delta}{2\lambda_w} \times \ln\frac{d_0}{d} + \frac{1}{\alpha_0} \times \frac{d_0}{d_1}$ (for annular channel), where α_0 is the convection heat transfer coefficient between external wall and surroundings, λ_w and λ are wall and liquid-specific thermal conductivity coefficients, respectively, $Nu = \alpha d/\lambda$ is a conventional Nu number, and α is the convection heat transfer coefficient between internal wall and liquid flow.

Very common physical considerations set a dependence between the laminar flow heat transfer intensity and wall thermal conductivity (i.e., on Bi number) over the initial thermal part of the channel. Such dependence is based on the fact that along the initial thermal part in which the external convection heat transfer coefficient α_0 is constant, the internal convection heat transfer coefficient α, temperature T_w, and heat flux q are variable. Consequently, Nu number will depend on Bi (or relative thermal conductivity of the wall) (see Figure 1.2).

At the counting coincidence between hydrodynamic and thermal initial sections, basic heat transfer dependencies stay the same, and only some quantitative parameters such as relative initial lengths and limit values of Nu numbers are subject to change. Hence, core flow inside annual tube is characterized by $l/d = 0.058Pe$ and $Nu_\infty = 4.94$.

The influence of roughness and curvature on heat transfer inside capillary channels is similar to conventional tubes, that is, the effect of roughness on heat transfer is absent at laminar flow, although it can effect transition from laminar to turbulent flow. It happens when the critical value of the relative roughness $(\Delta_{rh}/r)_{CR}$ will be exceeded, that is, at $(\Delta_{rh}/r)_{CR} \geq 5/\sqrt{Re}$. When this condition becomes invalid, the heat transfer intensity depends on roughness parameters, and the most important of these parameters is ratio Δ_{rh}/s, where s is the average pitch between roughness elements: $Nu_{rh} = Nu_{sm} f(\Delta_{rh}/s)$.

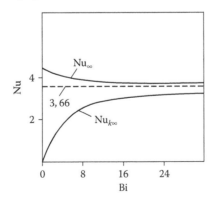

FIGURE 1.2 Dependency of Nu on Bi number for laminar flow at the initial part of annular tube (third-type boundary conditions).

When flow occurs inside a curved channel, secondary flows can appear to be caused by the action of centrifugal forces in curved zones. Secondary flows development appears at laminar regime, and the initial condition is characterized by parameter $K = Re\sqrt{d/R}$, where d is the tube diameter and R is the curvature radius. At $K < 18.9$, secondary flows are absent and heat transfer inside curved tubes is similar to that in plane channels. Secondary flows inside curved channels appear at $K > 18.9$ and enhance heat transfer up to the turbulent flow intensity. Further increase in Re number leads to the intensification of secondary flows, and starting from $Re_{CR2} = \dfrac{18000}{(d/R)^m}$ the more enhanced effect of curvature on heat transfer occurs at turbulent flow. At $Re \geq Re_{CR2}$, the heat transfer correlation should comprise the following correction multiplier: $\varepsilon_R = 1 + const(d/R); Nu = Nu_0 \times \varepsilon_R$.

1.2 HYDRODYNAMICS AND HEAT TRANSFER OF NONSTEADY SINGLE-PHASE LAMINAR FLOW

Analyzing Equation (1.6) while taking into account that $w_x = w_x(y, z, \tau)$ and $p = p(x, \tau)$, it is noticeable that $-dp/dx = f(\tau)$, that is, pressure gradient is a function of time. Once function $f(\tau)$ is known, the solution offered by the following equation

$$\frac{\partial w_x}{\partial \tau} = v\nabla^2 w_x + \frac{1}{\rho}f(\tau) \tag{1.14}$$

will depend on time conditions. The analytical solution for the velocity profile inside tubes with step changes in pressure gradient is

$$\frac{w_x}{w} = \varphi[E_i, Zh]$$

where $E_i = \left(i + \dfrac{1}{2}\right)\pi$ and $Zh = \dfrac{v\tau}{r_0^2}$ is a Zhukovsky number.

For pulsating flow, when

$$\frac{dp}{dx} = \left(\frac{dp}{dx}\right)_0 (1 + \gamma \cos \omega\tau); \quad \left(\frac{dp}{dx}\right)_0 = -\frac{3\mu\overline{w}}{r_0^2}$$

where $(dp/dx)_0$ is a steady term of pressure gradient, γ is a dimensionless amplitude of the pressure oscillations, and ω is an oscillation frequency. Equation (1.14) can be modified as

$$\frac{W_X}{W} = \frac{3}{2}((1 - y^2) + a\overline{u})$$

where $(3/2)a\overline{u}$ is a nonsteady pulsation velocity term.

The solution for the hydrodynamic problem in laminar liquid flow with constant physical properties can be applied in the analysis of nonsteady heat transfer. Since its theoretical solution is rather complicated, simplified models and approximation methods should be used.

The dimensionless equation of nonsteady heat transfer inside annular tubes is

$$\frac{\partial \theta}{\partial Fo} + (1 - R^2)\frac{\partial \theta}{\partial Z} = \frac{1}{R}\frac{\partial}{\partial Z}\left(R\frac{\partial \theta}{\partial R}\right)$$

The initial condition is $\theta(Z, R, 0) = 0$. Boundary conditions at the inlet, axis, and wall are taken as $\theta(Z, R, Fo) = 0$, $(\partial \theta/\partial R)_{R=0} = 0$, and $\theta(Z, 1, Fo) = 1$, respectively. The solution is

$$\theta = 1 - \sum_{n=0}^{\infty} A_n \psi_n(r) \times \begin{cases} \exp(-\gamma_n Fo), Fo \le a_n Z \\ \exp(-\varepsilon_n^2 Z), Fo \ge a_n Z \end{cases} \qquad (1.15)$$

Here, eigenfunctions ψ_n and constants A_n and a_n have the same values as in the steady solution.

Analysis of Equation (1.15) demonstrates that when the cross section of the tube at the present time (Fo) does not contain a liquid, which was initially located away from the heating zone, then heat transfer intensity is a function of time and radius, and is not length-dependent, that is, the heat transfer process is similar to transient heat conduction in the nonflow liquid.

For cross sections full of liquid that was initially located away from the heating zone, heat transfer depends both on time and length (coordinate Z). Such dependency exists until steady state is achieved. Moment of time of steady-state emergence is within the range of $Fo = Z$ to $Fo = 2Z$.

Similar regularities also occur in other channel cross section shapes, for example, in rectangular channel cross section at temperature or heat flux jump change on the wall. Therefore, at laminar flow inside tube, heat transfer intensity has to be estimated by transient heat conduction (contact mechanism) if $Fo < 1$, or by quasi-steady thermal regime if $Fo > 2$.

Pulsations of hydrodynamic parameters, including pressure gradients, are the most common ones for thermal pulsating regimes (e.g., two-phase flow with liquid slugs or slug flow). Pressure gradient is a periodic function of time.

$$\frac{dp}{dz} = \left(\frac{dp}{dz}\right)_0 \left(1 + \frac{a}{2}\cos \omega \tau\right)$$

The velocity profile for laminar flow under a pulsating pressure gradient is obtained by the superposition of the steady and oscillating terms of velocity that depend on Y, $\omega \tau$, and M ($\omega \tau = 2\pi \tau/\tau_0$, $Y = y/r_0$, $M = (\omega r_0^2/2\nu)^{1/2}$).

Comparison of steady and nonsteady relative heat fluxes for plain tube is shown in Figure 1.3 at different values of parameter M (A is a double value of dimensionless amplitude of pressure oscillations) [1]. At $M \le 0.1$, nonsteady heat fluxes are assumed to be quasi-steady ones. At $M = 1, 2$, heat transfer is not quasi-steady throughout the initial tube zone. However, when $Z = (1/Pe)(z/d) > 1$ (i.e., at sufficient tube length), even at large enough M values, variation between steady and unsteady heat fluxes becomes insufficient and can thus be neglected.

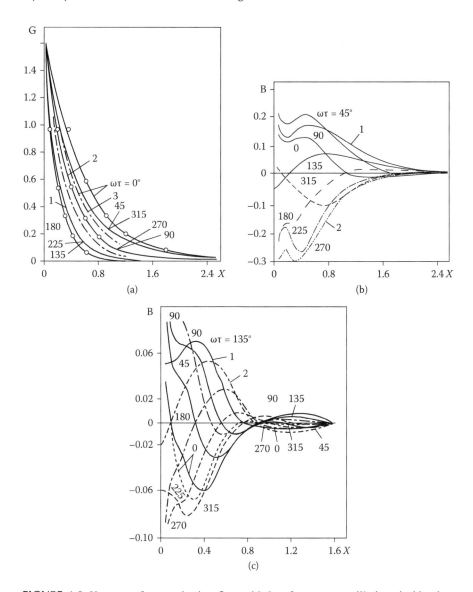

FIGURE 1.3 Heat transfer at pulsating flow with low-frequency oscillations inside plane $2r_0$-width tube. $Pr = 0.7$; $A = 1.0$. (a) $M = 0.1$; $\Gamma = q_w r_0/(\lambda(t_w - t_0))$; $X = \dfrac{8 \times x}{3Pe \times 2r_0}$; 1, calculation: $\omega\tau = 0–315°$; 2, quasi-steady process calculation; 3, the same for core flow. (b, c) Variation between steady and unsteady heat fluxes; $M = 1$ and $M = 2$, respectively; $\boldsymbol{\mathit{B}} = \dfrac{(q_w - q_{w0})r_0}{\lambda(t_w - t_0)}$; 1, calculation: $\omega\tau = 0–135°$; 2, the same: $\omega\tau = 180–315°$ [1].

A study [1] on transient heat transfer for laminar flow inside tube concludes that "velocity pulsations don't change sufficiently an average heat transfer in comparison to the steady-state flow." This conclusion, of course, is correct when pulsation amplitude is not large enough to initiate transient or turbulent flow regime.

Heat transfer for transient flow appears to depend on a number of factors, such as relative channel length L/d_E, Archimedean force (i.e., Gr number, including the very low Gr values), Re number, etc. The combined influence of these parameters is hard to predict. Flow instability is the basic distinction of the transition regime. Thus, an accurate assessment of transient flow can be done by using the "periodic factor," ω, which can account for the flow volume fraction occupied by turbulence.

Assumptions $\omega \to 1$ and $\omega \to 0$ correspond to the developed turbulent and laminar flow regimes, respectively. Since the value of ω increases as Re number increases and for developed turbulent flow we have $Nu \sim CRe^{0.8}$, it means that for transient flow regime $Nu \sim CRe^m$, where $m > 1$. Thus, at low Ra values ($Ra < 1$) and when Re slightly exceeds the critical value Re_c, an approximation of heat transfer correlation $Nu = \varphi(Re)$ comprises $m > 2, \ldots, 3$. At $Re > 4000$, $m = 1.6, \ldots, 1.7$, and at $Re = 10{,}000$, m is close to 1. At first glance the dependence of Nu numbers on Re values appears paradoxical, because it implies an increase in heat transfer coefficient as characteristic length increases. Indeed, such a correlation occurs because of the escalation in flow turbulence, and not just a result of particular changes in boundary layer width.

1.3 HYDRODYNAMICS AND HEAT TRANSFER OF SINGLE-PHASE TURBULENT FLOW IN CAPILLARIES AND SLITS

The Reynolds hydrodynamic analogy coincides well with experimental data on steady-state turbulent flow in channels and gives the following formula

$$Nu = \left(\frac{C_f}{2}\right) RePr\psi_T$$

where ψ_T is a function accounting for nonisothermity of the flow.

The viscosity coefficient of turbulent flow in smooth channels depends slightly on Re number as $C_f = const/Re^n$, where $n = 0.2, \ldots, 0.25$. The conventional formula for the heat transfer of droplet liquid turbulent flow in channels is

$$Nu = C_1 Re^{0.8} Pr^{0.4} \left(\frac{Pr_L}{Pr_W}\right)^n$$

The effect of nonisothermity in gas flow can be accounted for by using the temperature factor $(T_W/T_L)^m$ instead of the ratio of Pr numbers. Since the appearance of turbulent flow in capillary channels is less probable, distinctions of turbulent flow in small-diameter channels were not studied comprehensively.

It is supposed that regularities of single-phase turbulent flow appear the same for channels having a small cross-sectional area. Kalinin et al. [2] noted that in spite of the existence of noticeable secondary flows in slit channels "in wide range of Re numbers ($6000 < Re < 500{,}000$) correlation for the annular tube allows generalization to experimental data for all ratios of sides." Moreover, comparison of extensive research data with the Blasius formula for annular tube has confirmed the above-mentioned conclusion.

Kalinin et al. [2] suggested using a generalized equation to calculate the internal liquid turbulent flow heat transfer in plain channels at symmetric heating, developed by Petukhov and Roizen [3] for annular channels, assuming ratio $r_2/r_1 = 1$. Hence, the following formula for turbulent gas flow ($Pr = 1$) is obtained

$$Nu = 0.016Re^{0.8}$$

Under similar conditions for annular channels, the following correlations were recommended [3]:

- External surface: $Nu_2 = 0.0186Re^{0.8}[1 - 0.14 \, (r_1/r_2)^{-0.6}]J$
- Internal surface: $Nu_1 = 0.016Re^{0.8}(r_1/r_2)^{-0.16}J$

where $J = 1 + 7.5[((r_1/r_2) - 5)/Re]^{0.6}$ at $(r_1/r_2) < 0.2$; and $J = 1$ at $(r_1/r_2) > 0.2$.

The equivalent hydraulic diameter d_E should be used as a characteristic length for both plain and annular channels. Kalinin et al. [2] also included recommendations to account for length (for short channels) and asymmetric heating. Because the range of changes of nonstabilized turbulent flow relative lengths is $20 < d_E/l < 100$, in small-diameter channels ($d_E < 0.001$ m) length effect is observed in very short channels and usually has an insignificant contribution. Heat input asymmetry has noticeable influence on the process at $Pr < 1$ by lowering the Nu number by 1.2–1.5 times.

A review of more than 70 publications on hydrodynamics and heat transfer of forced single-phase and two-phase flows was presented by Peng and Wang [4]. However, directly related problems of single-phase heat transfer and hydrodynamics were considered only in about 10 papers. Peng and Wang [4] qualified the studies of Tuckermann and Pease [7, 8] as pioneering ones. These studies showed an approximate agreement between experimental data on single-phase flow hydrodynamics and conventional correlations. Peng and Wang [4] explained the critical differences for conventional channels revealed in the work of Wu and Little [5, 6] by specific near-wall effects on fluid properties. In spite of very small channel sizes, a major agreement with known hydrodynamic formulas for conventional channels was reported by Choi et al. [9], Harley et al. [10], and Harlow et al. [11].

Thus, most of the reviewed data do not reveal the critical differences between regularities of single-phase hydrodynamic flow in microchannels and conventional channel dimensions.

Hypothetically, near-wall effects should primarily influence the hydrodynamic characteristics, but this was not discovered by investigators. Hence, it was natural to suppose that single-phase heat transfer regularities in conventional channels are also valid. The review of Peng and Wang [4] shows that this is not true.

For example, the generalized correlation for turbulent flow region is presented by Wu and Little [5, 6] as

$$Nu = 0,00222 \, Re^{1.09} \, Pr^{0.4}$$

Transition to turbulent flow takes place at the condition $Re > 3000$. Wu and Little [5, 6] reported a significant enhancement in turbulent flow heat transfer in channels at two-side heating compared with that for one-side heating.

In addition, an atypical single-phase turbulent flow correlation was presented by Choi et al. [9]. Hence, the formula for Nu number gives

$$Nu = 3{,}28 \cdot 10^{-6} \, Re^{1,96} \, Pr^{1/3}$$

Zhang and Xin [12] demonstrated that turbulent flow heat transfer regularities obtained for tube diameters $d > 0.8$ mm match those of conventional ones. For tube diameters $d < 0.8$ mm, it was suggested to use a slightly changed generalized formula, as follows:

$$Nu = 0{,}021(RePr)^{0,7}$$

Contradictions between generalized correlations suggested by different authors and the odd character of the Nu dependence on the Re number (i.e., when correlation $Nu \approx CRe^m$ has an index $m > 1$) led Peng and Wang to conduct further experiments [4]. This study gives the following correlation for single-phase heat transfer in capillary channels,

$$Nu = 0{,}00805 \, Re^{4/5} \, Pr^{1/3}$$

However, the review [4] does not consider the following significant problems in heat transfer experimental studies:

- Reliability of the channel geometric parameter measurements, that is, diameter, width, and height of plain channel, etc.
- Checking of geometry constancy within the channel lengths
- Reliability of the local wall temperature measurements
- Reliability of the average wall temperature measurements
- Reliability of the heat flux and flow rate measurements

Overcoming of the problems cited above requires the application of special techniques and careful analysis of possible systematic and nonsystematic errors. Because Peng and Wang [4] did not provide details on experimental methods, evaluations of the quality of experimental data (e.g., [5–12]) reviewed and analyzed by Peng and Wang [4] is almost impossible. Experiments were performed using the experimental setup, consisting of four rectangular parallel channels, cut out in the groove shape having a depth of 0.7 mm in stainless steel plate (width, 18 mm; thickness, 2 mm; length, 125 mm). Groove width was in the range of 0.2–0.8 mm and groove pitch ranged from 2.4 to 3.8 mm [4].

Heating was done by using direct electric current. A wall temperature was measured by six thermocouples, located on the back side of the plate with grooves (three at the inlet and three at the outlet). Unfortunately, the authors did not describe the flow-rate measurement technique used. Suppose that the entire flow rate was measured for the whole channel set. The problem of average temperature determination was excluded, because only local heat transfer coefficients were determined in cross sections, where wall temperatures were measured. However, the effect of thermocouple insert on local heat generation was not discussed.

Direct measurement results presented by different authors demonstrate considerable scattering and essential contradictions between experimental data on laminar flow heat transfer regularities. Hence, experimental data [4–12] cannot be considered reliable.

It seems reasonable to suppose that the transition from channel cross section sizes of 0.1–1 mm to micron levels significantly complicates experimental methods. Therefore, at present, a statement on the existence of specific hydrodynamic and heat transfer regularities of single-phase flow in capillary channels cannot be proved. The same relates to recommendations for the determination of the limit of capillary channel size characterizing the switch to such specific regularities.

Analysis of the hydrodynamics and heat transfer of nonsteady single-phase turbulent flows is essential for understanding the nature two-phase flow. This problem has its own value, and a number of papers are devoted to its study. Two of the first attempts to generalize numerous experimental results in the field are presented in Refs. [13, 14].

Consider the major conclusions of these research activities:

- Transient heat transfer is characterized not only by steady-state process parameters (i.e., Re, Pr, dimensionless geometric ratios, such as l/d, etc.), but also by the time dependency of boundary conditions, mass flow G, wall temperature T_W, heat flux density q_W, etc.
- The ratio between turbulent flow transient heat transfer coefficient and quasi-steady heat transfer coefficient is defined by the boundary condition changing rate. It gives both new dimensionless variables

$$K_q = \frac{dq_W}{dz}\frac{d_E}{q_W}; \quad K_T = \frac{\partial T_W}{\partial \tau}\beta d\sqrt{\frac{\lambda}{c_p g G}}; \quad K_G = \frac{dG}{d\tau}\frac{d^2}{Gv}$$

and generalized correlations for transient convective heat transfer as

$$K = Nu/Nu_0 = f\ (Re_0, Pr_0, K_q, K_T, K_G)$$

- According to Koshkin et al. [13] and Burundukov [14], major effects of nonsteady state can be estimated by relative convection heat transfer intensity $K = Nu/Nu_0$ as follows:

> 1 at $(\partial T_W/\partial \tau) > 0$.
$K < 1$ at $(\partial T_W/\partial \tau) < 0$.
$K > 1$ at wall heat flux increase $(\partial q_W/\partial \tau) > 0$.
$K < 1$ at wall heat flux decrease $(\partial q_W/\partial \tau) < 0$.

Change in mass flow rate also influences the relative heat transfer intensity, K: $K > 1$ at heating and flow acceleration ($K_G > 0$), and $K < 1$ at decreasing velocity ($K_G < 0$). Lowering of the Re number and temperature factor increases the effect of nonsteady state.

In Refs. [13, 14], correlations, which connect $K = Nu/Nu_0$, K_T, K_q, K_G, and other dimensionless parameters that accounted for transient effects, determine such nonsteady regimes, which allow the treatment of transient heat transfer and viscosity friction coefficients as quasi-steady ones.

If we suppose at the first approximation that experimentally proved regularities of single-phase transient heat transfer in conventional channels can be used for estimations of non-steady-state effect in capillary channels, then transition to capillary channels should be accompanied by a reduction in non-steady-state effect under other similar conditions (i.e., identity of values $\partial T_W / \partial \tau$; $\partial q / \partial z$; $\partial G / \partial \tau$, etc.), because $K_T \approx C d_E$; $K_q \approx C d_E$; $K_G \approx C d^2$.

Thus, there is a strong ground that data on quasi-steady heat transfer and mass transfer coefficients can be used for analysis of transient processes in capillary channels.

2 Hydrodynamics and Heat Transfer at Single-Phase Flow through Porous Media

2.1 PHYSICAL FUNDAMENTALS, MODELS, AND EQUATIONS OF MOMENTUM AND HEAT TRANSFER AT SINGLE-FLOW THROUGH POROUS MEDIA

The momentum equation of single-phase flow through porous media based on the linear Darcy's law in the form $-(dp/dz) = (\mu w/K)$ or its modified form (for large Re numbers, $Re > 0.01$) gives

$$-(dp/dz) = (\mu w/K) + \beta \rho w^2$$

where K is the permeability coefficient.

The right time to use $\alpha = 1/K$ instead of K, then α and β, can be anticipated via the viscosity and inertia coefficients of Darcy's equation, respectively, and they should be determined experimentally for each type of porous structure.

There are various formulas for determining coefficient K (or α). The most common formula used to estimate permeability is the Kozeny equation

$$\alpha = const \frac{(1-\varepsilon)^2}{\varepsilon^3 d_0^2}$$

where ε is porosity, d_0 is an average pore diameter, and constant ($const$) depends on the method for manufacturing porous structure.

The flow through porous structure Re number essentially depends on the characteristic size choice. The most common models for flow through porous structures use a ratio β/α as the characteristic size; consequently, $Re = G \times \beta/\mu \times \alpha$ (where G is a mass flow rate related to the cross section of porous structure unit cell). The flow velocity w is a conventional value related to the entire cross-sectional area of porous structure and is called filtration velocity. The corresponding filtration equations are based on Darcy's law and mass conservation law for multidimensional single-phase flow (equations of fluid flow through porous media). These equations are commonly used for the conditions of the linear Darcy's law.

Solving various applied problems is usually connected with such empirical correlations as

$$K = C\varepsilon^m D^n$$

where D is the characteristic pore size or average diameter of particles forming porous structure.

Numerous researchers have reported a decrease in filtration flow with time at constant pressure drop. This phenomenon is called obliteration.

Belov [28] and Polyaev et al. [30] presented data on obliteration effect and its physical principles. It was shown that the known experimental facts of the decrease in liquid flow due to filtration in porous media can be characterized by gas leakage into liquid [30]. This significantly decreases the permeability and consequently, the mass flow rate. Thus, it was suggested that thorough liquid degassing be carried out and to increase liquid pressure level in order to raise the gas leakage limit.

Generally, heat transfer through porous structures occurs as heat conduction through the structure skeleton, and as convection and heat conduction in the fluid. These processes are interconnected because in common cases heat is transferred between skeleton and fluid.

Consider the porous media, which are characterized by the following thermophysical and geometric parameters: λ_{Ei}, ε_i, α, S. The energy conservation equation can be presented separately for skeleton and fluid. According to Cheiphez and Naimark [17] and Aerov and Todes [27], the skeleton energy equation is

$$\frac{\partial}{\partial x}(\lambda_{Ex}(1-\varepsilon))\frac{\partial T_{SK}}{\partial x} + \frac{\partial}{\partial y}(\lambda_{Ey}(1-\varepsilon))\frac{\partial T_{SK}}{\partial y} + \frac{\partial}{\partial z}(\lambda_{Ez}(1-\varepsilon))\frac{\partial T_{SK}}{\partial z}$$

$$+ q_{VSK}(1-\varepsilon) + \alpha S_v(T_0 - T_{SK}) = \rho_{SK} c_{SK}(1-\varepsilon)\frac{\partial T_{SK}}{\partial \tau}$$

Here, λ_{Ex}, λ_{Ey}, and λ_{Ez} are components of the skeleton effective thermal conductivity; T_{SK} and T_0 denote the local temperature of skeleton and fluid, respectively; ε, S_v, ρ_{SK}, and c_{SK} are porosity, specific surface of solid sintered capillary structure, density, and heat capacity of the skeleton, respectively; q_V is the heat generation volumetric density of the porous structure skeleton; and α is the heat transfer coefficient from fluid to the skeleton. In the absence of internal heat generation, the fluid energy equation is

$$\lambda_L \left[\frac{\partial}{\partial x}\left(\varepsilon\frac{\partial T}{\partial x}\right) + \frac{\partial}{\partial y}\left(\varepsilon\frac{\partial T}{\partial y}\right) + \frac{\partial}{\partial z}\left(\varepsilon\frac{\partial T}{\partial z}\right)\right] - \alpha S_v(T_0 - T_{SK})$$

$$= \rho c_p \varepsilon \left(\varepsilon\frac{\partial T_0}{\partial \tau} + W_x\frac{\partial T_0}{\partial x} + W_y\frac{\partial T_0}{\partial y} + W_z\frac{\partial T_0}{\partial z}\right)$$

When fluid flow in porous structures is absent or is at low liquid velocities, systems of two equations simplify to a single energy equation as

$$\frac{\partial}{\partial x}\left(\lambda_{Ex}\frac{\partial T}{\partial x}\right) + \frac{\partial}{\partial y}\left(\lambda_{Ey}\frac{\partial T}{\partial y}\right) + \frac{\partial}{\partial z}\left(\lambda_{Ez}\frac{\partial T}{\partial z}\right) + q_{VSK}(1-\varepsilon) = \rho_E c_E\left(\frac{\partial T}{\partial \tau}\right)$$

Here,

$$T_0 = T_{SK} = T..; ..\lambda_{Ei} = \lambda_{SK}(1-\varepsilon) + .\lambda_L \varepsilon; ..\rho_E c_E = \rho_{SK} c_{SK}(1-\varepsilon) + \rho_L c_L \varepsilon \quad (2.1)$$

One-dimensional steady-state heat transfer without internal heat generation is a common problem. In such cases, the corresponding system of equations is

$$\frac{d}{dx}\left(\lambda_E(1-\varepsilon)\frac{dT_{SK}}{dx}\right) + \alpha S_v (T_0 - T_{SK}) = 0$$

$$\lambda_L \frac{d}{dx}\left(\varepsilon \frac{dT_0}{dx}\right) - Gc_p \frac{dT_0}{dx} = \alpha S_v (T_0 - T_{SK})$$

In the absence of local temperature slip, we only have one equation instead of two:

$$\frac{d}{dx}\left((\lambda_E(1-\varepsilon) + \lambda_L \varepsilon)\frac{dT}{dx}\right) - Gc_p \frac{dT}{dx} = 0 \quad (2.2)$$

When λ_E, ε, and λ_L are constant, assumption $\lambda_E(1-\varepsilon) + \lambda_L \varepsilon. = .\lambda_{E0}$ simplifies Equation (2.2) as

$$\frac{d^2T}{dx^2} - \frac{Gc_p}{\lambda_{E0}}\frac{dT}{dx} = 0$$

The general solution for such an equation is

$$t = C_1 \frac{\lambda_{E0}}{Gc_{p_1}}\exp\left(\frac{Gc_p}{\lambda_{E0}}X\right) + C_2$$

At the optional boundary condition,

$$X = 0 : T = T_s; X = -\delta;.. - \lambda_{E0} \times \frac{dT}{dx} = q$$

the current temperature value is

$$T = T_s + \frac{q}{Gc_p \varepsilon}\exp\left(\frac{Gc_p \varepsilon \delta}{\lambda_{E0}}\right)\left[1 - \exp\left(\frac{Gc_p \varepsilon}{\lambda_{E0}}x\right)\right]$$

for heat conduction surface, at $x = -\delta$

$$T_W - T_s = \frac{q}{Gc_p \varepsilon}\left[\exp\left(\frac{Gc_p \varepsilon \delta}{\lambda_{E0}}\right) - 1\right]$$

When $Gc_p \varepsilon \delta / \lambda_{E0} \ll 1$, the temperature can be defined as $T = T_s + q(\delta/\lambda_{E0})$.

Thus, the effective thermal conductivity of capillary structure λ_E or thermal conductivity of porous structure wetted skeleton λ_{SK} and heat transfer coefficient α are the most important parameters determined for the heat transfer intensity of the single-flow through porous structure. Determination of λ_E, λ_{SK}, and α can be classified into two independent stages:

- Determination of dependencies on characteristic pore size for the local values λ_E, α, and other parameters of elementary (representative) cell.
- Averaging of λ_E and α over the porous structure with respect to actual pore size distribution.

2.2 EFFECTIVE THERMAL CONDUCTIVITY OF POROUS STRUCTURES

Dulnev and Zarichnyak [15] presented, discussed, and generalized results of numerous studies on thermal conductivity of complex structures. Basic principles of approximation of heat conduction models for such materials, including simple porous media, were also formulated. In contrast to porous structure heat conduction models [17], Dulnev and Zarichnyak [15] classify different heterogeneous structures as: (1) structures with inclusions, (2) structures with interpenetrated components, (3) combined (with inclusions and interpenetrated components), and (4) granular (bed).

According to Dulnev and Zarichnyak [15], modeling of the effective heat conduction comprises the following:

- Determination of the elementary representative cell.
- Formation of thermal models based on the generalized conductivity theory, and assembly of thermal resistance series.
- Determination of the entire thermal resistance terms via approximated or exact solutions.
- Examination of the adequacy of the effective thermal conductivity correlation.

The main terms of the entire thermal resistance of the representative cell are:

- Skeleton thermal resistance (dry, moist, or wetted), which has an inversely proportional dependency on value λ_{SK}.
- Thermal resistance of pore volumes, being filled with gas or liquid heat carrier.

Feasible models and correlations depend on the structure type, heat flux direction, and orientation of skeleton elements. Hence, in a limit case, when the orientation of skeleton elements coincides with the main direction of the heat flux, parallel resistance model is valid

$$\lambda_E = \varepsilon \times \lambda_L + (1-\varepsilon)\lambda_{SK} \qquad (2.3)$$

Equation (2.3) gives the maximum assessment λ_{Emax} for effective thermal conductivity value.

When the orientation of skeleton elements is perpendicular to the heat flux direction, then the effective thermal conductivity is minimal, and it is determined by the stratified model:

$$\frac{1-\varepsilon}{\lambda_{SK}} + \frac{\varepsilon}{\lambda_L} = \frac{1}{\lambda_{E\,min}} \tag{2.4}$$

Different correlations for the calculation of λ_E were suggested by Dulnev and Zarichnyak [15] at intermediate orientations of skeleton elements:

$$\log \lambda_E = (1-\varepsilon)\log \lambda_{SK} + \varepsilon \log \lambda_L \tag{2.5}$$

$$\lambda_E = \frac{[(1-\varepsilon)\lambda_{SK} + \varepsilon\lambda_L]^p}{[(1-\varepsilon)/\lambda_{SK} + \varepsilon/\lambda_L]^{(1-p)}} \tag{2.6}$$

where p and $(1-p)$ are probabilities of parallel and consecutive mutual positions of skeleton and liquid layers, respectively. If these probabilities are equal, then the effective thermal conductivity is

$$\lambda_{E\,p=0.5} = \sqrt{\lambda_{SK}\lambda_L \frac{(1-\varepsilon)\lambda_{SK} + \lambda_L\varepsilon}{(1-\varepsilon)\lambda_L + \lambda_{SK}\varepsilon}}$$

The presented correlations are related to the systems with interpenetrated components. The Maxwell formula [15, 18] can be applied for structures with circular inclusions:

$$\frac{\lambda_E}{\lambda_1} = \frac{\lambda_2/\lambda_1 + 2 - (2-\varepsilon)(1-\lambda_2/\lambda_1)}{\lambda_2/\lambda_1 + 2 + (1-\varepsilon)(1-\lambda_2/\lambda_1)}$$

where λ_2 is a thermal conductivity of the inclusions and λ_1 is a thermal conductivity of the continuous component.

The Odolevsky formula was suggested for the structures with cubic inclusions

$$\frac{\lambda_E}{\lambda_1} = 1 - \frac{3(1-\lambda_2/\lambda_1)\varepsilon}{2+\varepsilon+(1-\varepsilon)\lambda/\lambda_1}$$

If the thermal conductivity of inclusions is very small, that is, $\lambda_2 \ll \lambda_1$, then the following correlation is valid

$$\lambda_E = \lambda_1 \frac{2(1-\varepsilon)}{2+\varepsilon}$$

The basic initial data for calculating λ_E are the thermal conductivities of the porous skeleton λ_{SK} and the liquid or gas λ_L. In the case of liquid, λ_L is a conventional thermal conductivity. In the case of gas, λ_L radiation and convection heat transfer effects should be accounted for (if these effects were not accounted for by the initial energy equation).

The influence of radiation can be estimated from the recommendations of Dulnev and Zarichnyak [15]. For thin optical layers, the radiation component is

$$\lambda_{Ra} = 0.227 \varepsilon_{RE} L \left(\frac{\bar{T}}{100} \right)^3$$

where ε_{RE} is the darkness coefficient.

Similar to the general case, for thick optical layers, the radiation component is

$$\lambda_{Ra} = \frac{16n^2 \sigma_0 T^3}{\beta} f(\varepsilon_W, \tau)$$

where ε_W is the wall darkness coefficient, σ_0 is the Stefan–Boltzmann constant, β is the reduction coefficient, n is the media refractive index, $f(\varepsilon_W, \tau)$ is a function accounting for influence of wall darkness and the optical thickness, and $\tau = \alpha L$ (α is the volumetric spectral absorption coefficient and L is the physical thickness of the optical layer, through which radiation is transferred).

As a rule, convection heat transfer is absent in small size pores, excluding only the cases of high pressure flows (≥ 100 bar) and significant temperature drops, whereas continuous media approximation could be nonvalid for very small size pores. Therefore, according to the recommendations of Dulnev and Zarichnyak [15], wall temperature jump should be accounted for by the correction factor to the thermal conductivity of gas bulk flow λ_{L0} at pressure p_0 and temperature T as

$$\lambda_L = \frac{\lambda_{L0}}{1 + B/(p\delta)}$$

$$B = \frac{4k}{k+1} \frac{2-f}{f} l_n p_0 \frac{1}{Pr}$$

where k is the adiabatic index, δ is the pore size, f is the accommodation coefficient characterizing the energy transfer due to the molecule collisions with wall surface, and l_n is the length of the mean free path of molecules at given pressure and temperature values.

In most cases, the correction factor accounting for temperature jump only appears essential for very small size pores at very low pressures ($p < 0.1$–1.0 bar), that is, neglecting radiation term and temperature jump, λ_L can be determined as a flow thermal conductivity.

In contrast to λ_L, in regular cases, λ_{SK} cannot be simply defined by the skeleton thermal conductivity. The value λ_{SK} essentially depends on the skeleton structure, connections between the structure elements, their shapes, roughness, etc.

The main factors influencing skeleton thermal conductivity are as follows:

- Conditions of contact between the porous structure elements, such as contact type (mechanical pressing, soldering, welding, gluing, etc.), contact spot size, number of contacts within the elementary cell, thermal conductivity of the medium, filled the gap in the contact zone, surroundings, etc. These conditions determine the thermal contact resistance of the elementary cell.
- Deformation of the heat flux paths in the contact zones and related supplementary thermal resistance.
- Roughness of porous structure elements.

Contact spot area, S_C, and its relative value, $\mu_s = S_C/S_0$, have a major significance on the deformation of paths and in the calculation of connected thermal resistance, where S_0 is the cross section area of a granular bed single element ($S_0 = \pi r_0^2$ for bed and metal-fibrous structures, where r_0 is the radius of single particle or fiber). Values of μ_s range from 10^{-2} to 10^{-5} in granular bed structures [15]. S_C depends on skeleton mechanical properties such as elasticity module E_0, Poisson coefficient μ_s (in spherical structures without cross deformation $\mu_s = 0$), radius of element r_0, and pressure related to the single contact.

The Hertz formula gives S_C as

$$S_C = \prod \frac{3(1-\mu_s)^2 r_0 p_y}{4E_0}$$

The pressure on a single element p_y can be defined by the pressure in the given bed cross section and elasticity module of the bed layer E_b. When calculating for S_C, granular bed elasticity E_b should be applied instead of E_0. The experiments demonstrated that at low specific loads ($p_y \leq 3 \times 10^5$ Pa), the value $E_0 \sim 6.5 \times 10^7$ Pa is constant; at greater loads, the correlation between E_0, E_b, and p_y yields

$$E_b = K_E p_y^{1/3} E_0^{2/3}$$

The formulas present μ_s and other contact spot parameters as a function of porosity and pressure on the bed surface [15]. In such cases, the terms of thermal contact resistance are as follows:

- Thermal resistance of the actual contact spot that is identical to the plain-wall thermal resistance of the wall thickness being equal to the average half-height of microroughness δ_{RG}:

$$R_{1C} = 0.5 \frac{\delta_{RG}}{\lambda_L \mu_s} \pi r^2$$

- Thermal resistance caused by deformation of paths is

$$R_{2C} = \frac{1 - 0.5\delta_{RG}/r}{\lambda_L \mu_s \pi R} \Phi\left(\frac{b}{r}, \frac{r}{L}\right)$$

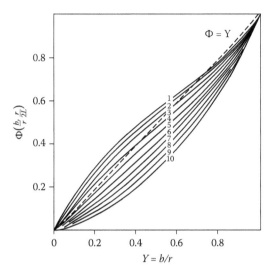

FIGURE 2.1 Function $\Phi = \Phi(X, Y)$ [15]. 1–10: $X = r/(2L) = 1, 0.9, 0.8, 0.7, 0.6, 0.5, 0.4, 0.3, 0.2, 0.1$, respectively.

where $\Phi(b/r, r/L)$ is a function depending on the ratio between contact spot size b and size of representative cell r, and relative extension of the representative cell r/L (Figure 2.1).

Function Φ for systems with interpenetrated components can be approximated by fourth-order polynomial as

$$\Phi - \Phi_1 = (1 - \Phi_1)\left(\frac{\lambda_L}{\lambda_m}\right)^2 \left[7,5 - 11\frac{\lambda_L}{\lambda_m} + 4,5\left(\frac{\lambda_L}{\lambda_m}\right)^2\right]$$

where $\Phi_1 = \sqrt{\dfrac{C}{2 - C}}$, $C = \Delta/L$ is the characteristic size of the representative cell, Δ is the characteristic size of the element, and λ_m is the skeleton thermal conductivity.

Within the granular materials, it is necessary to distinguish between first-order (dense pile of granules) and second-order structures, represented in the alternation of voids and elements of the first-order structures [15].

The relative void concentration in the second-order structure is

$$\varepsilon_{22} = \frac{V_3}{V_1 + V_2} = \frac{\varepsilon_2 - \varepsilon_{2C}}{1 - \varepsilon_{2C}}$$

where V_1 denotes the volume of solid particles; V_2 is the volume of pores in the first-order structure; V_3 is the volume of voids in the second-order structure; Δ_1, L_1 and Δ_2, L_2 are the characteristic sizes of the first- and second-order structures, respectively; and ε_2 is the total concentration of voids in the second-order structure (including voids in the first-order structure, their relative concentration in representative cell of the second order is equal to ε_{2C}).

The effective thermal conductivity of the granular structure with known parameters of the first- and second-order structures is

$$\lambda_E = \lambda_{SK}\left[C_2^2 + \frac{\lambda_L}{\lambda_{SK}}(1-C_2)^2 + 2\frac{\lambda_L}{\lambda_{SK}}C_2(1-C_2)\left(\frac{\lambda_L}{\lambda_{SK}}C_2 + (1-C_2)^{-1}\right)\right] \quad (2.7)$$

Figure 2.2 presents a comparison of experimental data with theoretical models [15]. The essential scattering in the λ_E experimental points for the free filling up granular beds appears predictable, because the structures and parameters for the free filling up are not well reproducible.

Comparison of λ_E calculations given by various authors are shown in Figure 2.2. Accordingly, Equation (2.6), which was suggested by Dulnev and Zarichnyak [15], gives the best concurrence with experimental data.

Sintering, welding, spraying, and other so-called "coupled" structures have better reproduction of characteristics than materials with free filling up. Comparison between the effective thermal conductivities of such coupled structures is shown in Figure 2.3 [15].

Extensive experiments on thermal conductivity of fiber–metal structures (FMS) were conducted in Kiev Polytechnic University [19]. It was shown that FMS thermal conductivity models can be created based on the generalized conductivity theory

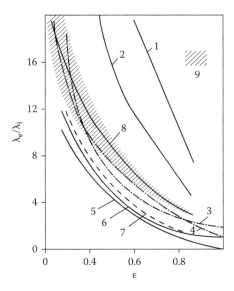

FIGURE 2.2 Dependence of quartz sand relative thermal conductivity on porosity under the normal conditions according to the experimental data and theoretical models of: 1, Lalikov; 2, Vasiliev; 3, Prasolov; 4, Dulnev and Sigalova; 5, Bogomolov; 6, Kaganer; 7, Cuney and Smith; 8, calculated by using Equation (2.7); 9, scattering between experimental data of different authors. $\lambda_m = 3$ W/(mK) and λ_L are thermal conductivities of quartz sand grain and air inside pores, respectively.

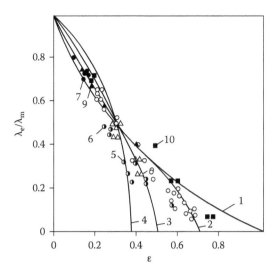

FIGURE 2.3 Comparison between thermal conductivities of coupled porous structures [15]. 1–4: calculating at $\varepsilon = 1.0$, 0.7, 0.5, 0.4, respectively. Experimental data: 5, Ni, $\lambda_m = 57$ W/(mK) at $T = 900$ K; 6, Ni, $\lambda_m = 85$ W/(mK) at $T = 293$ K; 7, Ni, $\lambda_m = 85$ W/(mK) at $T = 293$ K; 8, Fe, $\lambda_m = 78.5$ W/(mK) at $T = 293$ K; 9, Fe, $\lambda_m = 39.5$ W/(mK) at $T = 293$ K; 10, graphite, $\lambda_m = 1100$ W/(mK) at $T = 500$ K.

with accounting for defects in spaces of contact between fibers. Semena et al. [19] suggested calculating λ_E using the following equation

$$\lambda_E = \lambda_m \left((1-\varepsilon)^2 M + \varepsilon^2 \frac{\lambda_L}{\lambda_m} + 4 \frac{\lambda_L}{\lambda_m} \varepsilon(1-\varepsilon) \frac{1}{1-\varepsilon} \right) \qquad (2.8)$$

$$M = \mu_1 + \frac{2A\sqrt{1-\mu_1^2}(1-\mu_1)}{A\sqrt{1-\mu_1^2}+(1-\mu_1)}$$

where $\mu_1 = a/2r$ is a ratio between the contact spot size and the radius of fiber;

$$A = \sqrt{\pi} \frac{\lambda_L/\lambda_m}{1 - \lambda_L/\lambda_m} \left\{ \frac{1}{1 - \lambda_L/\lambda_m} \times Ln \frac{\lambda_L}{\lambda_m} - 1 \right\}$$

The dependencies of skeleton and effective thermal conductivities for FMS filled with liquid are presented in Figure 2.4 [19].

Data comparison shows that decrease in variations between skeleton and effective thermal conductivities was observed when porosity dropped to 0.65, that is, lowering the effect of heat carrier properties on effective thermal conductivity.

For steel porous structures with porosity of less than 0.5 (for copper structures, <0.65), there is no marked difference between effective thermal conductivity and skeleton thermal conductivity, because heat transfer mainly takes place through the

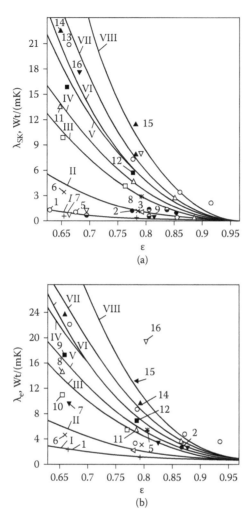

FIGURE 2.4 Dependence of (a) skeleton and (b) effective thermal conductivities of fiber–metal porous structure on porosity [19]. 1–16: experimental data. Calculating data: I, $d = 30\ \mu m$, $l = 3$ mm, stainless steel; II, $d = 50\ \mu m$, $l = 3$ mm, nickel; III, $d = 70\ \mu m$, $l = 3$ mm, copper; IV, $d = 40\ \mu m$, $l = 3$ mm, copper; V, $d = 70\ \mu m$, $l = 10$ mm, copper; VI, $d = 20\ \mu m$, $l = 3$ mm, copper; VII, $d = 40\ \mu m$, $l = 10$ mm, copper; VIII, $d = 20\ \mu m$, $l = 10$ mm, copper.

skeleton contact spots. Under highly porous conditions, the effect of liquid properties appears to be considerably less (e.g., the effective thermal conductivity of copper structures saturated with water increases comparatively to the skeleton thermal conductivity by 1.1–1.7 times, and in steel structures by 2.8–4.0 times).

Screen structures represent a partial case of FMS. Mistchenko [26] did not observe any essential differences in the effective thermal conductivities of sintered and screen structures under the same skeleton materials and heat carrier properties (Figure 2.5).

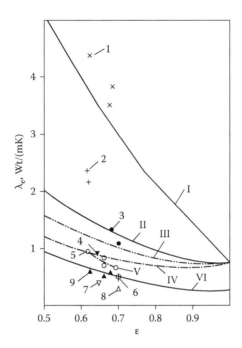

FIGURE 2.5 Comparison between experimental and calculation data on effective thermal conductivities of screen and metal–fiber structures [19, 26]. 1–9: experimental data. Calculation data: I, Equation (2.3); II and III, Equation (2.5); IV, Equation (2.5); V, Equation (2.6); VI, Equation (2.4). I–V and 1–5, structures saturated with water; VI and 6–8, structures saturated with acetone or ethanol.

Generally, in heat carrier flows through porous media, skeleton and liquid or gas temperatures are not equal to each other. Thus, conditions and correlations determining heat transfer between skeleton, porous structure, and flowing medium are of crucial importance.

2.3 INTERNAL HEAT TRANSFER IN POROUS MEDIA

Both heat transfer and mass transfer regularities are important when evaporation occurs in porous structures, which are the same or similar in various cases. The ideal model of grain bed represents the bed of equal spheres with the uniform piling. Hence, theoretical models of heat transfer in the grain bed are based on regularities of heat transfer between single sphere and passing flow as

$$Nu = const\ Re^m Pr^n$$

where indexes m and n are determined by the flow regime, connected to the corresponding Re number range. It is known that when Re increases, the value of m changes from 0.5 up to 0.9. Aerov and Todes [27] presented numerous experimental data on heat and mass transfer in grain beds, and gave the following recommendations on the calculation of steady-state heat transfer coefficients when there are contacts between grains:

$$\text{At } 2 < Re < 30 \qquad\qquad Nu_E = 0.725Re_E^{0.47}Pr^{1/3}$$

$$\text{At } 30 < Re < 30,000 \quad Nu_E = 0.4Re_E^{0.64}Pr^{1/3} \qquad\qquad (2.9)$$

When $0.01 < Re < 2$, the natural convection effect occurs and leads to an increase in m value:

$$Nu_E = 0.515Re_E^{0.85}Sc^{1/3};$$

$$Re_E = 4w/S_V v_L; \; Pr = v/a; \; Sc = v/D; \; Nu_E = 4\alpha\varepsilon/\lambda_L S_V$$

Comparison of typical empirical and semiempirical correlations for internal heat transfer in spherical grain beds is presented in Figures 2.6 and 2.7 [27].

A review on internal heat transfer and hydraulic resistance in flow through metallic and metallic–ceramic porous materials is presented in Mayorov [29].

According to Aerov and Todes [27], Belov [28], Mayorov [29], and Polyaev et al. [30], it is necessary to distinguish between the following regimes:

- Viscosity (Darcy regime), $Re < 0.01$
- Transient, $0.01 < Re < 100$
- Inertial, $Re > 100$

For the Re number given as $Re = G\beta_0/(\mu\alpha_0)$, the characteristic size is represented by the ratio β_0/α_0. Choosing such a characteristic size is advantageous, because β_0 and α_0 can be measured with higher reliability and lower scattering compared to

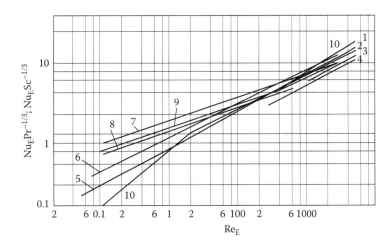

FIGURE 2.6 Heat and mass transfer in spherical granular bed at various porosity levels [27]. 1–9: $\varepsilon = 0.477, 0.463, 0.365, 0.33, 0.48, 0.33, 0.47, 0.488$, respectively; 10, calculated by Equation (2.9).

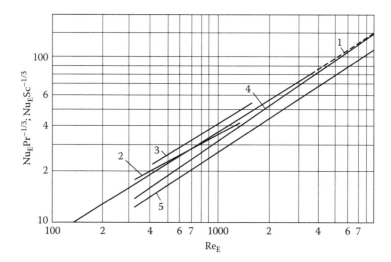

FIGURE 2.7 Comparison between experimental data on heat and mass transfer inside grain bed and calculations using Equation (2.9) [27]. 1, Equation (2.9). Experimental data: 2, d_p = 10.3 mm; 3, d_p = 12.1 mm; 4, in the bed center; 5, at the bed peripheral.

parameters S_v (specific surface), d_0, and d_p (pore and particle diameters, respectively). However, using β_0/α_0 as the characteristic size gives significantly lower values for Re numbers at flows inside porous media than in the case when the pore or particle diameter is used as the characteristic size. Thus, parameter β_0/α_0 reallocates results from the region where $Re > 1$ to the region with $Re < 1$.

Comparison of experimental data on internal heat transfer in metallic sintered porous materials is presented in Table 2.1 [29].

Table 2.1 shows that majority of experiments were conducted over the range $Re = G\beta_0/(\mu\alpha_0)$ from 10^{-2} up to 10^2. Analysis of the momentum equation demonstrates that this region is characterized by the transitional flow regime, that is, laminar flow is mixing with the turbulent one [28]. It is known that increasing the characteristic size in the transitional regime leads to an increase in the heat transfer coefficient, that is, when the value of Re increases, Nu number increases at a higher proportion. It means that $Nu \sim Re^m$, where $m > 1$. Considerable scattering of data on internal heat transfer coefficients under heat carrier flow inside porous structure is not only caused by a variety of characteristic size choices (β_0/α_0, d_0, d_p, etc.), but also by probable variation in laws of pore size distribution.

In reality, both experimental methods and theoretical models of the internal heat transfer do not consider pore influence on heat transfer intensity.

Current calculation methods for heat transfer inside porous media accounting for the pore distribution effect were reported by Cheiphez and Naimark [17]. These methods use the following concepts:

- Correlations between rates of mass flow, momentum, and concentration have a linear form:

At filtration mass flow rate inside porous structure: $G_{ij} = -F_{ij}\dfrac{K_{ij}\Delta p_{ij}}{\mu l_{ij}}$

TABLE 2.1

Experimental Conditions and Research Results on Internal Convection Heat Transfer in Porous Materials

Heat Transfer Criteria Equations	Nu Number	Re Number	Re Range	Heat Carrier	Porous Material	Structure Thickness, mm	Porosity
$Nu = 0.0286\,Re^{1.84}$	$\alpha_V(\beta_f/\alpha_0)^2/\lambda_L$	$G\beta_f/(\mu\alpha_0)$	0.016–0.15	air	A	0.72	0.30
$Nu = 0.006\,Re^{1.84}$	$\alpha_V(\beta_f/\alpha_0)^2/\lambda_L$	$G\beta_f/(\mu\alpha_0)$	0.016–0.1	air	A	4.22	0.52
$Nu = 0.038 \times Re^{1.34}\,(d_p/\delta)$	$\alpha_V(\beta_f/\alpha_0)^2/\lambda_L$	$G\beta_f/(\mu\alpha_0)$	0.03–1.15	air	A	2.0; 5.1; 8.1	0.42
$Nu = 0.005\,Re\,Pr$	$\alpha_V(\beta_f/\alpha_0)^2/\lambda_L$	$G\beta_f/(\mu\alpha_0)$	0.003–30	air, oil	A	1.3	0.30
$Nu = Z\,Re^{0.65}$	$\alpha_V(\beta_f/\alpha_0)^2/\lambda_L$	$G\beta_f/(\mu\alpha_0)$	0.01–5.0	air, helium	B	3.2; 6.4	0.6; 0.8
$Nu = 0.007\,Re^{1.2}$	$\alpha_V(\beta_f/\alpha_0)^2/\lambda_L$	$G\beta_f/(\mu\alpha_0)$	0.04–10	air	C	3.0	0.37; 0.45
$Nu \approx C\,Re^{0.9}$	$\alpha_V(\beta_f/\alpha_0)^2/\lambda_L$	$G\beta_f/(\mu\alpha_0)$	0.005–0.2	air	D	1.0–1.4	0.31–0.44
$Nu = f(\varepsilon,\delta)Re^{1.2}$	$\alpha_V(\beta_f/\alpha_0)^2/\lambda_L$	$G\beta_f/(\mu\alpha_0)$	0.007–0.7	air, nitrogen	A, E	1.0–5.6	0.3–0.5
$Nu = 9.2 \times 10^{-3}\,Re^{1.34}$	$\alpha_V(\beta_f/\alpha_0)^2/\lambda_L$	$G\beta_0/(\mu\alpha_0))$	0.3–7.0	air	C	7; 10	0.29–0.6
$Nu = C\,Re$	$\alpha_V d_p^2/(6\lambda_L(1-\varepsilon))$	$Gd_p/(\mu\varepsilon)$	2–160	air	F	1.6–9.0	0.33–0.39
$Nu = 0.0042\,Re^{0.9}$	$\alpha_V d_p^2/(6\lambda_L(1-\varepsilon))$	$Gd_p/(\mu\varepsilon)$	2.5–100	air	G	3.0	0.35
$Nu = 0.0175\,Re\,Pr$	$\alpha_V d_p^2/(6\lambda_L(1-\varepsilon))$	$Gd_p/(\mu\varepsilon)$	2.0–75	Ar, He	A	4.75	0.33
$Nu = 0.09\,Re$	$\alpha_V d_p^2/(6\lambda_L(1-\varepsilon))$	Gd_p/μ	3.0–15	air; He	A	1.3	0.31
$Nu = 1.25(d_p/\delta)Re$	$\alpha_V d_p^2/(6\lambda_L(1-\varepsilon))$	Gd_p/μ	0.7–5.0	air	A	1.3–8.1	0.31–0.55
$Nu = 0.8\,(d_p/\delta)Re^{1.34}$	$\alpha_V d_p^2/(6\lambda_L(1-\varepsilon))$	Gd_p/μ	50–250	air	A	1.3–8.1	0.31–0.55
$Nu = 0.1(Re\,Pr)^{1.25}$	$\alpha_V d_p^2/(6\lambda_L(1-\varepsilon))$	Gd_p/μ	0.56–95	Ar; He	A, F	3.0–16.9	0.23–0.31
$Nu = 0.028\,Re^{1.2}\varepsilon^{2.9}$	$\alpha_V d_0^2/\lambda_L$	$Gd_0/(\mu\varepsilon)$	0.4–5.0	air	H	0.5; 1.4; 5.0	0.25; 0.32; 0.4
$Nu = C\,Re$	$\alpha_V d_0^2/(6\lambda_L(1-\varepsilon))$	$Gd_0/(\mu\varepsilon)$	1.0–4.5	air	A, B, D	0.4–1.85	0.3–0.65
$Nu = 0.0014[(1-\varepsilon)/\varepsilon]^{0.9}$	$\alpha_V d_0^2/\lambda_L$	$G\delta/\mu_L$	14–600	air	A	0.79–3.18	0.35–0.58
$Nu = 0.0004 \times Re^{1.16}$	$\alpha_V K/\lambda_L$	$G\sqrt{K}/\mu_L$	0.0003–0.02	air	A	1.92–2.53	0.2–0.3

α_V, volumetric heat transfer coefficient; d_p, particle diameter; d_0, pore size; A, stainless steel powder structure; B, fiber stainless steel; C, fiber nickel structure; D, screen stainless steel; E, nichrome; F, powder bronze; G, powder copper; H, powder nichrome.

At mass flow rate of the ith component: $G_{ij} = -F_{ij}D_{ij}\dfrac{\Delta C_{ij}}{l_{ij}}$

where l_{ij} is the length of mass transfer path, m; D_{ij} and μ denote diffusion coefficient and dynamic viscosity, respectively; F_{ij} is the cross section of the porous medium through which mass flow j_{ij} [kg/(m²sec)], takes place, m^2; and K_{ij} is the permeability coefficient, related to length l_{ij} (m).

- The porous media can be described by the mesh model. Mesh structure is related to the real structure through the conformity of pore distribution, identity in a number of connections within a single mesh knot, coordination number in the porous structure, and basing on the description of the effective pore radius R_E on the self-conformed field model as

$$\int_0^\infty \frac{\left(R_E^4 - R^4\right)\varphi(R)\,dR}{R^2\left[R^4 + (N_K(1/2)-1)R_E^4\right]} = 0 \quad - \text{ Filtration process}$$

$$\int_0^\infty \frac{\left(R_E^2 D(R_E) - R^2 D(R)\right)\varphi(R)\,dR}{R^2\left[R^2 D(R) + ((1/2)N_K - 1)R_E^2 D(R)\right]} = 0$$

$\quad - \text{ Diffusion process (D is a diffusion coefficient)}$

- The effective parameters R_E, D_E, etc., determined by the cited approach, are used in the numerical calculations for the determination of integral characteristics based on the effective parameters of the representative cell.

The practice by which such methods are applied for modeling of heat and mass transfer processes in porous media is still unknown.

The internal heat transfer and hydraulic resistance inside nonsintered screen structures show none of the peculiarities observed in sintered metallic, and metal–ceramic porous materials. Essentially, it can be explained by the fact that the actual flow conditions in screen structures approach the flow conditions in dense tube (cylinder) bundles. It is known that for external problems a function of heat transfer intensity on the Re number does not contain zones with nonmonotonous Nu variation. For example, heat transfer in screens packed with a number of beds ranging from 20 up to 80 at Re number range of 20–400 can be described by the criteria equation $Nu = 0.09 + 0.49\,Re^{0.5}$ [31].

Mori and Miyazaki [31] did not observe any influence of stiffened flow conditions (including a decrease in cross section) on heat transfer intensity. Baumann and Blab [32] conducted an extensive experimental study on screens with different packing techniques. Using the analogy of flow in tube bundles for experimental data generalization yields the following criteria equation

$$Nu_d = 0.786\ (Re_d\, d_E\,/2d)^{0.54}\ Pr^{1/3}$$

where $d_E = 4\varepsilon/S_{vi}$ denotes the characteristic size and d is the wire diameter.

Baumann and Blab [32] paid special attention to the effect of turbulence on internal heat transfer regularities at flow in screen structures, and suggested determining the characteristic size as $L = 0.015/(\varepsilon^2 S_v)$. Such characteristic size allocates a Re number range of 0.01–10. Heat transfer in screen packages can be prescribed with maximum inaccuracy of 20% as

$$Nu_E = 0.11 Re_L^{0.54} Pr^{2/3}$$

Experimental studies on the internal heat transfer of gas flow in screen packing were conducted by Pritula and Zablotskaya [33, 34]. Eight screen packing types with longitudinal pitch values ranging from 1 up to 5.8 were tested. Mesh equivalent diameter was selected as the characteristic size. The heat transfer coefficient was related to the entire screen surface.

Comparison of research results [33, 34] with the data of other authors on internal heat transfer of gas flow inside screen structures led to the following generalization:

$$Nu_0 = \varphi_1(f) Re^{0.5} + \varphi_2(f) Re^{0.8}$$

where $\varphi_1(f)$ and $\varphi_2(f)$ are stiffening index functions. The stiffening index correlates with porosity as $f = 1 - \varepsilon$. Pritula and Zablotskaya [33, 34] did not define these functions but they suggested the following constrained correlations:

- For screen $Nu_0 = 0.98 Re^{0.5}$
- For screen packing $Nu_0 = 0.065 Re^{0.8}$

Another recommendation was to account for the stratification effect of screen packing on the heat transfer intensity by using a parameter $\varepsilon_L = S_L/S_0$, where S_L denotes the pitch between screen beds in the packing and S_0 represents the same characteristic but under the dense contact between beds. According to Pritula and Zablotskaya [33, 34], the generalized correlation for heat transfer of gas flow in screen packing porous structures is $Nu = \varepsilon_L^n Nu_0$.

2.4 HEAT TRANSFER ON THE POROUS STRUCTURE EXTERNAL SURFACE

Problems relating to heat transfer on the porous structure external surface can be classified into the following cases (Figure 2.8):

- Heat transfer caused by the effective heat conduction through the wetted porous structure from the one heat transfer surface to another at the small flow velocities (case a1).
- Heat transfer between wall and single-phase flow inside the porous structure (case a2).
- Heat transfer at condensation on the porous structure external surface (case b).
- Heat transfer at evaporation from the porous structure external surface (case c).

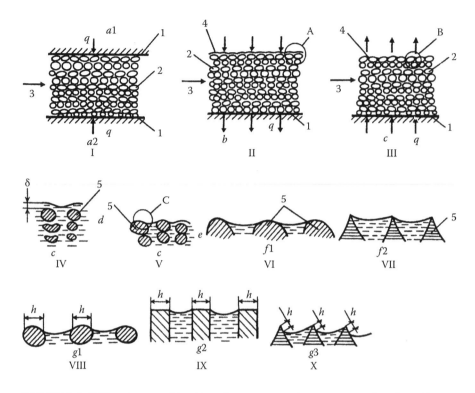

FIGURE 2.8 Different cases of heat transfer on porous structure external surfaces: I, convection heat transfer between wall and ingle-phase flow through the porous media (case a); II, heat transfer at condensation on the surface covered by porous structure (PS) completely filled with liquid (case b); III, heat transfer from wall through the filled PS to the interface at evaporation (case c); IV, PS surface element at which condensate film is forming with thickness d (case d); V, surface element from which evaporation takes place (case e); VI and VII, evaporation from the wetted PS surface without drying zones $f1, f2$; VIII, IX, and X, evaporation from the wetted PS surface with drying zones $g1, g2, g3$; 1, wall; 2, porous structure; 3, heat carrier; 4, heat transfer external surface of wetted PS (surface of condensation or evaporation); 5, elements of PS skeleton.

We shall consider the fundamentals of heat transfer corresponding to the above-mentioned conditions. The essential heat transfer mechanism for the regime marked as "a" is heat conduction, and its specific mechanism is associated with the determination of the effective thermal conductivity coefficients. This problem was outlined in Section 2.2.

Heat transfer between heat carrier flow inside the porous structure and the wall has been extensively studied for the grain bed case, whereas there are only a few analogous investigations on gas-sprayed, sintered, and other similar porous structures coupled with a wall.

In general, heat transfer between an unmoved bed and a wall represents a combined effect of convection inside the porous structure and heat conduction through the skeleton. The presence of a bed in the vicinity of the wall significantly dumps convection and affects the turbulence mixing mechanism. Moreover, local porosity values are essentially increasing, and the number of contacts is decreasing. Hence, it leads to the formation of specific film thermal resistance close to the wall.

The overall heat transfer coefficient in the vicinity of the wall α_c can be expressed as the sum of the local thermal resistances

$$\frac{\alpha_c d}{\lambda_L} = \frac{\alpha_1 d}{\lambda_L} + \frac{1}{\dfrac{\lambda_L}{\alpha_2 d} + \dfrac{1}{C_0 Re\,Pr}}$$

where α_1 is a heat transfer coefficient in the vicinity of the wall, which is independent of the liquid linear velocity. In the case of gaseous heat carrier, it is also necessary to account for a radiation term; according to Aerov and Todes [27], $4 < \alpha_1 d/\lambda_L < 15$.

If we suppose that the laminar boundary layer appears next to each point of grain element contact with the wall, and the incoming jet is ruining it, then α_2 can be estimated. The consequent heat transfer correlation is

$$Nu_2 = \alpha_2 d/\lambda_L = ARe_0^{0.5} Pr^{1/3}$$

$C_0 Re\,Pr$ characterizes the decrease in the heat transfer coefficient due to the changing of interpositions of bed elements, $C = 0.054$ [27].

The dependence of $Nu_E/Pr^{1/3}$ on Re_E is shown in Figure 2.9, where $Nu_2 = Nu_E = \alpha_2 d_E/\lambda_L$ and d_E is the equivalent diameter of the grain layer, $d_E = 4\varepsilon/S_v$. If the tube or apparatus size is compared to the grain diameter, then wall surface should be accounted for in the calculation of S_v.

If we consider the heat carrier flow, then grain bed temperature distribution is determined by the superposition of heat transfer between wall and bed, and heat conduction. Assuming that

$$k^{-1} = \frac{1}{\alpha_2 + \alpha_r} + \frac{1}{\alpha_0} + \frac{\delta}{\lambda_m}$$

where α_2 is the film heat transfer coefficient, α_r is the radiation heat transfer coefficient, α_0 is the heat transfer coefficient from the external surface, λ_m is the wall thermal conductivity, and δ is the wall thickness. Value $Bi = kD_a/\lambda_R$ can be used in the analysis of heat transfer through the wall (where $\lambda_R = \lambda_{SC}$ is the radial thermal conductivity of the bed and D_a is the diameter of the apparatus channel).

As a rule, Bi number is quite large (i.e., $Bi \to \infty$) and temperature distribution inside the bed is given as

$$\frac{T_2 - T_L}{T_1 - T_L} = \exp\frac{4kL}{jc_p D_a}$$

where T_L is the temperature of heat carrier.

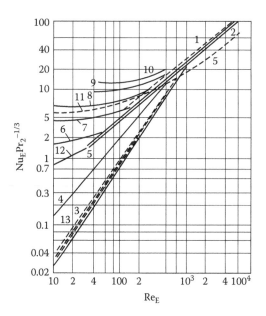

FIGURE 2.9 Dependence of $Nu_E/Pr^{1/3}$ on Re_E [27]. Experimental data: 1, Aerov and Umnik; 2, Aerov and Batischev (longitudinal heat conduction is not accounted for); 3, Aerov and Batischev (longitudinal heat conduction is accounted for in different values of D_d/d); 4, Gelperin and Kagan; 5, Yagi and Vakao. 6, Glass spheres, $d = 0.57$ and 0.94 mm; 7, glass spheres, $d = 2.75$ mm; 8, glass spheres, $d = 5.1$ mm, and brass spheres, $d = 3$ mm; 9, glass spheres, $d = 12.3$ mm; 10, steel spheres, $d = 11.2$ mm.

The experimental determination of heat transfer coefficients in the vicinity of the wall leads to dependencies $Nu_{E2} \approx f(Re_E, Pr)$. The most reliable correlations are

- If $1 < Re_E < 38$, then $Nu_{E2} = 0.253 Re_E^{0.5} Pr^{1/3}$.
- If $38 < Re_E < 10{,}000$, then $Nu_{E2} = 0.09 Re_E^{0.8} Pr^{1/3}$.

Heat transfer at condensation on the porous structure external surface can be considered as the heat conduction through the following two layers:

- Condensate thin film (film thickness is determined by liquid suction through the porous structure).
- Wetted porous structure (heat transfer is determined by effective heat conduction through the condensate film).

In a number of cases, it is possible to ignore the thermal resistance of condensate thin film and to consider only the effective heat conduction.

Evaporation from external surfaces of thin-layered porous structures with drying zones has several peculiarities (Figure 2.8, g1–g3).

The main peculiarity can be explained in the following manner: in most cases during evaporation, external surfaces of porous structure elements are drained out, that is, evaporation takes place from liquid meniscus inside pores; however, heat is transferred mostly by the metallic skeleton. Because skeleton thermal conductivity is considerably higher than that of the heat carrier, a significant deformation of heat flux streamlines appears. It is identical in appearance to the additional surface thermal resistance.

The influence of such a thermal resistance is especially important for the small specific height thin-layered porous and capillary-grooved structures. If the thickness of a porous structure exceeds 1.5–2 times the characteristic pore size, then the overall porous structure thermal resistance R_S can be expressed as the sum of the surface thermal resistance R_0 and thermal resistance of the major part of the porous structure R_1. R_1 can be determined by using the corresponding correlations for effective thermal conductivity, λ_E. Hence, the overall porous structure thermal resistance R_S is

$$R_S = R_0 + \frac{H - \delta_0}{\lambda_E} \tag{2.10}$$

where δ_0 is the thickness of the surface layer and H denotes the entire thickness of the porous structure.

The correlation suggested by Mistchenko [26] for the calculation of R_0 gives

$$R_0 = \frac{S}{\lambda_L} \frac{0.14}{\dfrac{S}{B}\sqrt{\dfrac{\lambda_L}{\lambda_m}}\left(1 + \sqrt{\dfrac{\lambda_L}{\lambda_m}}\right)} f\left(\frac{S}{R}\right) \tag{2.11}$$

where $f(S/R)$ is the function accounting for the influence of meniscus shape (curvature) changing on thermal resistance R_0 at intensive evaporation

$$f\left(\frac{S}{R}\right) = \exp\left[-1.16\left(1 + \sqrt{\frac{\lambda_L}{\lambda_m}}\right)\frac{S}{R}\right] \tag{2.12}$$

Naturally, at $H \gg \delta_0$ the influence of δ_0 can be neglected.

In the case of thin-layered structures ($H \sim d_0$), R_0 represents a complex function of major geometric, flow, and thermophysical parameters. By using electric analogy methods in their study of thin-layered structures with rectangular and triangular grooves, Burdo and Smirnova [21] obtained the dependencies of λ_E of these factors. For example, in the case of rectangular grooves, the following was obtained

$$\frac{\lambda_E}{\lambda_m} = \frac{S}{B}\left[1 + C\left(\frac{S}{B}\sqrt{\frac{h}{S}}\right)^m\right] \tag{2.13}$$

where $C = 1.58$, $m = 1.64$ at $\lambda_E/\lambda_m = 0.03$–$0.04$, and $C = 2.6$, $m = 1.5$ at $\lambda_E/\lambda_m = 0.02$–$0.04$. In the case of triangular grooves, it is

$$\lambda_E/\lambda_m = a - b \sin \alpha \tag{2.14}$$

where $a = 0.112$, $b = 0.0059$ at $\lambda_E/\lambda_m = 0.03$–$0.04$, and $a = 0.006$; $b = 0.0048$ at $\lambda_E/\lambda_m = 0.0008$–$0.001$.

Burdo and Smirnova [21] accounted for the influence of meniscus curvature as the function $f(S/R)$ in agreement with recommendations [26].

The approximation approach of accounting for heat and mass transfer peculiarities through the wetted groove capillary structures was suggested for triangular grooves [22] and trapezoidal grooves [23]. Later, Schekriladze [24] applied this approach to clarify various heat and mass transfer regularities of evaporation and condensation at the profiled surfaces. As the initial details of this method have already been thoroughly discussed in Refs. [22, 23, 25], this work will only consider several important correlations.

Various studies [22–25] have recommended using the following formula to calculate the heat transfer coefficient for both evaporation and condensation at the flooded triangular grooves:

$$\alpha = \sqrt{\frac{\lambda'\lambda_m}{\sin\Theta \, tg\,\varphi}} \bigg/ h$$

where λ' and λ_m are the liquid and grooved material thermal conductivities, respectively, φ is the half-angle of the groove ledge, Θ is the contact angle between the tangent to the liquid–vapor interface and wetted groove side surface, and h is the groove height.

The following formula can be applied in cases of flooded trapezoidal grooves and some deepening of the liquid layer [23–25]

$$\alpha = \frac{1}{h^*}\sqrt{\frac{\lambda'\lambda_m}{\sin\Theta}\frac{2h^*}{t}}$$

where h^* is the thickness of the liquid layer in the groove (for most cases, $h^* < h$; when $h^* = h$, the groove is flooded) and t is the groove pitch.

It is necessary to discuss the most essential drawbacks of the approach suggested and developed in several studies [22–25].

(1) Prof. Schekriladze suggested using the present model for both evaporation and condensation. This does not correspond with the physical reality, particularly for rectangular and trapezoidal grooves.

In the case of condensation on the types of grooves cited above, the whole heat transfer surface is covered by thin condensate film, in which thickness depends on both condensation intensity and the amount of liquid suction. The last one is essentially conditional on many factors such as condensation surface position within the gravitational field, the possible scale of the capillary forces, viscosity of liquid, etc. In contrast, during evaporation this surface dries out.

Such a distinction is less relevant to experiments with flooded triangular grooves. In the case of condensation, the absence of liquid film on the external surface and the impossibility of removing the liquid from grooves due to gravity or capillary forces represent only a limited number of cases. Thus, the efforts of Shekriladze and Rusishvili [24, 25] to compare results based on the above-mentioned simplified correlations with experimental data on condensation at the low-finned tubes or heat pipes with grooved surfaces failed because the cited approach does not account for hydrodynamic phenomena, which determine the film condensation regimes.

The presented approach is physically proved mainly for evaporation within flooded triangular grooves. Similar to other models in the first approximation, it accounts for the formation of the additional surface thermal resistance of the contact between the liquid–vapor interface and the high thermal conductivity skeleton. Such a thermal resistance is absent in the majority of actual condensation conditions.

(2) The next critical problem is related to the contradiction in the choice of the contact angle value. According to various studies [22–25], it is not a wetted angle, which approaches zero in the case of wetting liquids, that is, concurring with the correlations presented above, it means that $\alpha \to \infty$. Hence, if the current approach is correct, then other authors' experimental data on heat transfer at evaporation from grooved surfaces should be hardly reproducible and should have a large scattering; so far, however, this has not been reported in known publications, including [22–25].

An even stranger result relates to the assumption of Θ constancy (i.e., constancy of the interface curvature radius) under decreasing liquid level in the grooves, which creates one more contradiction to the physical reality. It is known that a decrease in the interface level leads to a substantial increase in hydraulic resistance and, consequently, requires an increase in the capillary potential, that is, lowering of the interface curvature radius. Therefore, parameter Θ has an empirical value without any special physical meaning.

(3) The presented approach does not account for the fact that the value of the equilibrium saturation temperature above the curved phase surface sufficiently exceeds the corresponding value above the plain surface (the variation is equal to the nucleation overheat $\Delta T^* = 2\sigma T_S / r \rho'' L$, where L is an interface curvature radius). It means that the temperature drop in consequent evaporation processes is different than that in the common case of the convection heat transfer (i.e., difference between saturation and wall temperatures, $\Delta T_S = T_W - T_{js}$), but as $\Delta T = \Delta T_S - \Delta T^*$.

(4) It is wrong to consider L as a steady parameter that is independent from the heat flux and that other factors influenced the hydraulic resistance of liquid transport to the evaporation surface. Curvature radius decreases sufficiently from $L = \infty$ (plain surface) to $L = R_{min}$ (inserted circle radius for ideal wetted liquid) with the increase in heat flux [20, 82]. This decrease in L is accompanied by the corresponding growth of heat transfer intensity. Such a growth was observed during evaporation on grooved surfaces in experiments [24, 25]. However, Shekriladze and Rusishvili [24, 25] described this phenomenon as the boiling initiation.

Direct comparison between results based on approximations suggested in Refs. [22–25] and data from other models using the electric analogy [20, 21] or numerical methods [82] has not been reported.

Borodkin [82] concluded that direct comparison between calculations based on the current approach [23, 24] and numerical calculations with two-dimensional model and electric analogy data shows poor agreement.

The necessity of taking into account the additional thermal resistance, R_0, could also be significant for massive porous structures, when heat transfer occurs in thin layers (comparing with pore sizes) under evaporation in several vapor channels inside internal volumes of porous structures.

3 Thermohydrodynamics at Vaporization inside Capillary-Porous Structures

3.1 VAPORIZATION CONDITIONS IN CAPILLARIES AND CAPILLARY-POROUS STRUCTURES

Regimes of vaporization in capillaries and capillary structures are restricted by liquid flow supply, heat input conditions, and geometry of capillary structure (Figures 3.1–3.4).

Liquid flow arrangement is assigned mainly to the physical nature of the forces maintaining the heat carrier circulation; the following cases could be specified:

- Forced liquid supply under the action of external forces, when the force volumetric density can be estimated by the following dimensionless parameters: $\rho w^2/L$, $\rho' w_0^2/d_e$, $\Delta p'/L$.
- Gravitational heat carrier supply (natural convection). The heat carrier flow is governed by the action of Archimedean forces. Their density can be evaluated as $F_g = (\rho' - \rho'')g \sin \beta$, where β is the angle accounting for the capillary channel orientation in the gravity field.
- Capillary liquid supply, when the liquid motion is maintained by the action of capillary forces. Their density can be estimated by as $F_\sigma = \sigma/(aL)$, where L is the length of the capillary channel, a is the characteristic length of the channel, and $1/a$ is an interface curvature).
- Inertial (pulsating) heat carrier supply, when a periodic liquid feeding of the heat transfer surface occurs under the action of inertial forces. These inertial forces appear due to the vaporization process. Density of the forces can be estimated as $F_{In} = (\rho'' w_0''^2)/L$.
- Combined heat carrier flow, when densities of two or more volumetric forces cited above have the same order of magnitude, that is, F_g and F_w, or F_g and F_σ, etc.

The a/D_0 ratio is an essential parameter determining the conditions of vaporization inside the capillary channels and porous structures. It represents the ratio between the characteristic size of elementary channel cross section and the bubble

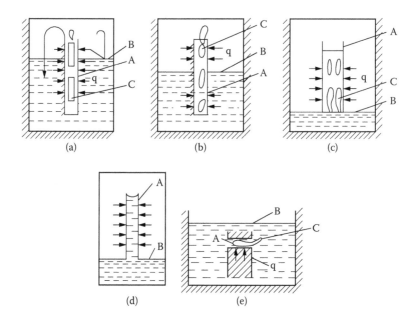

FIGURE 3.1 Mechanisms of the heat carrier supply at vaporization inside narrow slits and capillaries: (a) gravitational in the single channel, (b) combined (gravitational + inertial), (c) inertial, (d) capillary input, (e) gravitational supply to the horizontal slit. A, slit or capillary channel; B, liquid level; C, vapor phase, "slug" regime; q, heat input.

departure diameter. Excluding only the forced flow case, the heat carrier properties and the saturation pressure govern the value of such a diameter, D_0. It can be estimated as $D_0 = \sqrt{\sigma/(\rho' - \rho'')g}$, that is, inequality $a/D_0 \ll 1$ can be reformulated as

$$\sigma/\left\{(\rho' - \rho'')gD_0^2\right\} = Bo \gg C_1 \tag{3.1}$$

where Bo is the Bond number.

Equation (3.1) allows us to determine whether the channel can be considered a capillary. According to Equation (3.1), for many liquids, channels with diameters less than 1.0 mm are capillaries.

Weber number, We, yields a similar estimation in the case of forced heat carrier flow

$$\sigma/(\rho' w_0^2 d^2) = We \gg C_2$$

When $\rho' w_0'^2 \ll (\rho' - \rho'')gd$ (small velocities of the two-phase flow) for the forced liquid circulation, Equation (3.1) can be used as the estimation tool.

Because single bubbles are blocking the channel cross section in the capillaries, it means that some regimes of separated phase motion will be not present. Therefore, in the capillaries and capillary-porous structures, all feasible two-phase flows with separate phase motion do not exist and only such regimes as "slug" and plug are accessible. It is probable that "annular" and "dispersed" regimes are also obtainable under certain conditions.

In the case of the intensive heat carrier vaporization, the "slug" regime is not dominant because it is associated with the restricted range of the mass vapor quality change, Δx.

Thus, two-phase flow regimes in porous structures depend on the heat input conditions. There are several typical heat input conditions for vapor-generating porous structure (Figure 3.2):

- Heat input through the heat transfer wall to the porous structure wetted by heat carrier
- Heat generation inside the porous structure
- Convective or radiative heat input to the external evaporative surface of capillary structure (this is common in evaporative cooling, thermal protection systems, and drying technology)

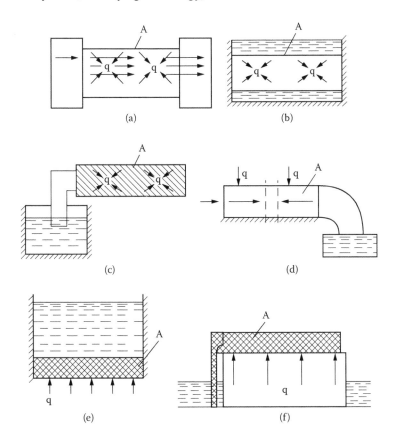

FIGURE 3.2 Heat input conditions at vaporization inside the porous structures: (a) volumetric heat generation, forced liquid supply; (b) volumetric heat generation, gravitational liquid supply; (c) volumetric heat generation, capillary liquid supply; (d) forced and capillary liquid supply, heat transfer from the evaporating surface; (e) gravity liquid supply, heat transfer from the wall to the wetted capillary structure; (f) capillary liquid supply, heat transfer from the wall to the porous structure.

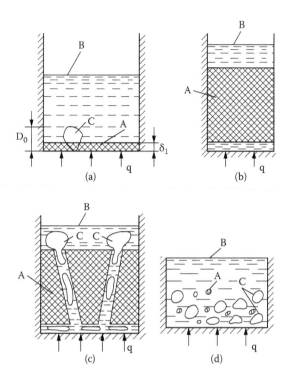

FIGURE 3.3 Mechanisms of vaporization on the heat transfer surfaces covered by moveable porous structures: (a) gravitational liquid supply, heat transfer from the wall to the thin layer porous structure; (b) gravitational liquid supply, heat transfer from the wall to the massive structure; (c) regime of the channels' formation; (d) thermal "pseudo-fluidization" regime. A, porous structure; B, liquid level; C, vapor phase; q, heat input.

Drying processes are directly connected with vaporization in porous structures. It is a very important technological process. Extensive data on the drying process are reported in several well-known publications [18, 37–39].

It is important to acknowledge the common features and main differences between the drying process and mechanisms of evaporation, and boiling in the capillaries and in the capillary-porous structures discussed in this book:

1. The drying process is essentially an unsteady process of moisture removal from the capillary-porous medium. The distinctive parameters of the process are time, moisture content, etc. The steady states are significant only as the boundary conditions.

   However, vaporization in capillaries and capillary structures considered in this work signifies a steady state. The governing parameters are the local and average heat and mass transfer coefficients, with heat and mass fluxes constraining vaporization regimes by the intensities of steady heat and mass transfer.

Unsteady processes are the subject of interest with respect to the analysis of transient condition from one steady state to another.

2. The main objective of the drying process is moisture removal up to the complete drying out of the body. Hence, in the given case it is necessary to account for entire types of connections between moisture and the porous structure, that is, chemical, physicochemical, and physicomechanical.

   Consideration of the steady evaporation and boiling processes in capillaries and capillary-porous structures requires us to account for mainly physicomechanical connections in the microcapillaries.

3. The variety of thermohydrodynamic phenomena accompanying the drying process maintains improvement of the phenomenological theory based on the application of such concepts as moisture transfer potential, moisture capacity, etc. Special experimental methods were developed for measuring consequent empiric coefficients in the fundamental heat and mass transfer equations.

In most cases, liquid transfer in capillaries and capillary-porous structures at the steady vaporization process was determined by another set of parameters and regularities in comparison with the drying processes.

The quantitative correlations for the two-phase hydrodynamics have indirect effect on the main heat transfer characteristics via the thickness variations of the liquid film at heat transfer surfaces, liquid film origin, arrangement, etc.

Therefore, application of the known single-phase heat and mass transfer, filtration, and capillary hydrodynamics correlations to porous media flow represents a rational approach for the analytical treatment of the steady-state heat and mass transfer at vaporization in capillaries and capillary-porous structures.

3.2 HYDRODYNAMICS AT VAPORIZATION IN CAPILLARY-POROUS STRUCTURES

By considering the entire capillary structure at the two-phase flow filtration, similar to the drying processes, it is possible to use the fundamental hydrodynamic equations for porous structures [18]. The main assumption of such an approach is that, for multiflow filtration, each flow is treated separately according to Darcy's law (single-phase filtration). A similar equation set with initial conditions is presented in by Charny [40] based on the phenomenological approach with the empirical correlations. The Buckley-Leverett theory is the most widely used. It uses the experimentally proven statement that each phase of the mixture is filtrating through the porous medium in accordance with Darcy's law.

Thus, neglecting volumetric force action, Darcy's law yields the following correlation for the elementary volume with the cross section S

$$Q_i = \frac{-K_i(W_1, W_2)}{\mu_i} S \frac{dp_i}{dz}$$

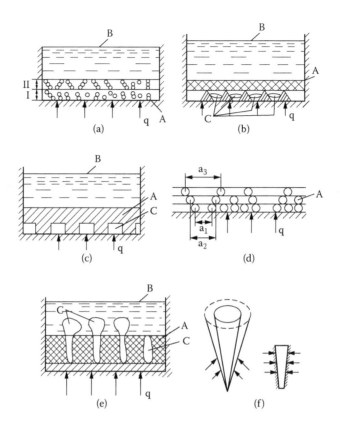

FIGURE 3.4 Forms of nonuniform capillary-porous structures: (a) double-layer structure with different pore sizes of external and near-wall layers; (b, c) double-layer structures with longitudinal vapor channels in the near-wall layer; (d) with the increasing pore size; (f) with the variable size of capillary channels. A, porous structure; B, liquid level; C, vapor channels; q, heat input.

Here, Q_i is the ith-phase volumetric flow rate, K_i is the permeability of the ith phase, μ_i is the Newtonian viscosity, W_i is the saturation of the entire cross section with the ith phase, and dp_i/dz is the pressure gradient within the ith phase.

As a rule, capillary forces establish the difference in pressure gradients at the steady-state regimes. It is also assumed that phase permeabilities K_i depend on porous media saturations values, W_i.

The problem of one-dimensional filtration of the gas–liquid two-phase flow through the porous media is the most studied.

Within the two-phase flow, $W_1 + W_2 = 1$, that is, K_1 and K_2 are functions of the subsequent values of saturations W_1 and W_2 only. Introducing relative saturations $\bar{K}_1$ $(W_1) = K_1(W_1)/K_0$ and $\bar{K}_2$ $(W_2) = K_2(W_2)/K_0$, characteristic curves $\bar{K}_1$ (W_1) and $\bar{K}_2$ (W_2) can be obtained (Figure 3.5 [40]).

These dependencies account for the fact that a fraction of the liquid phase W_0 is not involved in the filtration process (micropores, the deadlocked pores, etc.). It is known from the soil filtration experiments that $W_0 = 0.2$.

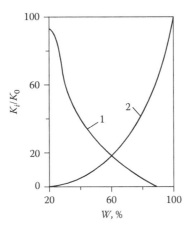

FIGURE 3.5 Dependency between relative phase permeabilities K_i/K_0 and phase saturations W [40]. 1, Gas; 2, liquid.

Assuming that values $\bar{K}_i$ represent simple functions of saturation W_i, and basing on experimental data on the gas–water two-phase system (W is the moisture content), the following empirical approximated correlations were obtained for the phase relative permeabilities of gas $\bar{K}_2$ (W) and liquid $\bar{K}_1$ ($1 - W$):

$$\bar{K}_1(W) = 0; \quad 0 \leq W \leq 0.2;$$

$$K_1(W) = \left(\frac{W - 0,2}{0,8}\right)^{3,5}; W \geq 0.2; \quad K_2(W) = \left(\frac{0,9 - W}{0,9}\right)^{3,5}; 0 \leq W \leq 0.8;$$

$$K_2(W) = 0; W > 0.8 \tag{3.2}$$

The capillary pressure depends on the saturation as

$$p_k(W) = J_0(W)\sigma\sqrt{\frac{\varepsilon}{K_0}}\Big/\cos\theta \tag{3.3}$$

where $J_0(W)$ is the Leverett function shown in Figure 3.6 for drying and wetting processes.

Equations (3.2) and (3.3), in conjunction with the filtration equations of each phase, phase generation, and boundary conditions, determine the governing regularities of the relative saturation W (moisture content in the porous structure) and $(1 - W)$ (vapor quality in the porous structure).

Charny [40] considered various underground hydro–gas–dynamic problems for air–water, gas–oil filtration, etc., based on the relative permeability model. The related approach was not concerned with evaporated flows and vaporization in the porous structures. Mayorov et al. [30, 41, 58–64] performed one of the first attempts to apply such an approach under the conditions of internal heat generation.

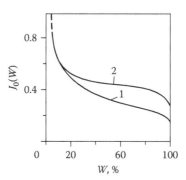

FIGURE 3.6 Leverett functions for the capillary pressure (Equation 3.3). 1, Wetting of the porous structure; 2, drying of the porous structure.

The homogeneous flow model and the relative phase permeability model of two-phase flow in the porous media were presented and discussed by Polyaev et al. [30] and Mayorov [41]. Results were presented as the dependency between the two-phase flow parameter, F, and the vapor quality, x. Comparison with experimental data was not presented, because such experiments were unknown to the authors.

Four values of exponents ($n = 1, 2, 3, 6$) in dependencies between the Newtonian viscosity of the two-phase mixture and the vapor quality were considered [30, 41]. The authors assumed that it allows accounting for the influence of liquid and gas concentrations on the corresponding relative permeabilities for each phase. In some cases, the values of F defined by the two above-mentioned models diverged from each other in 10 instances. On the other hand, under certain conditions the results obtained by the phase permeability model approximately coincided with those obtained via the homogeneous approach (Figure 3.7). However, the authors did not include an explanation for such an outcome.

Correlations used for the calculation of F are also different. In some cases, they represent a monotonous function; but sometimes there are essential extremums (maximums or minimums), etc. [41]. The approach suggested by Mayorov [41] requires the following concepts:

1. In the relative phase permeability model [40], the conventional correction for semiempirical dependency between the saturation and the liquid content has a power law form $\bar{K} = \bar{W}^n$ within the entire range of W values, that is, from 0 to 1. It significantly distorts the actual physical conditions when the definite range of low $\bar{W}$ values ensues in the absence of liquid flow. On the other hand, there is a specific range of low values of vapor concentrations, $(1 - \bar{W})$, at the deficiency of vapor flow. These deviations are essential for such specific vaporization phenomena in porous structures as formation of stable vapor cavities within the saturated area by liquid porous structure, the presence of residual liquid in capillaries at the steam flashing, etc.

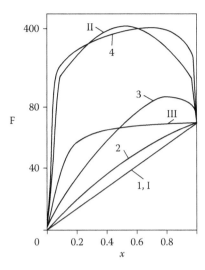

FIGURE 3.7 Dependency between the two-phase property parameter, F, and vapor quality, x. 1–4, calculations based on the relative permeability model at $n = 1, 2, 3, 6$, respectively; I, II, III, the same for the homogeneous model when the Newtonian viscosity of the two-phase mixture was calculated by different correlations.

2. Equations for describing the vapor quality dependence of thermal parameters (i.e., heat fluxes; temperatures, etc.) are missing from the models.
3. The authors assumed the x linear dependency of the coordinate that was in contrast with the physical essence of the process, even in the case of uniform volumetric internal heat generation, when the entire heat flux was transferred into the latent heat neglecting the sensible heat. In those cases when heat is transferred from the wall or from the evaporation surface, the assumed linear function of $x(z)$ is completely incorrect.
4. In the known problems of vaporization in porous structures, maintaining conditions of vapor–liquid mixture flow (filtration) are supposed to be given, whereas the distribution of W within the porous structure is not set. For example, in the case of heat carrier forced flow, the pressure gradient dp/dz is known; in the case of gravitational flow, the hydrostatic pressure on the external surface of the porous structure is also given too, and the capillary potential is accounted for. In addition, it relates both to the capillary feeding conditions and to heat carrier supply under the action of inertial forces. Therefore, neglecting the pressure drop between vapor and liquid phases is incorrect in many heat supply methods.

Thus, one-dimensional two-phase filtration model represents a set of fundamental hydro–gas–dynamic equations of filtration inside the porous medium and simplified equations of the internal heat transfer. The volumetric liquid and vapor flow rates j_1

and j_2 according to Darcy's law are related to the external forces (dp/dz) and volumetric forces $\rho_j Z$ as

$$j_1 = -\left(K_0 \frac{\bar{K}_1(W)}{\mu_1}\right)\left(\frac{dp_1}{dz} - \rho_1 Z\right); j_2 = -\left(K_0 \frac{\bar{K}_2(W)}{\mu_2}\right)\left(\frac{dp_2}{dz} - \rho_2 Z\right) \quad (3.4)$$

The pressure drop across the liquid–vapor interface is the capillary pressure, p_{CL}. When the interface is destroyed, Equation (3.4) accounts for the corresponding dynamic effects

$$p_1 - p_2 = \sigma\left(\frac{1}{R_1} + \frac{1}{R_2}\right) \equiv p_{CL}(W) \quad (3.5)$$

The one-dimensional energy equation of the evaporating flow yields

$$-rj\frac{dx}{dz} - j(1-x)c_{p'}\frac{d\vartheta_L}{dz} + \lambda_{SC}(1-\varepsilon)\frac{d^2\vartheta_{SC}}{dz^2} + q_V(z) = 0 \quad (3.6)$$

The heat transfer equation with internal heat generation term is

$$q_V(z) + \lambda_{SC}(1-\varepsilon)\frac{d^2\vartheta_{SC}}{dz^2} - S_F(z)\alpha'(\vartheta_{SC} - \vartheta_L) = 0 \quad (3.7)$$

where ϑ_{sc} and υ_L denote the local temperature of skeleton and liquid, respectively.

The equations set should be supplemented by the Equations (3.2) and, (3.3), and the corresponding boundary conditions.

The precise form of Equations (3.2) and (3.3) was obtained based on the experimental data on filtration of gas–water or oil–gas flows through the soils. Other types of porous structures (thin layers, high porosity metallic and ceramic porous structures, etc.) and heat carriers require subsequent experimental tests of the correctness of Equations (3.2) and (3.3). Equations (3.2)–(3.7) can probably be regarded as the first approximation method of the analysis of the above-mentioned porous structures.

When the internal heat generation is absent and heat is transferred from the wall to the wetted porous structure, the assumption that vaporization is restricted to very thin zone along the z axis (i.e., $(dx/dz) \approx 0$; $(d\vartheta_{sc}/dz) \approx 0$; j_1 and j_2 are constant) essentially simplifies the problem. In such a case, Equations (3.6) and (3.7) need not considered.

When thin layer porous structures (coverings, coatings, etc.) have such thickness H and pore size a values that the inequality $\rho'gH \ll \sigma/a$ is satisfied, then capillary pressure is dominant, and mass forces are neglected. It means that, according to the relative permeability model, conditions of the vapor–liquid steady-state filtration are defined as

$$jv_1 K_0\left(\frac{1-x}{K_1(w)} - \frac{x}{K_2(W)}\frac{v_2}{v_1}\right) = \frac{\sigma\sqrt{\varepsilon/K_0}}{\cos\Theta}\frac{dI_0(W)}{dW}\frac{dW}{dz} \quad (3.8)$$

According to Equation (3.8), since $[dI_0(W)/dW] < 0$, the sign in the left-hand defines the sign of the derivative (dW/dz) at fixed x value and given vapor and liquid viscosities

v_2 and v_1. When $(dW/dz) < 0$, liquid saturation W decreases with increasing z; $K_2(W)$ increases and $K_1(W)$ decreases, that is, (dW/dz) remains negative until W reaches its minimum value, W_{min}. It means that the entire internal volume is filled with vapor.

Similarly, condition $(dW/dz) > 0$ corresponds to the filling of the entire internal volume with liquid. It means that in case capillary forces prevail, a steady cocurrent flow of vapor and liquid inside the porous layer is unattainable. The cocurrent flow is evident only in the presence of external or mass forces, and when their effect exceeds the capillary action, a probability of such a flow increases. Thus, if capillary forces are dominant at vaporization inside porous structures, vapor and liquid flows are countercurrent or crosscurrent. The counter flow occurs in the porous structure when there is pool boiling from coated surfaces. The cross flow accompanied by vaporization inside the porous structure is representative of the heat pipe "boiling" regime. In addition, cross flow ensues from porous coatings applied in the vapor generation channels of high-power generators. It allows elimination of a heat transfer crisis of the second type.

Ovodkov et al. [42–45] conducted systematic investigations of the countercurrent two-phase flow inside porous structures driven by external pressures compared with the capillary pressure.

Proper understanding of experimental data [42–44] on the hydrodynamics of countercurrent liquid–gas flows requires consideration of a schematic design of the experimental setup presented by Kovalev and Soloviev [36] (Figure 3.8).

In the experiments conducted by Ovodkov [44], gas (nitrogen) flowed from left to right, and the gas pressure inside the porous structure was decreased in the same direction correspondingly. At specific conditions, the counter flow of liquid was

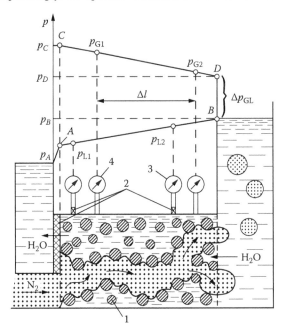

FIGURE 3.8 Schematic design of the experimental setup on the hydrodynamics of liquid and gas counter flows in the porous structure [36]. 1, Porous sample; 2, microporous screens; 3 and 4, gas and liquid internal pressures measuring devices.

observed in the porous structure zones filled by liquid, that is, from right to left, and the liquid pressure was reduced in the corresponding direction.

Typical experimental data on countercurrent filtration regimes of gas and liquid are shown in Figure 3.9 [36, 44].

Based on the analysis of these data, the following conclusions can be drawn:

1. Restrictions exist regarding the driven pressure drops in liquid and gas phases. Hence, exceeding of such restricted values is required to attain the directed gas or fluid flow. Such a value for gas was introduced by Kovalev and Soloviev [36] and Ovodkov [44] as the disruption pressure, ΔP_{CL}. It corresponds to the beginning stage of the gas minimal filtration, when pores of the maximum size were involved in the filtration process [17].
2. Curves 3 and 4 correspond to the liquid (water in Ref. [44]) single-phase filtration and validate Darcy's law for the given conditions.
3. At the two-phase countercurrent, the liquid phase filtration regularities remain linear (i.e., governed by Darcy's law). However, their inclination angles (curves 4–9, Figure 3.9) increase significantly as gas filtration flow rates and driven forces increase. This happens when both gas concentration in the porous volume rises and relative liquid permeability declines.
4. According to Figure 3.9, gas filtration regularities are essentially nonlinear.

This does not imply a violation of Darcy's law for gas filtration, because the increase in both driven pressure drop and gas flow rate is connected with the essential increase of the gas relative permeability. On the other hand, the influence of

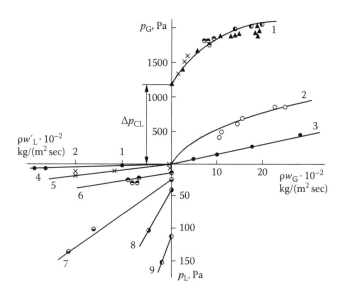

FIGURE 3.9 Liquid and gas counter flow inside the porous layer [36]. 1, Gas pressure drop due to gas flow rate growth; 2, gas pressure drop due to counter flows of the phases; 3, gas pressure drop in the dry sample; 4, gas pressure drop in the wetted sample; 5–9, gas pressure drop due to counter flows of the phases at different gas flow rates.

phase viscous interaction also appears possible. The correct understanding of the physical mechanism responsible for such an interaction is extremely important for the theoretical analysis of the thermohydrodynamic phenomena at vaporization inside porous structures. The phase interaction was quantitatively estimated via the friction pressure drop along the liquid–gas interface, ΔP_f.

Consider the mechanism of the viscosity phase interaction that is fitted to the condition $G_L = 0$ at $\Delta P_L \le \Delta P_{CL}$. Such a condition becomes evident as a result of the action of viscous forces causing the cyclical circulating flows [36]. However, Kovalev and Soloviev [36] did not present any physical explanation of the circulation mechanism. It is reasonable to assume that the gas flow inside the pores forms a cocurrent film liquid flow on the surface of the porous skeleton. The increase in gas mass flow rate intensifies the film flow, provides growth of the liquid flow rate in the film, and leads to an increase in the entire fluid flow rate (countercurrent gas flow).

The model explains both the nonlinear gas filtration regularity in the given conditions and the failure of the mechanism due to the increase in gas mass flow rate and the corresponding violations in the film flow continuity.

Apparently, Kovalev and Soloviev [36] did not analyze the probability of such a mechanism because their physical models of vaporization in porous structures assume the dry walls of vapor channels.

Consequently, the following two-phase filtration correlations were suggested [36]

$$G_G = K \frac{K_G}{\mu_G} \frac{\Delta p_G - \Delta p_{CL}}{\Delta l}; \quad G_L = K \frac{K_L}{\mu_L} \frac{\Delta p_L - \Delta p_{CL}}{\Delta l} \tag{3.9}$$

where Δl is the length of the filtration zone and K_L and K_G are the relative phase permeabilities.

$$K_L = \frac{\displaystyle\int_{R_{min}}^{R^*} R^2 f(R)dR}{\displaystyle\int_{R_{min}}^{R_{max}} R^2 f(R)dR}; \quad K_g = \frac{\displaystyle\int_{R^*}^{R_{MAX}} R^2 f(R)dR}{\displaystyle\int_{R_{MIN}}^{R_{MAX}} R^2 f(R)dR}$$

where R^* is related to the porous structure saturation $W(R)$ as

$$W(R) = \int_{R_{min}}^{R^*} f(R)dR$$

However, correlations of functions $f(R)$ providing reliable data for calculations of relative permeabilities $\bar{K}_L$ and $\bar{K}_G$ and ΔP_f remain unidentified. Unfortunately, pore size distribution calculations based both on the data of mercury porosimeter and statistical edge detection do not exclude the "closed pore" effect [44].

The approach suggested by Ovodkov [44] is completely an experimental one. Hence, data from this study are not applicable to other cases. When the testing sample represents porous coatings with thickness δ less than 1 mm, an application of the method leads to principal difficulties that are impossible to overcome at present.

However, besides the mentioned difficulties in the application of the present approach to hydrodynamic phenomena at the two-phase filtration in porous structures, it is important to note other important problems related to the two-phase filtration in porous structures [36].

When the condition $\Delta P_G \geq \Delta P_{CL}$ is realized for the gas phase, some vagueness is presented in correlations between parameters W, G_G, etc.

The problem becomes especially important when the vapor channel formation inside the porous structure is unrelated to the external mechanical action ($\Delta P_G \geq \Delta P_{CL}$), but is instead caused by vaporization. Here, some pores reach the nucleation limit, but it does not mean that they are occupied by the vapor, that is, the set of equations and boundary conditions is incomplete and requires additional correlations, for example, determining the porous structure saturation W [36].

This problem also exists for the phase permeability model. Thus, its solution appears to be important. Its significance becomes essential for the analysis of the thermal and hydrodynamic phenomena at vaporization inside porous structures, particularly when boiling occurs on coated surfaces. This process is essentially different from the conditions of its experimental modeling [36] for the following reasons:

1. Total countercurrent liquid and vapor mass flow rates in steady regimes are approximately equal to each other. The term "total" is mostly applicable to the liquid because the total liquid mass flow rate could be presented as the algebraic sum of the two flows. One of these flows, G_1, occurs inside the porous channels with the relative phase permeability $\bar{K}_L = K_1$, and the second one, G_2, flows in films covered in the vapor channel walls either countercurrent or cocurrent to the vapor flow, G''.

2. Existence of the vapor phase in the different vapor channels is restricted by the temperature conditions. According to such constraints, nucleation appears feasible for the different pore diameters only after overcoming the "nucleation limit."

3. Driven pressure drops limiting the achievable regimes of the mutual gas and liquid flows depend on the "capillary capacity" of the porous structure (pore sizes and their distribution). On the other hand, they are restricted by the thermal regimes and pore nucleation conditions, that is, the driven pressure drops at the vaporization on the coated surfaces depend on the parameters of the external actions. Therefore, such driven drops could be excessive at some sets of thermophysical, structural, and regime parameters. Consequently, it will not only compensate for energy losses for liquid supply to the heating surface (basement of the porous structure) and for vapor removal, but will also ensure the intensive liquid escaping.

   In the other cases, the driven pressure drops are insignificant, and then stable vapor layers ("vapor blankets") appear inside the porous structures near the wall (the so-called "interlayer crisis"), etc.

4. The internal heat transfer processes (evaporation and condensation) could compensate for each other under certain conditions, and the porous volume filling by the vapor could not be supplemented by external hydrodynamic phenomena.

3.3 FUNDAMENTAL PRINCIPLE OF "IRREVERSIBILITY MINIMUM" IN TWO-PHASE FILTRATION MODELING

Uncertainty in the choice for "saturation" W (moisture content) or "vapor quality" $(1 - W)$ (vapor content) leads to the incompleteness of two-phase filtration equations at vaporization inside porous structures. This uncertainty is essential in the following cases:

1. Porous structure contains a wide range of open pores ($R_{min} \ll R_{max}$).
2. Thermal and hydrodynamic filtration regimes are both of theoretical and practical interest. Vapor can partly penetrate the "open" pores, that is, $(1 - W_{max}) < (1 - W) < (1 - W_{min})$. The values W_{max} and W_{min} correspond to porous structure liquid flooding condition $G_1 = G_2 + G''$ at co-current flows of the vapor and liquid film or $G_1 + G_2 = G''$ at countercurrent flows.
3. Driven pressure drop that stimulates liquid supply to the wall and removal of vapor flow occurs because of vaporization inside those porous channels, where overheating exceeds the "nucleation" limit. Hence, the following steady-state condition is essential: the driven pressure drop Δp_0 exceeds the sum of hydraulic liquid and vapor resistances ($\Delta p'$ and $\Delta p''$): $\Delta p_0 > \Delta p' + \Delta p''$ (similarly to heat pipes). Cold model correlations between the driven pressure drops and the hydraulic resistances appear to be different.

Consider the rational approach to the two-phase hydrodynamic modeling when vaporization occurs in the internal volume of porous structure or when it is completely drying out. If either condition 1 or condition 2 is not valid, the problem of choice uncertainty for value W does not exist. Actually, if the porous structure consists of open pores of approximately equal sizes (the open pore distribution practically absent), then the system of "active" open pores will be defined as soon as a stable thermal regime is reached (the temperature drop exceeds the "nucleation temperature border"). The values W and $(1 - W)$ and the corresponding relative phase permeabilities $K_1(W)$ and $K_2(1 - W)$ can be determined independently based on the thermal regime's condition.

Naturally, there are such cases when the simultaneous steady vapor and liquid filtration through the porous structure is impossible (e.g., low permeability microporous structures). Subsequently, it makes no sense in the problem of W definition.

First, consider the possible approach to the problem, assuming both conditions 1 and 2 are met. One should take into account that the driven pressure drop at vaporization inside the porous structures is a capillary pressure (CP). Thus, the whole sum of hydraulic resistances in the vapor and liquid channels should not exceed CP. Once the growth of W encourages lowering of $\Delta p'$ and increasing of $\Delta p''$, it means that some optimal saturation value (moisture content or vapor quality), W_{opt}, corresponds to the minimum value of the sum of hydraulic resistances ($\Delta p' + \Delta p''$). From the physical point of view, it means that such a regime corresponds to the minimum level of dissipation losses, that is, the minimum irreversibility of the given steady irreversible filtration process.

Therefore, the condition of the total hydraulic losses minimum can be regarded as the particular case of the entropy generation minimization principle. This principle is known in irreversible thermodynamics as the condition of stable stationary states of nonequilibrium systems. In the considered conditions, the stationary two-phase filtration regimes at vaporization inside porous structures represent stable stationary states. First, this approach was suggested by Smirnov et al. [46–48] for modeling of embracing "new" vapor-generating open pores into the porous media occupied by the vapor phase depending on the wall overheat, ΔT, or the heat flux, q. Such an approach was developed and applied for two-phase filtration hydrodynamic models at vaporization inside porous structures [49].

It is necessary to grade the hydrodynamic phenomena that occur during vaporization on the surfaces covered by porous coatings at the submerged conditions (porous structure completely covered by liquid) or by "capillary feeding" (liquid supply is attributable to the action of capillary forces appearing in the sequence of vaporization inside porous structures) [49].

In the first case (Figures 3.2a,e and 3.4a,d,e), two-phase filtration inside the porous media is a result of the phase countercurrent flow of liquid and vapor. Liquid is suctioned to the wall due to the action of capillary forces and vapor flow is removed from the vapor-generating surface because of the pressure drop developed by the heat input.

In the second case (see Figure 3.2, "capillary feeding"), two-phase filtration inside the porous structure is characterized by a crosscurrent phase flow. The liquid appears to flow along the wall and the vapor is filtered in the direction normal to the wall.

The fundamental hydrodynamic equation of two-phase filtration at vaporization on submerged porous structure is

$$\frac{4\sigma}{D_E} - \frac{v''q\delta_0 k_1}{K_0 K_2 (1-W)r} - \frac{q^2}{[r(1-W)]^2} \times \frac{1}{\rho''} - \frac{v'q\delta_0 k_2}{K_0 K_1 (W)r} \geq 0\text{", besides it} \quad (3.10)$$

Here, δ_0 is the porous layer thickness; K_1, and K_2 are the coefficients accounting for the irregularities of vapor and liquid mass flow rates, for example, as a result of heat flux irregularity; D_E is the equivalent diameter defining the capillary potential of porous structure; $K_2(1-W)$ and $K_1(W)$ are the relative vapor and liquid phase permeabilities, respectively; K is the total porous structure permeability; and r is a latent heat. Inequality (3.10) represents a necessary condition to define the saturation, W. The sufficient criterion is delivered by minimizing total hydraulic losses at the two-phase filtration. In the scope of one-dimensional model of the phase permeability, this condition yields

$$\frac{v''k_1}{K_2^2(1-W)}\frac{\partial[K_2(1-W)]}{\partial W} + \frac{v'k_1}{K_1^2(W)}\frac{\partial[K_1(W)]}{\partial W} - 2K_0\left(\frac{q}{rp''}\right)^2 \frac{p''}{(1-W)^3 \delta_0} = 0 \quad (3.11)$$

The set of formulas comprising Equations (3.2), (3.10), and (3.11) defines the two-phase structure inside the porous media at vaporization on the submerged porous heat transfer surface.

The same system determines the "internal layer crisis" when the stable vapor film ("vapor blanket") occurs near the wall inside the porous structure. This is what happens when, at some value of input heat flux q for given conditions, a spatial consideration of Equations (3.10) and (3.11) is unable to satisfy inequality (3.10).

Subsequently, a spatial analysis of the above-mentioned equations allows us to find such a value of thickness of the porous structure wetted part, δ^*, that renders inequality (3.10) valid.

Naturally, this leads to the case $\delta^* < \delta_0$, and the difference $(\delta_0 - \delta^*)$ allows the estimation of the near-wall vapor layer thickness.

A set of hydrodynamic one-dimensional equations of the steady two-phase filtration at the capillary feeding conditions can be obtained in the same manner. In such a case, similar to inequality (3.10), we have

$$\frac{4\sigma}{D_E} - \frac{v''qk_1\delta_0}{K_0K_2(1-W)r} - \left(\frac{q}{r(1-W)}\right)^2 \times \frac{1}{\rho''} - \frac{v'qk_1(L_0)^2}{K_0K_1(W)r\delta_0} - \Delta p_0 \geq 0 \quad (3.12)$$

Here, L_0 is the liquid transport length (e.g., for heat pipe, $L_0 = (L_{EV} + L_C)\,0.5 + L_{TR}$, where L_{EV} and L_C are the heat input and output lengths, respectively, and L_{TR} is the transport zone length).

$$\Delta p_0 = \Delta p' + \Delta p'' - \rho' g h_0 \sin\beta$$

Here, $\Delta p_{TR}'$ accounts for the liquid hydraulic resistances inside the porous structure at transport and condensation zones, $\Delta p''$ is the vapor flow hydraulic resistance inside the vapor channel, and β is the angle accounting for the relative location of heat input and output zones in the gravity field (when $\beta > 0$, heat input zone is located lower the than heat output zone, and vice versa).

The correlations for Δp_0 could be essentially different depending on the heat pipe position and condition of heat carrier filling (in case of the presence of heat carrier overload). However, for submerged structures and capillary feeding conditions, inequality (3.12) should be supplemented with the following equation, defining the value of porous structure saturation, W,

$$\frac{v''}{K_2^2(1-W)}\frac{\partial[K_2(1-W)]}{\partial W}k_1\delta_0 + \frac{v'k_1}{K_1^2(W)}\frac{\partial[K_1(W)]}{\partial W}\frac{L_0^2}{\delta_0} - 2K_0\left(\frac{q}{r}\right)^2\frac{1}{\rho''(1-W)^3} = 0$$

$$(3.13)$$

Equations (3.12) and (3.13) combined with Equation (3.2) and known formulas for the calculation of Δp_0 at various specific thermohydrodynamic regimes define the hydrodynamics of two-phase filtration at vaporization on the porous coated surfaces for the capillary feeding conditions.

Analysis of Equation (3.13) proves the existence of δ_0 corresponding to the minimum of hydraulic losses when boiling occurs at the heat input section. The following equation gives the value of $(\delta_0)_{ext}$

$$(\delta_0)_{ext} = L_0\sqrt{\frac{v'}{v''}\frac{K_2(1-W)}{K_1(W)}\frac{k_1}{k_2}} \quad (3.14)$$

Equation (3.14) is obtained under the condition of neglecting the last terms in Equations (3.11) and (3.13).

When conducting spatial consideration of Equations (3.12) and (3.14), we obtain $\delta_{ext} < \delta_0$, and inequality (3.12) is subject to change:

$$\frac{4\sigma}{D_E} - \frac{v''qk_2\delta_{ext}}{rK_0K_2(1-W)} - \frac{v''qk_2(\delta_0 - \delta_{ext})}{rK_0} - \left(\frac{q}{r(1-W)}\right)^2 \frac{1}{\rho''} - \frac{v'qk_1L_0^2}{K_0K_1(W)r\delta_{ext}} - \Delta p_0 \geq 0 \tag{3.15}$$

Subsequently, Equation (3.14) transforms to

$$(\delta_0)_{ext} = L_0\sqrt{\frac{v'}{v''}\frac{k_1}{k_2}\frac{K_2(1-W)}{K_1(W)}}\frac{1}{\sqrt{1-K_2(1-W)}} \tag{3.16}$$

The transformation of Equation (3.14) to the form (3.16) is attributable to the appearance of the second and third terms in formula (3.15) instead of the second term in inequality (3.12). Such a change is connected with the physical explanations relating to the case when $\delta \cong \delta_{ext}$.

It is known that intensive liquid throwing flow occurs at specific regimes of the boiling stage on surfaces covered by thin porous structures in the capillary feeding conditions. Such a phenomenon leads to the significant decrease of liquid filling in the porous structure ($\delta_{ext} < \delta_0$). Then, vapor filtration occurs when the filtration over two-layered porous structure consists of one layer partially filled with liquid (its thickness is equal to δ_{ext}) and the other one is a dry layer with thickness ($\delta_0 - \delta_{ext}$).

Such a model allows us to account for the probability of partial drying of porous structures as a result of liquid throwing without contradictions with initial assumptions.

Kovalev and Soloviev [36] applied the current approach in the analysis of two-phase filtration hydrodynamics at vaporization inside porous structures, particularly for modeling of the heat pipe evaporator based on the inverted meniscus principle. They [50, 51] suggested calculating the interphase position inside the porous coating by vapor filtration hydraulic resistance minimization principle. However, the authors actually applied the condition of vapor flow cross section minimization, that is, $J_Y = \int y(x)dx = min$ (where $y(x)$ is the local thickness of the vapor zone and x is the coordinate determining the vapor filtration direction). The stipulation $J_Y = min$ does not match the minimization of hydraulic losses because it corresponds to decrease in the cross section, that is, augmentation of hydraulic losses.

Moreover, minimization of J_Y at fixed temperature drop indicates the maximization of the total heat flux value, Q, because, according to the model of Kovalev and Soloviev [36, 50, 51], $Q \approx (A + B/y)$, where A and B are empirical coefficients. The problem of the total heat flux maximization at fixed temperature drop can be treated as equivalent to the problem of temperature drop minimization at a given Q. From the physical viewpoint, it corresponds to the aforementioned approach based on the minimum irreversibility principle for the stable steady states.

Furthermore, the inverted meniscus model [36] uses some doubtful assumptions based on the physical explanations of vaporization inside the porous structure (e.g., model of heat transfer from the evaporating meniscus, assumption on the wall isothermality, $T = const$, etc.). A thorough discussion of the mentioned model is presented in Chapter 5.

Let us analyze the most important results of the research [52] devoted to the two-phase filtration and heat and mass transfer in the evaporators based on the inverted meniscus principle. In this study, the two-dimensional interconnected problems of the two-phase filtration and heat and mass transfer with respect to the representative cells of such evaporators were handled appropriately. Figure 3.10 presents the typical schematic layout of the evaporators performing the principal design. It is based on the bidisperse structure, consisting of the high-disperse (HD) and the micro-disperse (MD) media. In addition, the representative cell in the conjugation zones HD and MD can be selected. The geometry of such a cell corresponds to the conventional design of the inverted meniscus (Figure 3.10a and b), but it has notably smaller sizes.

Figures 3.11–3.13 [52] present typical theoretical results obtained by adding the corresponding boundary conditions to the combined numerical solution of the set of the filtration and convective heat transfer two-dimensional differential equations for the vapor and liquid zones, respectively. The results demonstrate powerful thermal and hydrodynamic irregularities accompanying joint two-phase filtration and convective heat and mass transfer processes. Such a phenomenon has determinative influence on the regularities of the overall process.

A simplified approach assumes the appearance of the two dominant additional resistances at the contact zones between HD and MD structures, that is, hydraulic resistance in the vapor filtration and thermal surface resistance.

Yatscenko [52] demonstrated good agreement between his numerical model and the available experimental data.

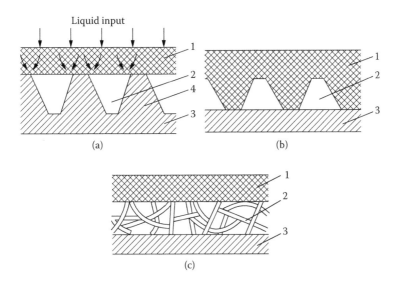

FIGURE 3.10 Typical schematic layout of the evaporators based on the inverted meniscus principle [52]. (a and b) Evaporator design (1, capillary structure; 2, vapor removal channels; 3, heating wall; 4, heat removal element); (c) principle of the inverted meniscus [1, microdisperse layer (the sintered metallic powder); 2, high-disperse layer (the metallic felt); 3, heating wall].

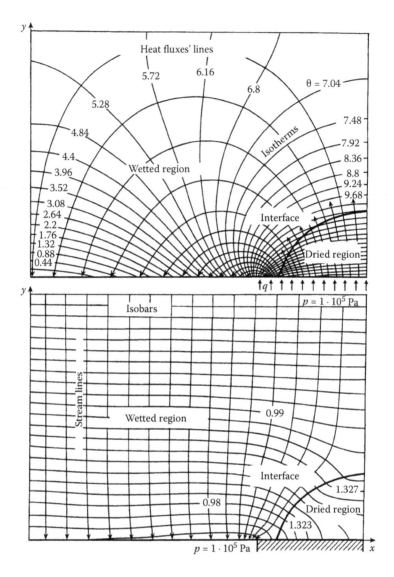

FIGURE 3.11 Results of numerical analysis for heat and mass transfer inside microdispersed layer (MDL) at appearance of vapor bubbles [52]. $\Theta = (T - T_{SO})/(T_0 - T_{SO})$; T, local temperature; $T_0 - T_{SO} = 1$ K.

Consider the physical nature of the numerical model [52] (Figures 3.11–3.13) of vaporization inside the wetted porous structure when heat is transferred to this part of such a structure.

Each figure consists of two parts. The upper part shows numerical data on temperature fields and heat flux distribution within the representative elementary cell of the porous structure. The Y axis corresponds to the normal direction of the evaporation surface coordinate (depth of the porous structure), whereas X axis coincides with the evaporation surface.

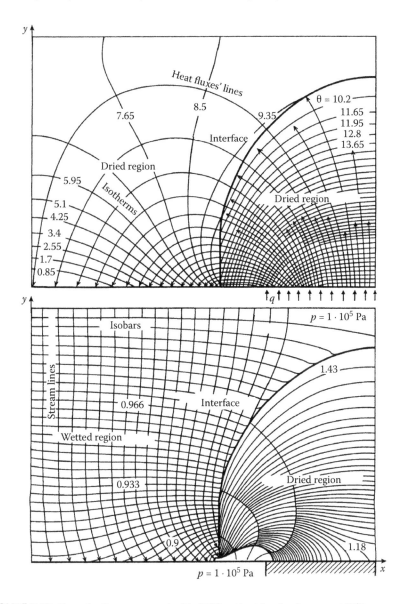

FIGURE 3.12 Growth of vapor zones inside MDL at heat load enlargement [52].

The unshaded part in the lower portion of the figure determines the contact-free wetted porous structure zone. Vaporization accompanied by exhaustive heat removal occurs mainly from this surface. The pressure at the external surface is $P = 0.1$ MPa. As shown, even at moderate heat fluxes, part of the wetted porous structure contact surface appears to be dried. The dried regions of the porous structure are shown in all figures located between the internal interphase boundary and the heating wall. It can be seen that the primary heat transfer occurs near the wall's left edge (see lower parts of the figures). There is also a certain interphase deepening.

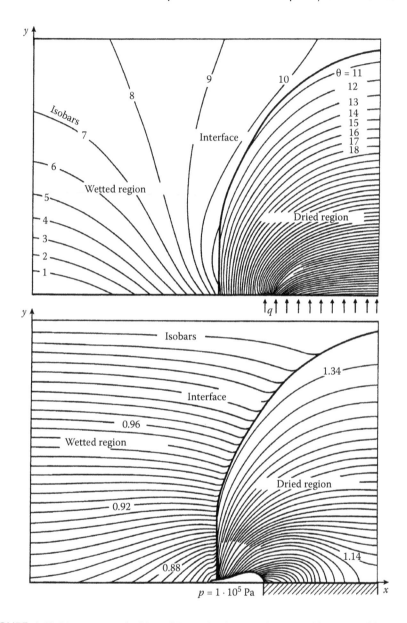

FIGURE 3.13 Vapor zones inside MDL at the heat load approaching the critical value (burnout) [52].

At low heat loads, the heat is transferred from the wall mostly through the wetted porous structure zone attached to the left edge of the wall. The rest of the heat is moved throughout the dried skeleton of the porous structure to the internal interphase and then through the wetted porous structure to the external evaporation surface.

The dried region increases with the growth of heat fluxes and finally it separates completely the evaporation surface from the heating surface. The heat is transferring through the dried element to the evaporation surface inside the porous structure. Then, the ensuing vapor is filtering through the dried skeleton to the external surface of the porous structure.

The main heat and mass losses take place in the zones adjoining the left edge of the heating surface. The increase in heat flux density is accompanied by the intensification of thermal and hydraulic irregularities and an increase in dried region volume. Finally, it leads to the complete filling of the porous structure by vapor and to the development of crisis regimes.

The basic calculations [52] were completed for copper micro- and bidisperse structures with water as the heat carrier. Results revealed that skeleton heat conductivity has a major significance at vaporization inside the porous structure.

It should be noted that Yatscenko [52] comprehensively analyzed a supplementary condition required for the reliable fitting of interphase position, and regarded that these correlations defined the phase equilibrium with respect to the corresponding pressure drop associated with the capillary pressure.

Without a thorough analysis of the specific boundary condition set and the numerical integration procedure [52], it should be noted that a correct physical interpretation of the hydrodynamic and thermal processes and good agreement with experimental results were obtained mostly in regard to the successful selection of values of the initial state parameters.

The combined development of the thermal and hydraulic processes at internal (volume) heat generation or the external heat input to wetted porous structures saturated by liquid (Figure 3.2a–d) has a number of specific peculiarities.

3.4 INVESTIGATIONS OF VAPORIZATION IN POROUS STRUCTURES WITH INTERNAL HEAT GENERATION

The analytical and experimental studies on the vaporization process inside porous structures were presented in several reports [41, 58–66] and generalized by Polayev et al. [30].

The stability of the two-phase flow through the porous structure represents an essential factor for the effective arrangement of the porous evaporation cooling; however, studies devoted to this problem are rather limited. Professor Vasiliev and his scientific team from the Luikov Heat and Mass Transfer Institute studied such a phenomenon for many years, and generalized their major results in their report [30].

The physical explanations of Mayorov et al. [58–66] were considerably transformed for the duration of the investigations. Sometimes, these changes were principal in character. The proper treatment of the analytical approach to the two-phase vapor flow inside the porous structures requires consideration of the above-mentioned results according to the sequence of their development [58–61].

The external heat flux q affects the uniform porous plate with thickness δ (Figure 3.14). Liquid with the initial temperature T_0 is used for the plate's cooling. During liquid filtration, its enthalpy increases due to the heat input and the

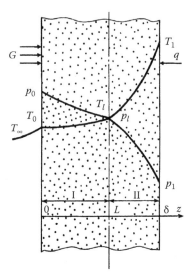

FIGURE 3.14 Physical model of two-phase porous cooling [58–61]. I, liquid flow; II, vapor flow.

temperature increase. Simultaneously, the saturation temperature decreases because of the filtration pressure drop. It was assumed that heat transfer intensity between the skeleton surface and the heat carrier is so high that the skeleton and the heat carrier are in equilibrium. For specific conditions at some distance from the inlet, the heat carrier temperature becomes higher than the saturation temperature. It initiates boiling of the heat carrier.

The initial model [58, 61] supposed that evaporation occurs in the thin region with very small thickness, located at distance L from the inlet. The pressure drop between phases due to the action of capillary forces and the phase interaction phenomena was neglected. It was also assumed that vapor overheating occurs from the interphase to the outlet. Vapor and liquid physical properties were supposed to be constant, corresponding to pressure from the surroundings. In addition, thermal equilibrium between the skeleton and the heat carrier is assumed for the overheating vapor region.

Such a model of two-phase porous cooling comprises a set of differential equations for the steady regime including the mass and momentum conservation equations for each ith zone (liquid and vapor).

Calculation results [58, 59] for the stabilized region are presented in Figure 3.15.

Because of the lack of experimental verification for the presented calculation results, the assumptions and conclusions arising from the model raise many doubts:

- The most suspicious hypothesis is an infinitesimal thickness of the evaporation zone in comparison with the whole structure thickness, that is, the size of the liquid and vapor heating zone. Taking into consideration that the latent heat is noticeably larger than the overheating values of liquid and vapor at real temperature drops and pressures, such an assumption appears quite doubtful.

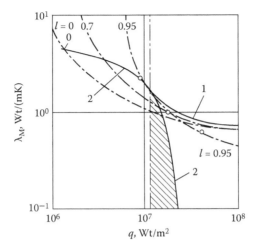

FIGURE 3.15 Numerical results of stability and safety operation for the two-phase porous cooling system ($p = 10$ bar and $Re = 0.1$) [58, 59]. 1, Stability margin; 2, safety margins.

- When the overheating vapor region embraces a significant part of the porous structure, then at specific values of heat flux the temperature drop between the skeleton and the heat carrier appears to be essential, and the assumption on their approximate equality is not valid. In such a case, the heat flux transferred due to the porous structure's thermal conductivity could be very significant.

Some of the above-mentioned assumptions were later withdrawn by Mayorov et al. [62–65]. The physical model accounting for the internal heat generation was developed (Figure 3.16), and the assumption neglecting the evaporation zone sizes was retracted [62]. The authors applied a conventional approach limited to the calculations based on equations of hydrodynamic and heat balances.

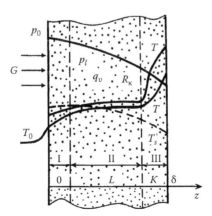

FIGURE 3.16 Physical model of the heat generating element (HGE) in two-phase porous cooling [62]. I, Liquid flow; II, evaporation region; III, vapor flow.

Several typical calculation results for dependency of the regions' margins on the surrounding pressure at internal heat generation (Figure 3.16) prove that the region of the phase transition is a dominant zone.

Further development of the same model was presented by Mayorov [63]. Figure 3.17 demonstrates the essence of the model when heat is applied from the side of the evaporation surface. As a supplement to the model [62], in the first approximation, the modified approach [63] accounts for several principal peculiarities of heat transfer on the evaporation zone. It was assumed that the evaporation zone consists of two parts. In the part where the skeleton surface has a secure contact with the liquid, mainly in the microfilm form, the evaporative heat transfer inside the pores is distinguished by an extremely high intensity, and this maintains until the skeleton temperature reaches some limit value corresponding to the metastable state of the substance.

In the part where the skeleton temperature exceeds the metastable state temperature, the contact between the liquid and skeleton appears to be impossible. Thus, the intensity of the heat transfer of the heat carrier with the skeleton decreases quickly.

In contrast to the model [58–60], Mayorov [63] suggested accounting for the porous structure pressure drop by using the phase permeability model [40]. The energy conservation equations were considered only within both parts of the evaporation region [63]:

$$\lambda \frac{d^2 T_1}{dz^2} = \alpha_1 (T_1 - T_s(z)) = G \frac{di_2}{dz}, L < z < z^*$$

$$\lambda \frac{d^2 T_2}{dz^2} = \alpha_2 (T_2 - T_s(z)) = G \frac{di_2}{dz}, z < z < K$$

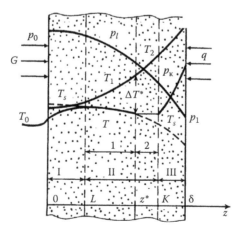

FIGURE 3.17 Physical model of the two-phase cooling [63]. I, Liquid flow; II, evaporation zone; III, vapor flow. 1, Contact zone between skeleton and liquid; 2, zone of the worse heat transfer.

at the following boundary conditions:

$$z = L: \quad T_1 - T_0 = \Delta T_0, \quad \lambda_1 \frac{dT_1}{dz} = G(i' - c'_p t_0);$$

$$z = z^*: \quad T_1 - T_0 = \Delta T, \quad \lambda_1 \frac{dT_1}{dz} = \lambda_m \frac{dT_2}{dz};$$

$$z = L: \quad \lambda_m \frac{dT_2}{dz} = G(i'' - c''_p t_0)$$

Mayorov et al. [41, 58–65] considered the evaporation cooling conditions when the thermal conductivity of the skeleton sufficiently exceeds its value for liquid. Therefore, the axial heat transfer through the liquid can be neglected at the actual porosity values. Calculations [63] were limited to the specific conditions: λ_m = 10 W/mK, δ = 10 mm, $T_0 - T_K = 2$ K, and $P = 0.1$ MPa. Calculation results corresponding to the above-mentioned conditions show that:

- The temperature drop between the skeleton and the heat carrier in the high-intensity heat transfer evaporation zone increases up to 200 K;
- The fraction of the high-intensity heat transfer zone for numerous thermal regimes is relatively small (lower than 0.1) [63].

What was the influence of such essential changes in the physical model of evaporative cooling of the porous structure on the process stability? Mayorov and Vasiliev [62–65] did not provide an answer to this question. Moreover, the results were not compared to experimental data, and estimations of heat transfer coefficients have not been included. These concerns were addressed later by Mayorov et al. [66], whose calculations were based on the analytical approach [41, 63].

Consider the experimental data shown in Figure 3.18 [66]. It allows for easy analysis based on the consideration of the elementary cell via the heat balance equation at the cooling parameters, specified in Figure 3.18. The heat flux required for the heat carrier evaporation in regime #5 (curve 5, Figure 3.18a) presents approximately up to 80% of the total heat input, and it is about 65% in regime #6. The temperature curves illustrates strong and continuous growth of the wall temperature up to 250–300°C. This could probably mean that the heat transfer of the skeleton with the evaporating heat carrier is decreasing, the contact between the skeleton and the liquid phase is failing, and, in spite of the extremely high growth of the vapor flow velocity, the temperature drop between the wall and the evaporating flow is increasing.

The same approach seems to be correct for the external heat input to the evaporation surface by radiation (Figure 3.18b). At the same time, the total heat flux required for the evaporation of the liquid at mass flow $G = 0.191$ kg/m² s is equal to 4.5–5.0 × 10^5 W/m², that is, from 80% to 90% of the maximum value of the total experimental heat flux. Therefore, the evaporation region embraces a significant fraction, whereas Mayorov [63] declared that the extension of such a zone is quite small. Mayorov and Vasiliev [62–65] assumed that the local heat transfer

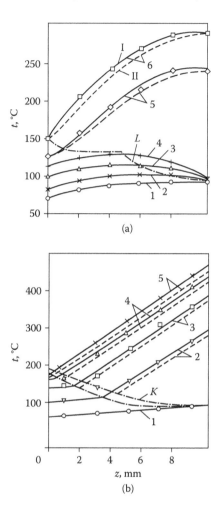

(a)

(b)

FIGURE 3.18 Thermal regimes of the porous vapor generated surfaces [66]. (a) Temperature variations of the porous HGE ($\varepsilon = 0.237$ porosity, sintered stainless steel powder with grain size in the range 63–100 μm); heat carrier flow rate, $G = 0.158$ kg/m^2 s; volumetric heat generation rate for curves from 1 to 6: q_V 10^{-7} = 1, 1.16, 2.24, 3.67, 3.95, 5 (W/m^3), respectively. (b) Temperature variations of the porous wall ($\varepsilon = 0.221$; sintered stainless steel powder with the grain size in the range 63–100 μm); $G = 0.191$ kg/m^2 s; values of the external radiation heat flux for curves 1 to 5: q 10^{-5} = 2.8, 5.63, 5.96, 6.12, 6.23 (W/m^2), respectively. I, Approximation of the experimental data; II, numerical data; L and K, conventional phase transition margins.

coefficients in the zones of intensive evaporation are defined by the liquid micro-film thickness, which could be determined in the simplest case by assuming the uniform distribution of a liquid over the skeleton. For the specified values of moisture content ($1 - W$) and grain size of the porous structure, d_0, a simple estimation of the microfilm thickness appears as $\delta = 0.5d_0[1 - \sqrt{1 - W}]$. Such a correlation delivers inaccurate values especially at the large W values, when the liquid stick-ing in micropores, dead ends, etc., should be accounted for. The strong influence of the interrelation between δ and W or X on the model accuracy could be caused by the existence of the liquid microdrops in vapor streams and jets. It is probable that these circumstances have a vital significance for the consequent inaccuracies within the numerical data obtained under the experimental conditions.

Khvostov and Marinichenko [72] attempted to analyze the stability of evaporation from the wetted porous structure throughout the two-phase flow under the action of capillary forces. Their model [72] uses the same initial assumptions as those used by Mayorov et al. [58–62]:

- Extension of the evaporation zone is infinitely small.
- Infinitely high intensity of heat transfer both within the liquid flow and the vapor zones.
- One-dimensional flow with constant thermophysical properties, etc.

In addition, it was assumed that two-phase flow occurs due to the action of capil-lary forces. Hence, external and gravity forces were neglected. Equalizing capillary pressure ΔP_k with friction pressure losses yields the initial integral equation

$$\Delta P_k = \frac{h}{K_0}(v'\bar{l} + v''(1-\bar{l})) \tag{3.17}$$

where h is the thickness of the porous structure and K_0 is the permeability coefficient of the dry porous structure.

By using the one-dimensional energy equation for two zones (liquid and vapor), Khvostov and Marinichenko [72] obtained a correlation between the heat flux q transferred to the external evaporation surface of the porous structure, dimensionless coordinate l of the evaporation zone, and the flow rate of the heat carrier, G

$$q = G(i'' - i_0)\exp\left(\frac{G(1-\bar{l})c_p''h}{\lambda_2}\right) \tag{3.18}$$

Here, λ_2 probably represents the effective thermal conductivity λ_E of a vapor-filled porous structure.

The combined analysis of the mentioned correlations yields the following inter-relation between the capillary pressure in the porous structure and the heat carrier flow rate G

$$\Delta p_k = G\frac{v'h}{K_0} + \frac{v'\lambda_2}{c_p''K_0}\left(1-\frac{v'}{v''}\right)In\frac{q}{G(i''-i_0)}$$

Supplementing this equation with the stability conditions [58–62],

$$\Delta p_k = \Delta p_{DR}; \quad \left(\frac{\partial\Delta p_k}{\partial G}\right)_q > \left(\frac{\partial\Delta p_{\Pi B}}{\partial G}\right) \tag{3.19}$$

(where Δp_{DR} is the driven pressure, i.e., a capillary pressure in the given case), allows a reasonable conclusion that in case of constant capillary pressure ($\Delta p_k = const$), that is, it does not depend on G, the stability condition (3.19) is not satisfied. Furthermore, the authors assumed the existence of the boundary layer in the plate depth from the side of the heating surface. This is in contradiction to the important initial assumption of the model (i.e., thickness of the evaporation zone is equal to zero).

Thus, even within the scope of the approximate model, this approach [72] is not proved. The authors assumed that, for specific capillary potential distribution and some combination of the parameters, a stability condition appears valid and has the following form

$$K = \frac{K_0\left(\dfrac{v'' - v'}{v'} - \dfrac{hGC''}{\lambda_2}\right)}{\dfrac{\Delta p_{0MAX}}{\sigma n}\left(1 + Ln\dfrac{q}{G(i'' - i_0)}\right)} < 1$$

As a result, application of the above-mentioned correlation as the porous structure vaporization stability criterion is not evident, particularly because a comparison with experimental data is missing. Experimental studies on stability restrictions are unidentified at the porous evaporative cooling for the liquid capillary feeding and the external heat input from the side of the evaporation surface.

Therefore, it can be concluded that regularities of thermal regimes in the evaporating heat carrier flow through the porous structures, including different phenomena of the thermohydrodynamic instability, require further (analytical and experimental) study.

Two publications by Nakoryakov et al. [67, 68], as well as their other studies on condensation and vaporization on surfaces submerged in the particle bed, can be considered as an example of such experimental and theoretical investigations.

Nakoryakov et al. [67, 68] solved the analytical problem of film boiling on the surface submerged in the particle bed under the assumption that the film flow appears analogous to the flow within the laminar boundary layer

$$Nu_s = \sqrt{2Ar_s\, \text{Pr}_s\, Ku_s\left(1 + \frac{2}{\pi}Ku_s\right)}$$

where $Ku_s = c_p'\Delta T_s/r$ is the Kutateladze number, $Ar_s = g \sin \varphi K_0 L((\rho'/\rho'' - 1)/v^2)$ is the Archimedes number, L is the characteristic size of the heating surface (the plate's length in the direction of the vapor film flow), K_0 is the permeability of the particle bed, and the subscript "s" denotes the saturation state.

The other variant of the analytical solution appears to be more accurate. It yields the following form of the heat transfer regularities at the film boiling on the submerged heating surface [67]

$$Nu_s = \sqrt{2Pe_s Ku_s\left(1 + \frac{0.5}{Ku_s}\right)}$$

where the characteristic velocity w in Peclet number is determined by the formula for hydraulic losses at the single-phase flow filtration within the porous structure

$$\frac{\Delta p}{L} = 180 \times (1-\varepsilon)^2 \times \frac{\mu'' \times w}{\varepsilon^3 \times d_0^2} + 2.2(1-\varepsilon) \times \frac{\rho w^2}{\varepsilon^3 \times d_0}$$

The numerical and the experimental data [67] were qualitatively and quantitatively correlated in a satisfactory manner.

Unfortunately, Nakoryakov et al. [67] considered only such experimental conditions that are completely different from the vaporization cases shown in Figures 3.2–3.4. They [67, 68] do not discuss the details of the formation of the liquid–vapor structures and the grain beds in a near-wall zone, including conditions of the grain bed removal from the heating wall, the abrupt changing of the specific effective parameters of the bed (viscosity, thermal conductivity, etc.), etc. Customization in the grain sizes (about 1 mm) in the experiment corresponds to the relatively large pore diameters. The above-mentioned and other peculiarities of the studies [67, 68] do not offer a successful allocation of the results and physical explanations for further models of the two-phase filtration at vaporization inside the porous structures.

Therefore, the systematic and focused experiments on filtration and heat transfer mechanisms at vaporization inside the porous structures are essential for the proper understanding of the physical nature of the thermohydrodynamic phenomena.

Only a few such experimental studies are known. Hence, their most important results will be considered in detail.

3.5 EXPERIMENTAL INVESTIGATIONS OF INTERNAL CHARACTERISTICS AND MECHANISMS OF THERMOHYDRODYNAMIC PHENOMENA

The experimental results presented by Bau and Torrance [69, 70] showed the existence of some specific pulsation thermal regimes at vaporization in porous structures. These regimes were determined as the consequence of thermal oscillatory instability phenomena.

Bau and Torrance [69] likewise proposed a thermohydrodynamic model for the determination of the critical heat flux, which is also called the maximum heat flux, q_{max}. The model was compared with the data of other authors.

Consider the conditions of the experiments [69, 70] and their results.

The experiments were performed with a special test setup (Figure 3.19). Bottom heating was realized in a cylindrical vessel. The main part of the vessel volume situated between the bottom and the movable, perforated plug lid was filled by the porous structure consisting of glass spheres of different diameters (the average sphere diameter was changed in the experiments from 0.1 to 1.2 mm).

The permeability of the porous structure and the effective thermal conductivity were defined in the special experiments at the reliable exclusion of the natural convection effect. The permeability ranged from 18×10^{-10} to 1400×10^{-10} m^2 and the effective thermal conductivity ranged from 0.92 to 2.6 W/mK. The porous structure was completely filled by the degassed, distilled water in such a manner

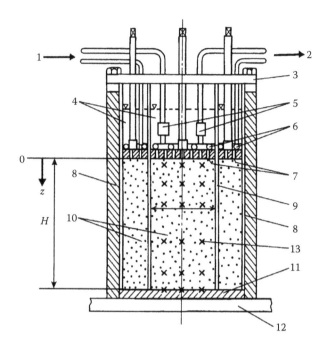

FIGURE 3.19 Schematic design of the experimental setup [69, 70]. 1 and 2, Inlet and outlet of the cooling water; 3, plugging lid; 4, water layer over the porous structure; 5, cooling water mixing chamber; 6, cooling serpentine tube; 7, perforated plate; 8, thermal isolation; 9, glass tube; 10, porous structure filled by the water; 11, vessel bottom; 12, isothermal hot plate; 13, thermocouples.

that the upper level of water exceeded the upper movable plugging lid by several centimeters. This allowed compensating for liquid volume expansion and heat removal to the water cooling serpentine tube.

Temperature measurements were made within the cross section and the height of the structure. These measurements allowed the operators to obtain the heat fluxes transferred through the submerged porous structure and to record both the beginning of the phase transition and its margins, that is, two-phase zone restrictions. The typical experimental data [70] demonstrated the existence of the following heat and mass transfer regimes in the submerged porous structure, when it was attached to the wall and heat transfer ensued from the wall to the external liquid layer through the porous structure (Figure 3.20):

- At low heat fluxes (about 0.5 kW/m²), the heat transfers through the submerged porous structure by the heat conduction.
- At the augmented external heat flux values, natural convection occurs in the porous structure. Its intensity depends on both the internal parameters of the porous structure (porosity, permeability, pore sizes, etc.) and the external parameters (layer width, height, contact conditions, etc.).

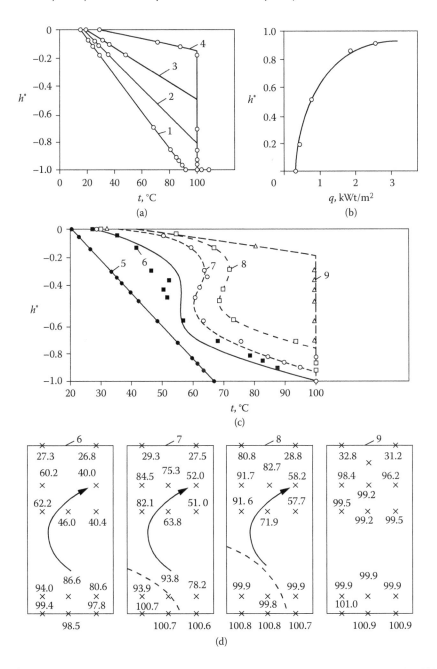

FIGURE 3.20 Experimental data on vaporization at submerged disperse layer [69, 70]. (a) Temperature distribution at the porous layer [permeability, $K = 8.5 \times 10^{-12}$ m^2; particle size, $d = 0.11$ mm; $h^* = z/H$; H, the layer height (thickness) [69, 70]; z, local coordinate (Figure 3.19)]. 1–4: $q = 0.40$, 0.53, 0.97, 2.37 kW/m^2, respectively. (b) Dependence of the water–vapor mixture relative height on heat flux. (c, d) Temperature distributions in submerged porous layer at varying heat flux; 5–9: $q = 0.5$, 1.8, 5.5, 6.5, 7.0 kW/m^2, respectively.

- The regime of the formation of the two-phase heat transfer zone was observed in the experiments [70] at heat flux = 1.8 kW/m². With the growth of the external heat flux, the height of the two-phase heat transfer zone increases and the size of the natural convection zone decreases at the same time. Two-phase zone height becomes stable at a certain heat flux value and heat transfers in the single-phase zone by heat conduction. Such a regime appears at $q = 7$ kW/m² [70]. In low-permeability porous structures [69], the two-phase boundary layer formation corresponds to a significantly narrow range of heat fluxes.
- With further growth of q, the stable two-phase thermal regime develops at extremely low temperature drops in the stabilized two-phase zone. Hence, it was impossible to register such temperature drops in the experiments. Temperature fields in the two-phase zone cross sections become uniform. It means that the natural convection was depressed. This regime was quite stable and reproducible until the heat flux reached $q < 11$ kW/m² [70].

At $q \geq 11$ kW/m², the regime of stable low-frequency thermal oscillations of local temperature and heat flux was developed in the two-phase heat transfer zone. The amplitude of temperature pulsations in the core of the two-phase zone was in the range of 2–5°C. The pulsation period increases with the heat flux growth up to 120 seconds (Figures 3.21 and 3.22).

Bau and Torrance [69] observed the same thermal regimes for low-permeability structures. However, they were characterized by the lower frequencies (pulsation period was about 10–20 minutes). Moreover, the regimes with varying pulsation period were examined. Such temperature regimes were qualified as emergence of the oscillating thermal instability with uncertain physical mechanism. One can

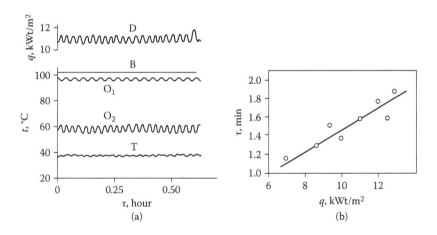

FIGURE 3.21 Pulsation regimes at vaporization in the porous structure [69, 70]. (a) Pulsations of the heat flux and the average value of heat flux, $q = 11$ kW/m²; D, B, O_1, O_2 are the thermal regimes with temperature pulsations. (b) Dependency of the pulsation period on heat flux.

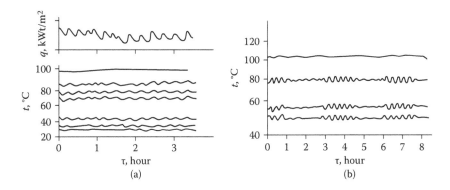

FIGURE 3.22 Pulsation regimes at vaporization in the low-permeability porous medium [69, 70]. ($K_0 = 64 \times 10^{-12}$ m²). (a) $q = 7.5$ kW/m². (b) $q = 8.7$ kW/m².

suppose that the phenomenon occurs as the sequence of the complex interaction between the heat transfer processes in two-phase and single-phase zones. Bau and Torrance [69, 70] suggested the following physical model, which explains the observed thermal regimes and allows analytical determination of the basic regularities at $q < 11$ kW/m².

The critical Raleigh number, Ra_{CR}, determines the development of the natural convection in the porous structure as

$$Ra_{CR} = \frac{g\beta\Delta TKH}{v'\left(\dfrac{\lambda_3}{p'c'_p}\right)}$$

Depending on the relative height of the porous layer H/d, the critical Raleigh number was in the range 55–90 in the specified experimental conditions. It was also assumed that in the two-phase zone the heat transfers inside the porous structure due to the vapor–liquid counter flow. The vapor flows in the upper part of the two-phase zone, where it condenses, and liquid flows down. Hence, the mass flow rates of liquid m_L and vapor m_V are equal to each other in the two-phase zone, $m_V = -m_L$.

Using the phase permeability model yields [69]

$$m_L = -\frac{\bar{K}_1 K_0}{v'}\left(\frac{dp}{dz} + p'g\right) \quad m_V = -\frac{\bar{K}_2 K_0}{v''}\left(\frac{dp}{dz} + p''g\right)$$

The relative phase permeability dependence on saturation was accepted in the simplest form: $\bar{K}_1 = W$ and $\bar{K}_2 = 1 - W$. The specific mass flow rate circulated inside the porous structure flow is

$$m = \frac{W(1-W)K_0 g(p' - p'')}{(1-W)v' + Wv''}$$

Assuming that the entire heat was transferred in the latent heat form, the authors provided the dependence between q and m via the dimensionless heat flux:

$$\bar{G} = \frac{qv'}{K_0 rg(p' - p'')} = \frac{W(1 - W)}{(1 - W)\dfrac{v'}{v''} + W}$$

The maximum heat flux values q_{max} correspond to the maximum values of the dimensionless heat flux $\bar{G}$ as a function of saturation W or the vapor content $(1 - W)$. The values q_{max} are defined by the limit equilibrium states of the two-phase zone. According to Bau and Torrance [69], q_{max} and G_{max} are independent of the porous layer height, H, because the influence of such important factors as capillary pressure, nonlinear character of the dependency between $\bar{K}$ and W, etc., were omitted. Nevertheless, simplicity represents the important advantage of the model and it is appropriate in some cases.

Using data from other sources, Bau and Torrance [69, 70] demonstrated that the matching order between the experimental and numerical data of q_{max} corresponded to the dry-out state and recommended the following formula:

$$q_{max} = \frac{K_0 rg(p' - p'')}{\left(\sqrt{v'} + \sqrt{v''}\right)}$$

Because the model does not account for the possible influence of capillary forces, it is unacceptable to the thin-layer porous structures working under zero-gravity conditions or in the horizontal position for the terrestrial cases, for example, the typical working conditions for heat pipes. Therefore, such a model does not fit many important applications of vaporization in porous structures. However, the initial concepts of the model mostly concurred with the above-mentioned general set of correlations with respect to the acknowledged limitations and assumptions. Thus, the consequent improvement of the approach and physical model presented by Bau and Torrance [69, 70] could be useful for the description of the vaporization process in the submerged porous layer attached to the heating wall.

Bau and Torrance [69, 70] did not suggest the model of heat transfer at the pool boiling on the surface within the submerged porous structure. However, they presented the experimental data as the dependency of the heat flux q on the temperature drop: $q \approx C\Delta T^m$ (Figure 3.23) — which is well matched to the model presented by Smirnov [71], based on the assumption of the countercurrent circulation of vapor and liquid flows through the capillary channels of the porous structure.

Ogniewez and Tien [73] applied the phase permeability model. However, the dependence of the relative phase permeabilities K_1 and K_2 on saturation was not shown. On the basis of the one-dimensional model, the integration of the momentum equations was completed without accounting for the dependency of the effective thermal conductivity of the capillary structure on saturation.

To develop the one-dimensional model, Ogniewez and Tien [73] considered conditions of vaporization inside the porous structure, when the heat flow and vapor

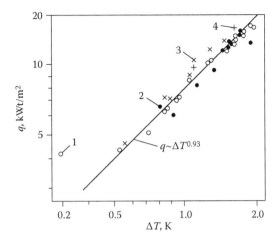

FIGURE 3.23 Dependency of heat flux, q, on temperature drop, ΔT [69, 70]. 1, 3, increase of heat flux q; 2, 4, decrease of heat flux q.

filtration occur in the opposite directions (conditions [69, 70]). Because a comparison with experimental data is lacking, assessment of the model reliability and practical application outlook is not feasible.

Wang and Groll [74] applied the phase permeability model to the analysis of the conditions affecting the maximum heat flux at vaporization in the bidisperse porous layer, when the dry-out regime was attained within the heating wall.

Analysis of the hydrodynamic phenomena, which was based on the one-dimensional equation of the two-phase filtration, accounted for the capillary effects via the capillary pressure defined with the Leverett function, $I(W)$:

$$p_k = \sigma \sqrt{\varepsilon/K} \cdot I(W)$$

where $I(W) = (W^{-1} - 1)^{0,175} / \sqrt{5}$.

Hydrodynamics of the two-phase filtration was analyzed by using the following assumptions:

1. Relative phase permeabilities were depicted by the simple power functions: $K_2 = (1 - W)^3$, $K_1 = W^2$.
2. The linear and quadratic terms of the total vapor and liquid hydraulic resistance of filtration were accounted for.
3. The linear terms are proportional to q/r, K_0^{-1}, K_1^{-1} when the quadratic terms are proportional to $(q/r)^2$ and η^{-1} (η denotes a conductivity [74]).
4. In the case of the grain bed layer, the authors used the following correlations for the values K and η:

$$K = \frac{d_P^2}{150} \frac{\varepsilon^3}{(1-\varepsilon)^2} ; \quad \eta = \frac{d_P^2}{1,75} \frac{\varepsilon^3}{(1-\varepsilon)^2}$$

According to the two-phase filtration model, the distribution of saturation was calculated numerically in layers consisting of top and bottom parts at the different parameters of the particle bed layer parameters (dp, ε, ρ', ρ'', etc.). The layers differ by particle size and, consequently, by permeability. The authors considered cases when the layer with higher permeability was located above the less permeable layer and vice versa.

It was observed that the saturation jump occurs at the border between the two layers. Moreover, if the more permeable layer is located over the less permeable one, it leads to an increase in maximum heat flux. It is accompanied by the drying out of the heating wall. In contrast, if the less permeable layer is located over the more permeable one, it leads to a significant decrease in critical heat flux. Information on the experimental validation of numerical data is absent. Wang and Groll [74] noted the qualitative concurrence of their data with two experimental studies, but a direct comparison with experimental data is lacking.

Therefore, the study of regularities of vaporization regimes and two-phase filtration inside the porous structures could not be limited to the experimental investigation of temperature fields alone. The direct visualization and observation of two-phase filtration through porous structures, including the typical hydrodynamic regimes of the two-phase flow in the separate pores, capillaries, etc., are required. Only a few investigations have been conducted on separate capillary flows, but they assisted significantly in helping other researchers to understand and to discover two-phase flows principal hydrodynamic phenomenon and the main driving mechanisms in the separate capillaries. First, it is necessary to mention works of Smirnov together with Dr. B. Afanasiev. These research activities cover the boiling mechanism within the screen capillary-porous structures attached to the wall. Dr. M. Berman and O. Chulkin completed another significant experimental research on movable grain bed structures. These experimental results first confirmed the existence of some hydrodynamic regimes that were analytically predicted earlier [71]. Moreover, these studies facilitated the discovery of several new unexpected vaporization hydrodynamic regimes. In addition, their results helped in the development of accurate physical representations and fundamental models of vaporization inside porous structures.

Consider the methods and the most important conclusions presented in the above-mentioned studies.

The study of thermohydrodynamic phenomena at vaporization in screen capillary-porous structures with the liquid capillary supply was accomplished by means of photo and high-speed video capture on the experimental setup developed by Afanasiev et al. [75–79].

The authors performed the following:

1. Recorded observations on hydrodynamic phenomena inside the porous structure at boiling on coated surfaces covered by screen wicks.
2. Studied the internal processes at the interface variations.
3. Observed vapor bubble and vapor channels inside the structure.
4. Investigated liquid droplets formation and throwing out structure characteristics.

The experimental setup is presented in Figure 3.24.

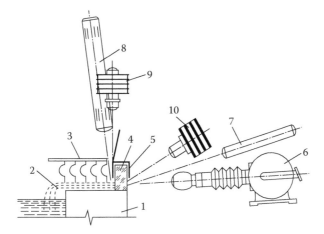

FIGURE 3.24 Schematic layout of the video capture of vaporization inside the screen capillary-porous structure. 1, Heat transformer made from the high thermal conductivity material (copper); 2, screen porous structure; 3, fixing device; 4, glass plate; 5, screen; 6, camera; 7, 8, gas lasers (LG-53 and LG-79) used as power light sources; 9, 10, light lamps (OI-24 and OI-27).

Photo and video capture were performed from the back and top sides. The last position was selected to study the hydrodynamic phenomena that occurred at the external surface of the porous structure. The backside video capture was completed in the reflected light over the optical glass. The glass was attached to the edge of the heating surface. It allowed the authors to observe the internal vaporization processes that developed inside the porous structure. The screen wick layers were used as the porous structures. They were attached perpendicularly to the glass plate.

The scale-up version of vaporization pictures allowed the selection of the depth of field lower than the cell size. High-speed video capture was performed at the steady thermal regimes. The video capture speed depended on the duration of the elementary vaporization process at different thermal regimes and it was changed from 250 to 3500 frames/sec.

The high-speed video capture was completed at the boiling of water and ethanol inside structures of various widths (the layer number was varied from 1 to 24, and the cell size ranged from 40 to 1600 µm at pressure 0.1 MPa).

The hydromechanical characteristics of the liquid droplet throw at boiling on coated surfaces covered by screen wicks (layer numbers 1–3) were investigated. The schematic layout of the measurement of the liquid droplet throwing rate is presented in Figure 3.25.

The distance between the liquid level and the center of the heating surface covered with the screen wick was 55 mm. The measuring device for the liquid flow rate consisted of the trap (the glass cylinder with separator) and the liquid collector having the branch pipe at the bottom. The jets and droplets were drawn in the separator trap and then into the liquid collector.

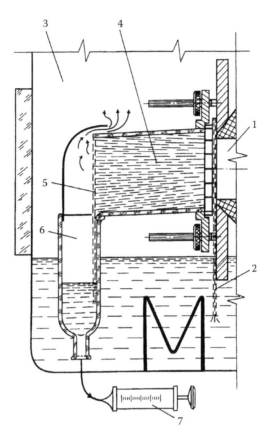

FIGURE 3.25 Schematic layout of the measurement of the liquid droplet throw flow rate at vaporization. 1, Heat transformer; 2, screen capillary structure; 3, working chamber; 4, liquid trap; 5, separator; 6, liquid collector; 7, syringe.

Analysis of research results [75–79] drew the following conclusions.

1. At low heat fluxes determined by the wall overheats $\Delta T = \Delta T_W - \Delta T_s \leq$ 1°C (water at the atmospheric pressure), vaporization occurs on the external liquid surface inside the porous structure, and it is characterized by the interface curvature. Such a phenomenon maintains the flow of liquid to the evaporation surface (Figure 3.26).

2. When the wall overheat increases and exceeds the nucleation temperature drop $\Delta T^* = 4\sigma T_S / r\rho''a$ (it is essential for the existence of the stable vapor phase in specific elementary cells of the porous structure), the interface moves from the external surface of porous structure to the heat transfer surface. Because thin screen layers were used as the porous structure [79] and each screen layer had approximately identical cell size, the transition from the external surface evaporation regime to the regime of evaporation from the deepened interface meniscus occurs simultaneously within all elementary cells at the specific heat flux value.

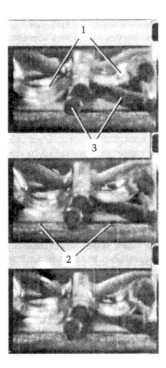

FIGURE 3.26 Photograph of vaporization at the two-layer screen structure [75–79], at the video capture speed of 200 frames/sec; heat flux, q =10 kW/m^2; cell size, a = 0.5–1.0 μm. Evaporation regime: 1, curvature of the interface; 2, heat transfer surface; 3, cell cross section of the screen porous structure.

Figure 3.27 presents a sharply deepened meniscus interface with the significant curvature in the adjoining regions with the wall. High-speed video observation shows that the oscillations of the interface in the liquid film covered the structure skeleton. It appears as the liquid flows from the external surface to the heat-generating wall.

Vapor cavities with the deepened meniscus were blocked near the external surface, and they can be considered the "micro-heat pipes" (Figure 3.27). Such cavities were properly stable at low heat flux values, whereas in some visualizations [75–79], periodical breaking in the cavities of the external interface were observed. Augmentation of the heat flux was accompanied by an increase in the frequency of external interface breaking. Lastly, the cavities were detached continuously at specific conditions (heat flux values).

3. The appearance of vapor bubbles in the liquid film on the heating wall at contacts between the skeleton and the wall, when temperature drop growth has reached a certain value, ΔT_1 (Figure 3.28), was also observed. It means that centers of nucleation were activated on the heat transfer surface at such values, ΔT_1, that is, boiling was developed in the microfilm-covered heat transfer wall. Both the heat flux growth and the increase in thickness of porous structure contradict vapor removal.

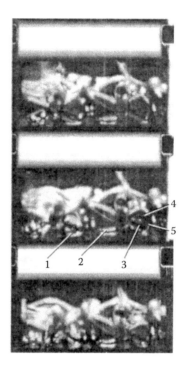

FIGURE 3.27 Photograph of vaporization at the three-layer screen structure [75–79], at the video capture speed of 2000 frames/sec; heat flux, $10 < q < 100$ kW/m²; cell size, $a = 130$ μm. Evaporation regime with the "deepened meniscus" adjoined to the heating wall: 1, 2, vapor-generated cells with "deepened meniscus" and free vapor removal; 3, the cell with the "deepened meniscus" with the vapor volume blocked from the top (internal microheat pipe); 4, upper interface of the microheat pipe.

As a result, liquid paths between the neighboring cells were destroyed and they formed large vapor zones that were periodically removed through some vapor channels.

4. The vapor zone covers the major portion of the structure at a certain heat flux value, that is, it dries out the structure. Such a regime can be defined as the critical or limit regime. High-intensity liquid removal from the porous structure was observed during capillary feeding in the regime accompanied by boiling in thin liquid films covering the heating surface. Typical liquid removal characteristics are given in Figures 3.29 and 3.30. The following observations were noted.

5. Under the certain conditions, the momentum source was created inside the capillary-porous structure near the wall. It passed the momentum to the liquid that is throwing it out from the capillary structure. The momentum source power was in the same order of magnitude as the capillary pressure, $4\sigma/a$. In some average thermal regimes, the liquid throwing flow rate exceeded the evaporated liquid flow rate, which was equal to $G'' = qS/r$ (Figure 3.30), by 4–5 times. These results represent the fundamental experimental fact that proved the development of a significant excess capillary potential inside the near-wall porous structure under certain vaporization conditions.

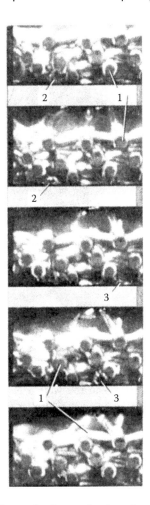

FIGURE 3.28 Photograph of vaporization at the three-layer screen structure [79], at the video capture speed 3500 frames/sec; heat flux, $q = 30$ kW/m^2; cell size, $a = 90$ μm. The evaporation regime is boiling within the near-wall microfilm; 1, wire cross section as the screen structure element; 2, 3, acting boiling nucleation centers in the cell core and in the contact zone (wall structure).

Therefore, experimental data [75–79] established that at vaporization inside the screen structures attached to the heating wall by the mechanical press, the following steady thermal regimes are feasible:

- Evaporation
- Microheat pipes
- Evaporation with the deepened meniscus
- Boiling in the liquid film covering the heating surface
- Vapor film (blanket) formation inside the structure or burnout

(a)   (b)

FIGURE 3.29 Boiling inside the coated heating surfaces with droplet liquid throwing [79]. (a) The entire picture; (b) enlargement of the same picture. Water at $p = 0.1$ MPa; $q = 3 \times 10^5$ W/m².

Development of the "vapor blanket" layer inside the wetted screen porous structure was reported in one of the pioneering studies examining heat pipes [80]. The regime appeared at very small specific heat loads, when q was close to the zero. The thickness of the "vapor blanket" increased with heat flux growth.

Note that Moss and Kelly [80] did not give particular attention to uniform and reliable contact between the wick and the wall. Hence, one can assume that the observed phenomenon was the result of the complete or partial screen wick layer

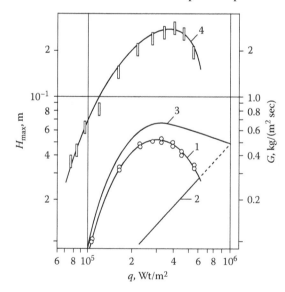

FIGURE 3.30 Energy characteristics of the droplet liquid throwing [79]. 1, Liquid throwing flow rate, G; 2, evaporated liquid flow rate, $G'' = qS/r$; 3, total flow rate of heat carrier, $G + G''$; 4, maximum height of liquid throwing.

detaching from the heating wall, which could be interpreted as the formation of the high-permeability layer near the wall of the porous structure. In such a case, a vapor–liquid intensive filtration along the heat transfer surface accompanies vaporization while it is in the single capillary channel. Therefore, at vaporization in wetted porous structures, when heat is transferred from the wall to the structure, the contact conditions between the wall and the structure have a fundamental impact on the development of thermal regimes.

Managing the reliable thermal contact between the wall and the structure represents an extremely important and complex technological problem for the practical realization of such devices as loop heat pipes (LHP) and capillary pumped loops (CPL). Therefore, Maidanik and his team conducted extensive studies on the influence of thermal resistance (mechanical contact between the wall and the structure of the LHP evaporator) on the evaporative heat transfer intensity in low-permeability porous structures.

The research results, together with the experimental methods and visual observations, were presented by Veshnin et al. [81].

Consider these results [81] apart from the detailed theoretical analysis devoted to the vaporization in LHP evaporators, which was based on the physical representations and numerical analysis developed by Kovalev and Soloviev [36].

It is known that the maximum intensity of heat transfer in the LHP is mainly due to the separation of liquid and vapor flows, and the considerable decrease in the liquid flow path inside the porous structure. It allowed the use of porous materials with extremely small pore sizes (≤ 1 μm) in the LHP design.

However, the problem relating to vapor removal from the near-wall zone appears more complicated than usual, and its solution is connected with the design of the system of vapor removal channels. The vapor removal flow rate increases with the extension of the vapor channel zone. On the other hand, the contact area between the wetted structure and the wall becomes smaller and maintains corresponding escalation of thermal resistance at the surface of the heat pipe evaporator.

Rational design of the LHP evaporator consists in forming a system of grooves at the internal contact surface of the wall or at the external porous structure contact surface of the evaporator. In addition, maintenance of the reliable surface contact is extremely important. A similar structure was used by Vershnin et al. [81] (Figure 3.31b). A schematic diagram of the experimental setup is shown in Figure 3.31a.

The experimental setup comprised a sealed chamber with the condenser chilled and thermostabilized by water, and the auxiliary heater maintained the saturation pressure (10 MPa) of the heat carrier. The heat carrier used in the experiments was acetone. The porous cylindrical sample with a diameter of 25.5 mm and thickness of 10 mm was placed into the frame and pressed by the spring to the heater with the force of 200 N. The heat flux of the main heater was determined by the "thermal wedge" method. Temperatures of the vapor zone and heater's wall were measured by copper–constantan thermocouples. The capillary feeding of the evaporating porous structure was maintained at the height equal to the sample thickness via the bellows system.

Vapor removal channels with a rectangular cross section were created by milling at the top sample surface. Two types of sample grooves with a width of 0.3 and 0.6 mm and a depth of 0.5 mm were made. Geometric parameters (fin width, etc.) were

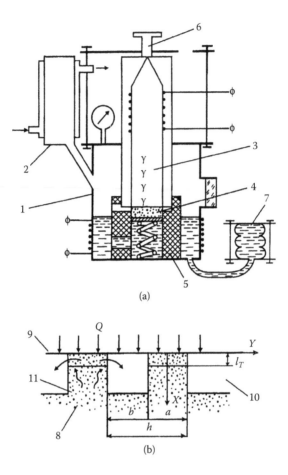

FIGURE 3.31 Schematic diagram of (a) the experimental setup and (b) the characteristic cell within the near-wall porous structure [81]. 1, Sealed chamber; 2, condenser; 3, heater; 4, sample; 5, frame; 6, pressing screw; 7, bellows ; 8, porous structure; 9, wall; 10, vapor removal channel; 11, porous fins.

varied for every sample type, that is, dimensionless parameter a/h was also varied (Figure 3.31). Titanium powder (PTOM) and nickel powder (PNK-OT1) were used as the porous structure materials. The titanium and nickel structures had porosity rates (ε) of 57% and 46%, and a maximum pore radius (R_{max}) of 4.5 and 2 µm, respectively.

The experimental data for the titanium samples are presented in Figure 3.32. The experimental points were found to be qualitatively and quantitatively consistent with the numerical data at $q = 1.6 \times 10^5$ W/m^2 and $R_C = 3 \times 10^{-5}$ m^2 K/W.

The experimental and numerical curves are equidistant at the whole range of the measured heat fluxes, whereas when the heat flux values were less than 1.6×10^5 W/m^2, the experimental points were located above the numerical ones. Similar results were obtained for the porous nickel samples. Analysis of the curve shapes demonstrated that experimental data on heat transfer coefficients and the corresponding

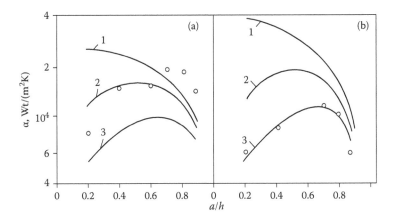

FIGURE 3.32 Dependencies of heat transfer coefficients on the relative part of the contact, a/h, of heat pipe surfaces [81]. h, groove depth; porous structure, titan powder; groove width, $b = 0.6$ mm; $a - q = 0.8 \times 10^5$ W/m^2; $b - q = 1.6 \times 10^5$ W/m^2; points, experimental data; lines, calculations by the model [36]. 1–3, $R_C = 0$, 10^{-5}, 3×10^{-5} m^2 K/W, respectively.

calculations coincided within an accuracy of some multiplier depending on the heat flux value. If such a statement is correct, then by using experimental and calculation data at $R_C = 3 \times 10^5$ m^2 K/W, one can estimate the value of the specific thermal contact resistance accounting for the multiplier determined at the concurrence of these data.

Calculations showed that increasing heat flux from 0.4×10^5 to 2.0×10^5 W/m^2 leads to an increase of the specific thermal resistance from 1.7×10^{-5} to 4.2×10^{-5} m^2 K/W. The calculation model [81] allows the determination of the total thermal resistance of the vaporization zone as

$$R_0 = \frac{1}{\alpha} = \frac{1}{\alpha_V} + \frac{h}{a} R_C$$

The heat transfer coefficient in the vapor (blanket) layer formation zone, α_V is defined as

$$\alpha_V = \frac{\lambda_E}{L_V} \frac{a}{h}$$

Here, λ_E and L_V denote the effective thermal conductivity and the thickness of the dry part of the porous structure, respectively (heat is transferred from the wall to the interface through this dry part), h is the pitch between the vapor channels, and a is the fin width.

L was determined by using the corresponding hydrodynamic problem of vapor filtration through the dry part of the porous structure to the vapor channels. Discussion of the experimental data presented by Veshnin et al. [81] requires the following explanation.

It was mentioned that the model and the experimental data coincided only in the case of accounting for the auxiliary specific thermal resistance of the contact that notably depended on the heat flux. Unfortunately, the model [36] used by Veshnin et al. [81] ignores the presence of the special surface thermal resistance caused by significant temperature irregularities at the evaporation surface. Such resistance augments the escalation of temperature irregularities. Assessments of the thermal resistance can be based on the recommendations and formulas presented by Burdo and Smirnova [20, 21] and Mistchenko [26]. These recommendations were developed, verified, and confirmed qualitatively and quantitatively by Borodkin [82].

Visualizations of vaporization in the near-wall zone obtained via high-speed video capture [81] represent a specific interest from a scientific viewpoint. Vershinin et al. [81] applied glass as the heating surface and obtained pictures of vaporization mechanism (Figure 3.33).

It should be noted that low resolution of the pictures casts some doubt on the authors' conclusion that the light zones of the pictures denote contact with vapor, whereas the dark regions are the liquid coverings. Hence, the existence of some wave surface elements can be observed in the light zones of the picture (Figure 3.33). It could happen, but only in the presence of the liquid film.

A similar concern relates to the light zones in places where the microporous fin tips are in contact with the glass heating surface. It is probable that the appearance of the liquid film has a periodic nature. A clear answer to this problem can only be based on further experimental investigations.

Vershinin et al. [81] also presented model experiments on the measurement of pressing force for the porous structure, and investigated its influence on heat transfer. Experimental measurements of heat transfer coefficients were completed for titanium samples at identical mechanical pressurization values (N = 3.1, 6.2, and 10 MPa). It was supposed that equal pressures create the same surface deformations, and identical thermal resistances occur in the samples with different a/h values. The pressing force was measured by the dynamometer DOSM-3-1 placed between the pressing screw and a heater, while a spring in the frame was replaced by the rigid support (Figure 3.31).

The experimental data on the measurement of heat transfer coefficients at q = 11 W/sm^2 and for different pressing forces are presented in Figure 3.34. It is clear that an increase in the value of N augments the heat transfer. The heat transfer coefficient increases as the a/h ratio increases. This effect is substantial when a/h is less than 0.7. Consequently, it may be supposed that thermal contact enhancement (e.g., the porous structure solid-phase sintering to the wall) allows attaining the maximum values of heat transfer coefficients at lower a/h values, similar to the case of the ideal thermal contact. In the experiments, the contact thermal resistance was 0.9×10^{-5} m^2 K/W at N = 10 MPa. This value was estimated under the assumption of concurrence between the experimental data and the calculations.

Hence, the long-standing practice of Vershinin et al. [81] and other researchers developing experimental and industrial LHP and CPL devices confirmed the essential influence of the hydrodynamic and thermal processes in the near-wall porous structure on total heat transfer intensity. In addition, the significant — and sometimes even crucial — effect of contact thermal resistance on heat transfer was established.

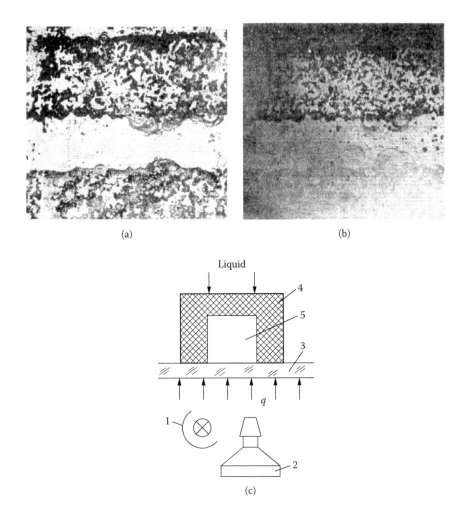

(a)   (b)

Liquid

(c)

FIGURE 3.33 Vaporization on the surface of loop heat pipe evaporator [81]. (a, b) Process photos (view from the bottom); dark zones, capillary structure elements; light zones, vapor-generating channels. (c) Schematic diagram of visualizations: 1, lighting source; 2, camera; 3, glass heating surface; 4, porous structure; 5, vapor removal channels.

At the same time, specific cases are promising and attractive, when the thermal contact between the wall and porous structure is poor or even absent. Such designs use partial or complete separation of the porous structure from the wall, (Figure 3.3b–d).

From this point of view, the following designs are notably different:

1. Thin layer capillary structures connected to the wall
2. Massive porous structures connected to the wall
3. Capillary structures separated from the wall by design or due to augmentation of the vaporization process

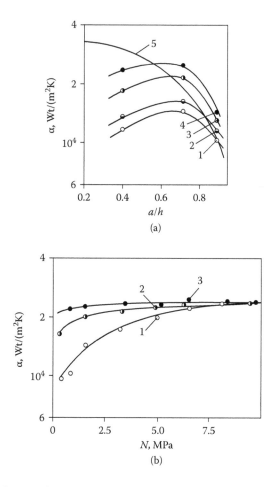

FIGURE 3.34 Influence of contact conditions on evaporation heat transfer intensity at the wall–structure contact zone of the loop heat pipe (PTOM, acetone) [81]. $b = 0.6$ mm; $a/h <$ 0.7; $q = 11$ W/sm². (a) Experimental data: 1–4, $N = 0.5, 3.1, 6.2, 10$ MPa; 5, calculations: $R_C =$ 0. (b) 1–3: $N = 0.5, 6.2, 10$ MPa.

Particle bed layer is a common case of the disjoined porous structure. Vaporization in the submerged dispersed bed layers represents a promising process in certain technological apparatuses (e.g., due to the significant heat transfer enhancement or at the evaporation of water solutions). Therefore, regularities of the vaporization heat transfer under such conditions were intensively investigated in recent years. The main experimental results, physical explanations, and existing analytical models are discussed in Chapter 5.

In this chapter, we consider only the basic principal hydrodynamic regimes that could be realized at vaporization inside particle bed porous structures.

As previously noted, the major problem associated with disjoining porous structures is the capability of gravity forces to secure a safe contact between the particle bed layer and the heating surface. When these forces are enough to provide secure contact conditions, the heat transfer and hydrodynamic regimes appear similar in conventional mechanical connections between the structure and the wall.

The pressure providing the contact between the layer's porous structure and the wall per surface area unit is equal to $f_{sc} = (\rho_{sc} - \rho'')(1 - \varepsilon)gH_0$ [if $H_0 < L$ (where L is the characteristic size of the particle bed), then H_0 is the height of the layer, and when $H_0 \gg L$, $H_0 = 2L$]. When the gravity force pressing the particle bed to the wall sufficiently exceeds the dynamic action of the vaporization separating the layer from the wall, then heat transfer in the porous structures is similar to the conventional case of mechanical contact between the capillary structure and the wall.

The dynamic effect of vaporization on the contact between the layer and the wall is defined as the following:

- Thermal (i.e., liquid flow from the heating surface is initiated as the consequence of the boiling process)
- Hydrodynamic (liquid forced filtration is developed through the porous layer with velocity rates ranging from the low values up to the velocities of fluidization)
- Combined (i.e., filtration appears as a combined result of the thermal and hydrodynamic interaction)

Hence, two limit regimes are feasible when the layer's porous structures are separated from the wall:

- Vaporization inside the motionless porous layer pressed to the wall by gravitational forces ($f_{sc} \gg \rho'w^2$)
- Vaporization inside the porous layer under thermal, hydrodynamic, or combined fluidization conditions

$$Hg(\rho_{SC} - \rho_L)(1 - \varepsilon) \ll \rho'w^2$$

In transient regimes, when

$$\rho'w^2 \approx H_0 g(\rho_{SC} - \rho_L)(1 - \varepsilon)$$

a specific density of the porous layer increases, and it is accompanied by the layer's separation from the heat transfer surface and by the formation of vapor removal channels (the so-called "channels formation regime"). The existence of "separation," "channels formation," and "fluidization" regimes was observed by Gorbis et al. [84–87]. Vaporization conditions were significantly dissimilar at boiling inside the layer connected to the wall, and in cases of separated layer or pseudo-fluidization. However, Mankovsky and Fridgant [88, 89] ignored such a distinction in the comparison of experimental data.

Analysis of current trends in the experimental and analytical studies and perspective applications of vaporization processes inside the capillary-porous structures demonstrated that the majority of researchers classified heat transfer at boiling on coated surfaces covered by wetted or submerged porous structures into two main cases:

- Porous structure is connected to the heating wall via thermomechanical forces, that is, soldering, sintering, etc.
- Porous structure is movable and separated from the wall at boiling, including the pseudo-fluidization regime.

In both cases, the hydrodynamic and the heat transfer peculiarities in capillaries and slits affect the vaporization process. Hence, further investigations are required not only to achieve a correct understanding of the physical nature of these phenomena, but also because of their importance in modern applied trends in the miniaturization of heat exchangers and other heat transfer devices.

Chapter 4 presents the theoretical analysis, experimental data, and summarized discussion of these processes.

4 Thermohydrodynamics at Vaporization in Slit and Capillary Channels

4.1 EXPERIMENTAL STUDIES ON HEAT TRANSFER AT BOILING IN SLITS AND CAPILLARY CHANNELS

The study conducted by Vasiliev and Mayorov [90] can be considered as the pioneering experimental research on boiling heat transfer in slits. They studied the influence of slit thickness on heat transfer intensity at water, ethanol, and 10% NaCl water solution boiling in vertical loop slits.

Experiments were performed on two working parts at one-side heating. The internal wall was heated in one part, and the external wall on another. The working element length used was 0.48 m; slit gap varied from 7.75 to 1.25 mm. The heat flux density, q, ranged from 2×10^4 to 1.1×10^5 W/m^2 under atmospheric pressure. The liquid input was realized by 20% filling of the working zone. With respect to the data [90], the maximum heat transfer coefficient could be reached.

The typical experimental dependencies conducted under internal heating conditions [90] are given in Figure 4.1 and generalized via the correlation $\alpha = 23.1q^{0.6}s^{-0.514}$ for slits with small gaps ($s \leq 3$ mm) and $\alpha = 15.6q^{0.6}s^{-0.15}$ at $3 \leq s \leq 10$ mm. At gap size $s \leq 3$ mm, the mentioned dependencies show that the slit's boiling heat transfer process was characterized by the considerable heat transfer enhancement with the decrease in gap size (with the law $1/\sqrt{s}$). The gap size increased (1.5–2 times greater than the vapor bubble departure diameter), leading to a decline in heat transfer enhancement.

The boiling heat transfer enhancement with the accompanying decrease in slit width could be attributed to the increase in two-phase flow turbulence. Visual observations show that a number of small bubbles increased the vapor quality growth as the slit's gap decreased. When heat flux (q) growth leads to a significant decrease in α, thermal regimes in which the heat transfer coefficient abruptly changed due to the increase in heat flux were observed. This was attributed to the essential destruction of the liquid film in the heat transfer surface [90].

Improvements in research [90] were achieved via experiments [91] conducted in the vertical annular slit at the internal surface heating and the working surface submerging into the vessel filled with the heat carrier. The height of the annular slit was 300 mm, the gap width size ranged from 0.2 to 4.0 mm, and the pressure varied from 1 to 10 bar. Water and oleate sodium–water solution were used as heat carriers.

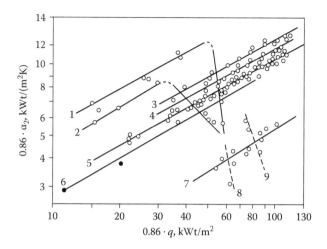

FIGURE 4.1 Dependency $\alpha = \phi(q)$ at boiling of water, ethanol, and 10% NaCl water solution inside the annular slits on the steel heating surface. Water and 10% NaCl solution boiling (s): 1, 1.25 mm; 2, 1.6 mm; 3, 2.7 mm; 4, 3.2 mm; 5, 6 mm. Ethanol boiling (s): 7, 4.7 mm; 8, 5.6 mm; 9, 7.75 mm.

Thermal boiling regimes were studied at the vertical annual slit by video and photo capture along the slit's height. It was observed [91] that there were different two-phase flow zone pictures: the economizer's zone with the single bubbles and the sticking vapor bubble zone. At the later zone, the liquid film covered the heat transfer surface on the initial part, and separate parts of the heat transfer surface were dry. When wall drying is absent, the local heat transfer correlations of the mentioned thermal regimes yield

$$\alpha = \text{const } q^{2/3} s^{-2/3} p_s^{-0.36}$$

Wall temperature oscillations were measured during the experiment. They appeared as the sequence of the vapor and liquid slugs or separate bubbles periodic heat transfer surface washing. The determined evaporation in the slits' boiling heat transfer physical mechanism was any "contact" heat conductivity mechanism because the time scales of the temperature pulsations were extremely small ($\tau \approx 10^{-2}–10^{-3}$ sec). On this basis, experimental data were generalized in the next stage: $Nu = 40(Fo \times Pr)^{-2/3}$. The Fo number was used as the determined time scale in the temperature pulsation period. These values were not independent variables and did not come into the simple conditions. The mentioned generalization confirmed the hypothesis regarding the determined role of the contact heat transfer mechanism in the entire boiling heat transfer inside the vertical slits. However, it was not convenient for many other cases when information about temperature pulsation times was not available.

Experiments involving the wide representative circle of refrigerants and ammonia boiling heat transfer inside the vertical slit channels with respect to the refrigeration machine conditions were conducted, and the results were presented by Azarskov [92] and Gogolin et al. [93]. One of the first experimental results presented by Azarskov [92] was done on the working element, which was as plate slit channel at a height of 300 mm with one transparent wall and one-side heating; the slit's gap ranged from 0.5 to 4.0 mm. As the main heat carrier, R-113 was used at atmospheric pressure.

To study the main regimes and the influence of geometry on boiling heat transfer, refrigerants R-12 and R-22 were conducted on the plate slit's channel at a height of 400 mm and with two-side heating. The boiling (saturation) temperature ranged from +20°C to −30°C, the heat flux density ranged from 10^3 to 20×10^3 W/m^2, and the flooding initial level ranged from 33% to 100%.

Similar experiments were also conducted for other heat carriers (e.g., ammonia, oil–refrigerants mixtures, etc.) on different types of corrugated surfaces. The details and general analysis were reported by Gogolin et al. [93], which contains physical concepts about slit channel boiling. It was shown that by boiling refrigerants in vertical slits, the following zones were observed:

• Single-phase flow (economizer) zone
• Subcooling boiling with single vapor bubbles on the heating surface zone
• Bubble flow regime with the boiling on the heating surface pressed
• Sticking vapor bubbles that were analogous to the slugs zone
• Annular slug flow regime zone

The parts occupied by the different zones were strongly changed depending on the conditions — heat flux densities, saturation pressure, the initial flooding level of the liquid at the borders of the transit from one zone to another, etc. The presence of the different two-phase flows led to significant and nonmonotonous changes in local heat transfer coefficients over the channel's height. The typical local heat transfer distribution over the channel's height by the refrigerant R-22 boiling inside the slit channel with respect to the experimental data in Ref. [92] is given in Figure 4.2. This complex and changing characteristic of the local heat transfer coefficient distribution along the height of the slit channels was the sequence of the coexistence of different boiling heat transfer mechanisms.

The boiling within the slit heat transfer mechanism can be considered as the conditional division of the whole heat flux, removed from the wall, into three parts [92, 93]: Q_{con}, Q_{boil}, and Q_{ev}, which denote the heat fluxes, removed from the heating wall by the corresponding different mechanisms, by single phase (liquid) convection, vapor bubble convection, and evaporation from liquid film in the bubble base, respectively.

It should be noted that it is not sufficient to divide the heat fluxes for the analysis; it is also important to determine or estimate the parts of local heat transfer coefficients (specific thermal resistances), corresponding to the mentioned mechanisms. The last mechanism (evaporation from liquid film in the bubble base) offered a more accurate way to separate, limit, and classify the complex heat transfer by the slit boiling parts. From a practical viewpoint, the diversity and complexity of the separate mechanisms and the strong dependence on most factors influencing their input on refrigeration machines' "working conditions" made the treatment of experimental

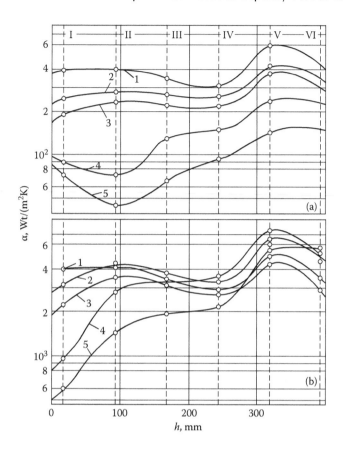

FIGURE 4.2 Local heat transfer coefficient changing over height h of the vertical slit channel by the refrigerant R-22 boiling on data from Ref. [92]. (a) $h_H = 100\%$, (b) $h_H = h_{OPT} = 33\%$. For parameter q: 1, 20 kW/m²; 2, 10 kW/m²; 3, 6 kW/m²; 4, 2 kW/m²; 5, 1 kW/m²; I–VI, places of measurements.

data on the empirical dependencies base and using them in practice more rational. Thus, Azarskov [92] suggested the vertical slit boiling heat transfer of the refrigerants R-12, R-22, and R-113 by whole slit flooding in the next correlation:

$$\alpha = 1.05 p_{CR}^{0.4} \times (p/p_{CR})^{0.2} \times (H/s)^{0.45} \times q^{0.6} / \left(T_{CR}^{0.8} \times M^{0.2} \right)$$

where M denotes the molecular mass.

This equation can be used when $75 < H/s < 400$, $1 < q < 20$ kW/m², $0.024 < p/p_{CR} < 0.184$, and also when $H/s = 800$ and $1 < q < 6$ kW/m². By the optimal flooding level, when recirculation was absent and $h = 1/3H$, it was recommended to use the following formula:

$$\alpha = 2.4 \frac{p_C^{0.6}}{T_{cr}^{0.7} M^{0.7}} \left(\frac{\sigma}{g\rho' s^2} \right) q^{0.4}$$

This equation can be recommended in the next range of q and p/p_{CR} changes: $100 < H/s < 400$ and $0.048 < (\sigma/g\rho's^2) < 1.25$, and also for $H/s = 800$ and $0.048 < (\sigma/g\rho's^2) < 5.0$.

Boiling in the slit channel heat transfer coefficient complex view was associated with the conditions described in Refs. [92–94], in which circulation in these channels can be taken into account for gravity forces. This is why from one side, flow velocity w_0 was dependent in a complex manner on the two-phase flow structure, vapor quality, and other important factors (slit's gap, heat flux density, etc.); but from the other side, w_0 has its own impact on heat transfer intensity, and consequently on vaporization and vapor quality inside the slit channels.

The heat transfer intensity increased as a result of the decrease in w_0, and also in the convective heat transfer zone part. Heat transfer growth was observed as the mass flow rate vapor concentration X and heat flux q increased. The heat flux has strongest influence on heat transfer coefficient α, when the surface part occupied by the boiling zone was sufficiently large and the vapor bubble density mechanism input is particularly substantial. The typical dependencies of the local and average heat transfer coefficients α on the main factor are shown in Figure 4.3.

The majority of the slit channel's gaps changing range can be related to the conditions of previous experiments [93, 94], that is, "the natural conditions," as soon as the vapor bubble "departure diameter" for refrigerants reaches $D_0 \leq 10^{-3}$ m.

This is why the slit's influence on the heat transfer intensity by R-12 and R-22 boiling appeared when the slit's gap dropped to less than 2 mm ($s < 2$ mm). Then, this dependency was prescribed by the next law: $\alpha \approx const \cdot s^{-m}$; where $m = 0.5$. When $s > 2$ mm, "the slit effect" was lessened, which made it possible to consider that boiling heat transfer law was similar to that for boiling under the two-phase lifting flow conditions.

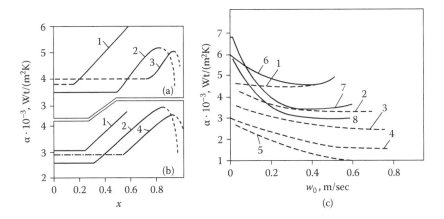

FIGURE 4.3 Refrigerant boiling heat transfer coefficients when they flow inside slit channels by natural convection [93]. (a) $q = 10$ kW/m². (b) $q = 5$ kW/m². For parameter w_0: 1, 0.25 m/sec; 2, 0.14 m/sec; 3, 0.07 m/sec; 4, 0.04 m/sec. (c) R-22 (dashed lines); $H = 0.8$ m, $s = 1$ mm, $t_0 = 20°C$, $x_1 = 0$. For parameter q: 1, 20 kW/m²; 2 kW/m², 10 kW/m²; 3, 5.6 kW/m²; 4, 2.8 kW/m²; 5, 1.5 kW/m². Ammonia (solid lines); $H = 1.4$ m, $t_0 = -30°C$, $x_1 = 0$. For parameter q: 6, 7 kW/m²; 7, 5 kW/m²; 8, 3 kW/m².

Experimental researches on the hydrodynamic and heat transfer of water and ethanol at flow boiling inside vertical plate slit channels with heating surfaces of 0.15×0.15 and 0.15×0.09 m when slit's gap was changing from 0.4 mm up to 1.3 mm and heat fluxes were varied between 2.10^3 and 2.10^4 Wt/m^2 at liquid flow rate equal to $G = 1.2\ G_{EV}$ (G_{EV} is the whole vaporization flow rate at the given geometry and heat loads) are presented in Refs. [96–98]. Experimental results [96] are shown in Figure 4.4. Data on experiments involving two-phase flow hydrodynamics by the slit's channel boiling are provided in Refs. [97, 98]. The hydrodynamic experimentation data were treated in the next nondimensional form:

$$Fr = u/\sqrt{g\Delta\rho s} \quad \text{and} \quad Re = \sqrt{(g\rho'\Delta\rho s^2/\sigma)}\Delta\rho = (\rho' - \rho'')/\rho'$$

by $Re < 0.23$; $u = 6.2g\sqrt{s^3(\rho' - \rho'')/\sigma}$; at $Re > 0.23$; $u = 1.5\sqrt{g\Delta\rho s}$; where u is the liquid relative velocity. These parameters were used in the model's [96] next correlations [97, 98]:

$$\varphi = \frac{1}{1.25 + 2.2\dfrac{r\rho''s}{qhk}u_\infty} \qquad u_\infty = 1.5\sqrt{gs} \qquad u_{CR} = \frac{qk}{r\rho''}\frac{h}{s}$$

Here, coefficient k accounted for the heat input conditions (for one-side heating, heat input $k = 1$; for two-side heating, heat input $k = 2$; h denotes the channel height local coordinate).

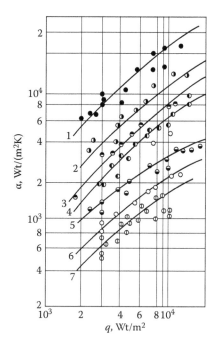

FIGURE 4.4 Water and ethanol boiling in plate vertical slit channel heat transfer based on data from Ref. [96]. Experimental data: $b = 90$ mm, $H = 150$ mm; $G = 1.2G_{EV}$; $p_s = 0.1$ MPa. Water (s): 1, 0.4 mm; 2, 0.7 mm; 3, 1.0 mm; 4, 1.3 mm; ethanol (s): 5, 0.5 mm; 6, 0.7 mm; 7, 1.0 mm.

The boiling heat transfer regularities in the slit's channels by some conditions caused the same result as in simple capillaries. The number of experiments conducted on heat transfer boiling in the capillaries is extremely limited. One of the first of such efforts was reported by Antipov et al. [98]. The experiments were conducted with capillaries whose length ranged from 96 to 175 mm and with an internal diameter of 0.6 mm. The heat carrier used was nitrogen in the liquid state. An empirical correlation was obtained in the form: $\bar{\alpha} = 6q^{0.6}l^{0.36}$. It was noted that temperature drops essentially had a nonmonotonous character along the capillary height. An experiment on water boiling heat transfer was performed on a plate nickel capillaries system with a height (H) of 0.04 m and an internal cross section of 4.9×0.5 mm. Results of the experiments [96, 99] were realized by extremely low heat fluxes ($q < 2 \times 10^3$ W/m²). Under these conditions, the researchers discovered the presence of extremely low average heat transfer coefficients that were even lower than the values found for the stabilized laminar liquid flow. The appropriate explanation for this can be done on the capillary flow boiling heat transfer model, which was suggested by Grigoriev et al. [96]. A version that was considerably different from the model presented in this study [96] was proposed by Labuntsov et al. [100]. Unfortunately, the calculations based on this the model for the short slit channels did not concur well with the corresponding experimentation data [99] for the forced flow.

It is possible to propose that a shift toward short channels will lead to the two-phase flow regime number essentially decreasing, and will allow simplifying the problem of generalizing experimentation data. This was recently confirmed by research on processes in the short slit channels.

The original experimental data were published by Leontiev [101] for water boiling at the slit channel with transparent walls. The heat input was realized at the channel's base. A special means was formulated to attain the thermal regimes with a single nucleation vapor bubble center action to initiate. The slit's height was 100 mm; the gap value was changed from 0.5 to 15 mm; the heat flux densities ranged from 0.2 to 4 kW/m², and the heat carrier used was water by atmospheric pressure. The heat transfer intensity dependency on the heat flux was obtained, and it was observed to be the same as for pool boiling without any dependency on gap size. It led to the conclusion that the heat transfer intensity by the boiling inside slit channel did not depend on the slit's gap size. However, it should be noted that as the boiling was caused only on the slit's foundation, that is, the conditions of vapor phase removal were approximately the same as by pool boiling, the possible "effect of the slit" was absent.

The heat input conditions were changed in the next set of experiments [102–104]. One-side heat input to the lateral slit's surface was realized, with the corresponding temperature and heat flux measurements by simultaneous visual observations for the boiling process over the nonheated transparent wall. The experiments were conducted for water, ethanol, and water + ethanol mixtures. It was discovered that the slit's boiling heat transfer coefficient increased by the slit's gap decreasing from 2.5 to 0.5 mm, and the coefficients increased by 2–4 times.

In the investigation [102–104], a weaker dependency of the heat transfer coefficient on the heat flux was noted ($\alpha \approx Cq^{1/3}$). It was weaker than in previous experiments [90, 91, 98, 99] performed for the same heat carriers. However, the main heat fluxes range was considerably higher (from 10^4 to 10^5 W/m²) in this series of tests [102–104], and with q growth, it was noted by Gogolin et al. [93] that the influence q on α weakened. It allowed an explanation of the mentioned regularities.

Unfortunately, visual observation results [102–104] were not included in the report. Comparison with the data of other authors was likewise lacking. The figure presented in Ref. [103] allows observers to propose that the "isolated bubbles" regime was observed taking into account that due to the heat flux level and the dependency $\alpha(q)$ view it had to be either "the sticking bubbles" thermal regime or "slug" one. The slit's boiling heat transfer model is developed and suggested in Refs. [103, 104]; unfortunately, this model is compared only with its own experimental data.

The experimental data of boiling in the inclined slit channels heat transfer are presented by Leontiev [103] as the data of boiling by weak gravity conditions. This lead to the following general conclusion: the relative heat transfer coefficient did not depend on heat flux, the slit's inclination angle (that, is the gravity force). This conclusion, associated with the independency of the slit's boiling heat transfer regularities, showed that the dependency of $\alpha(q)$ was not on the inclination angle changing.

It could be the sequence of many other factors, such as complex influence over the natural circulation conditions and change in parameters. The working element was included into the loop of the natural circulation. The number of these factors was extremely high. However, Leontiev et al. [103] did not analyze their influence on the process. This explains why, by the experimentation of the "slit channel boiling" heat transfer, it was preferable to organize research efforts to study independently the influence of various factors, including such factors as inlet channel parameters (W_0, X, etc.) [93].

From this point of view, it was of interest to conduct research on boiling heat transfer associated with the horizontal slit's channel, submerged into the pool boiling heat transfer. It could be considered an independent study of the clear "slit's effect" and could also be regarded as any limit case of the corresponding study for the inclined slit channels. These experiments could be related to the experimentation without any gravity conditions. However, even in the horizontal channels, gravity forces have some important influences on the different parts of process.

The combined experimentation of this process heat transfer, internal boiling characteristics, and critical heat fluxes (CHF) for the boiling inside the horizontal slits were stated and incorporated into the auspices of the Odessa Academy of Refrigeration (OSAR) and into the Institute of Applied Physics of Moldavian Academy of Sciences (APIMAS).

The experiments were conducted by the author's team at OSAR. The experiments were done in the plate horizontal slits, submerged into the pool of the heat carrier (water or ethanol). The slit's gap was changed from 0.15 to 5 mm. The experimental working elements: electric heating was provided via the direct electric current; sizes were varied in the following range — 0.13×0.015 and 0.1×0.03 m; heat fluxes densities were changed from 1000 W/m² to the CHF values; saturation pressure was varied from 0.036 to 5 bar. During the investigations, the heating surface type (copper, nichrome) was also changed as well as the heater's thickness (from 0.05 to 0.20 mm), but the influence of the heater's material thermophysical properties on boiling heat transfer regularities were not observed.

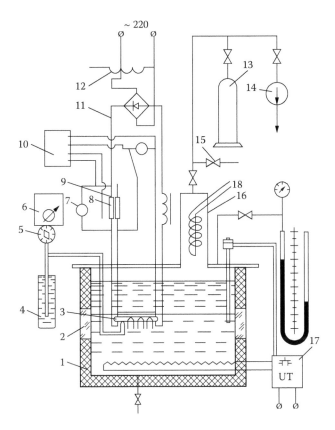

FIGURE 4.5 Schematic diagram of the experimental setup for the combined experimentation of boiling heat transfer, critical heat fluxes (CHF), and internal process characteristics inside the horizontal slits. 1, Chamber; 2, illuminator; 3, working element; 4, cold solder; 5, switch; 6, potentiometer; 7, millivoltmeter; 8, sample of electric resistance; 9, additional electric resistance; 10, double electric bridge; 11, electric current source block; 12, electric tension controller; 13, ballast capacity; 14, vacuum pump; 15, atmospheric valve; 16, condenser; 17, electric energy source of thermal-control heater; 18, refrigerator.

The working element design and experimental set-up schematic are given in Figures 4.5 and 4.6. The experiments were conducted with the working elements in the vertical position. The results for the horizontal position of the working elements are shown in Figures 4.7–4.10.

It was also observed that the boiling heat transfer enhancement takes place in the entire volume, when the slit's gap was lower than any size value. The heat transfer coefficient growth began when the gap size approached the vapor bubble departure size. Simultaneously, the CHF began to decrease. By the boiling in slit's channel, a different type of the dependency $\alpha \approx Cq^m$ was discovered. The values m by the atmospheric and increased pressure coincided with their own values in the pool boiling ($m = 0.6$–0.8). The thermal regimes with $m > 1$ appeared in the vacuum. These regimes, as it is known, have to be related to the unstable regimes in the thermal state.

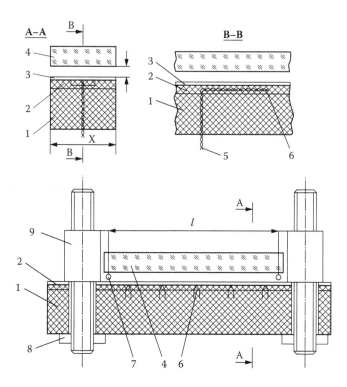

FIGURE 4.6 Working elements of the experimental setup [105–109]. 1 and 2, heat-insulating lining; 3, heater; 4, glass; 5, thermocouple output wires; 6, thermocouple head; 7, calibrating lining; 8, jamb nut; 9, current collector.

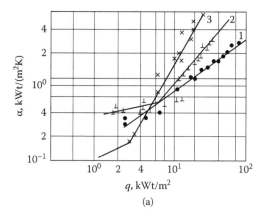

FIGURE 4.7 Influence of the horizontal slit's gap on ethanol boiling heat transfer, based on data from Refs. [105–109]. For parameter p_s: (a) 0.1 bar; (b) 0.2 bar; (c) 0.5 bar; (d) 1.0 bar. 1, Pool boiling; 2, $s = 1.0$ mm; 3, $s = 0.5$ mm.

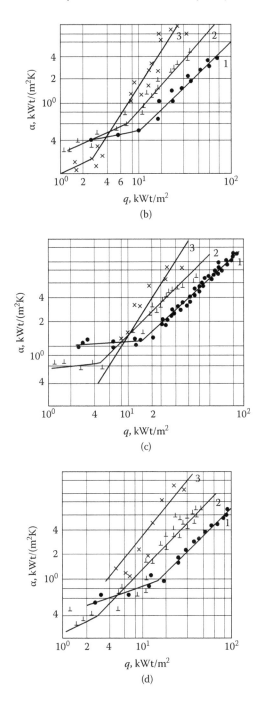

FIGURE 4.7 *Continued*

Research was conducted on the thermal regimes and the internal process characteristics of boiling inside the horizontal slits [105–109] for physical imaginations about the slit's boiling heat transfer mechanism development and the corresponding model working out.

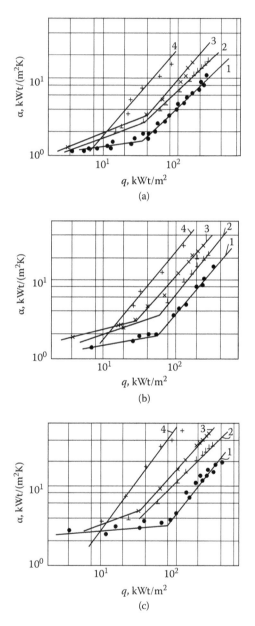

FIGURE 4.8 Influence of the horizontal slit's gap on water boiling heat transfer, based on data from Refs. [105–109]. For parameter p_s: (a) 0.06 bar; (b) 0.1 bar; (c) 0.2 bar; (d) 0.5; (e) 1.0 bar. 1, Pool boiling; 2, $s = 1.0$ mm; 3, $s = 0.5$ mm; 4, $s = 0.25$ mm.

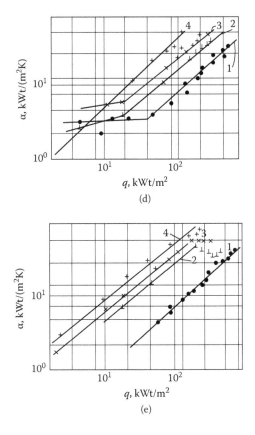

(d)

(e)

FIGURE 4.8 *Continued*

The APIMAS investigations were associated with the heat transfer, CHF, and internal characteristics (of such dielectric liquids as refrigerants, diethyl ether, pentane, etc.) boiling. These experimental results were published in several reports [113, 115, 116, 264]; they reflected the study on the influence of the electric field on the horizontal slit's boiling heat transfer. In addition, by this experimentation, separate data for the boiling without the electric field application were also obtained.

Next, the following exprimental conditions were used. The horizontal slits were submerged into the pool. These slits were organized from one side by the special heating surface and from the other side by the transparency plate. The last one was covered by electric resistive transparency layer. This allowed its use as the surface for the visual observation and for the electric field application. These were used in the working elements for direct and indirect heating. The working elements with direct heating were analogous ones, and were used in OSAR experiments. The metal heat transfer surface (copper, foil) was used in the thermoresistor, and it had the following dimensions: $b \times l = 10 \times 70, 20 \times 70, 30 \times 70$ mm.

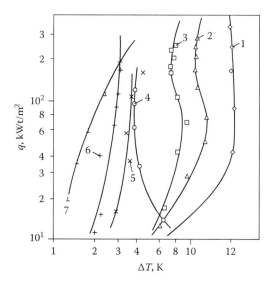

FIGURE 4.9 Boiling temperature curves for horizontal plate slits with width (b) = 15 mm, based on data from Refs. [105–109]. p_s = 0.1 bar; 1, pool boiling; 2, s = 1 mm; 3 and 4, s = 0.5 mm; p_s = 1.0 bar; 5, pool boiling; 6, s = 1 mm; 7, s = 0.5 mm; $\Delta T = T_w - T_s$.

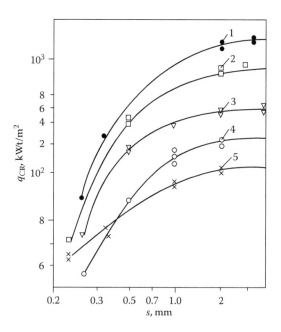

FIGURE 4.10 Influence of the slit's gap (s) on CHF by boiling inside horizontal plate slits, width (b) = 15 mm [105–109]. Water (p_s): 1, 5.0 bar; 2, 1.0 bar; 3, 0.5 bar; 4, 0.036 bar. Ethanol: 5, p_s = 0.1 bar.

The massive heater was used as the heat source for indirect heating conditions. The principle of the "thermal edge" schematic was used, with the heating part length $L = 250$ mm and diameter $d = 32$ mm. The CHF were determined by the abrupt temperature growth and were fixed with the help of a special automatic recorder. It automatically turned off the source of current if local heat transfer surface began to overheat; the goal was to preserve this surface from destruction.

The experimentation was performed by the atmospheric pressure. The heat fluxes q changed in the range 10^4 to 6×10^5 W/m²; the slit's gap changed from 0.5 to 4.0 mm. The typical experimentation results, which show the slit's gap influence on the boiling heat transfer intensity, are presented in Figure 4.11.

Here we can see the experimentation data that was obtained for the condition without the external plate. It was the pool boiling heat transfer data. The last ones had a good agreement with known data of other researcher's experimental results, including the known refrigerants' generalized correlations.

It follows from these experimental data that the heat transfer enhancement action of the slit was also caused by the slit gap's lower departure vapor bubble diameter (Laplace constant). However, it causes (Figure 4.11) an essentially different character of the dependency between heat transfer coefficient and heat flux for these liquid dielectrics with low thermal conductivities in comparison to other heat carriers boiling at the similar conditions (Figures 4.4 and 4.7). Index m in the dependency a $\approx Cq^m$ coincides or surpasses the corresponding value for the pool boiling case. When the slit's gap decreases the value of index m reduces too to 0.5...0.6. and even $m < 0$ at slit's gap $s = 0.5$ mm.

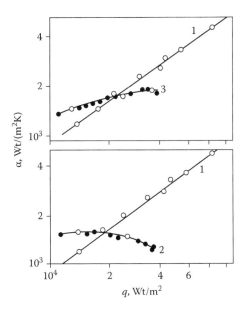

FIGURE 4.11 R-113 boiling heat transfer in horizontal slits' dependency on heat flux, based on data from Refs. [110–113]. 1, Pool boiling; 2, boiling in the slit, $s = 0.5$ mm; 3, the same, $s = 1$ mm.

It was especially interesting to note that when q was increased, the slit's enhancement action was not only lower, but it was a considerably lower heat transfer intensity for slit's boiling than for pool boiling even though similar conditions were maintained (liquid, pressure, etc.) It was especially clear for the following condition: $s = 0.5$ mm, $p = 1$ bar, R-113, and $q > 2 \times 10^4$ W/m^2. However, as long as the view of the dependency did not contain some abrupt change, it was observed that it was not associated with any crisis phenomenon.

This explanation and understanding became possible only after a series of special studies were conducted, accompanied by visual observations of the slit's boiling inside the horizontal slits. These observations showed that the significant change in the heat transfer coefficients' dependencies were associated with "drying spots" formation and their development, beginning from the very small heat fluxes (lower than 10^3–10^4 W/m^2).

The slit's influence on the CHF by the data [111, 115, 116] was observed to be the same as for the other heat carriers, which were reported in other studies [105–107, 109, 114]. It was confirmed that CHF decreased with the slit's gap decrease and, in particular, when the gap became smaller than the vapor bubble departure diameter.

The vertical and horizontal slit boiling heat transfer and CHF regularities have been of special interest in association with the study on boiling on finned surfaces. The heat transfer condensation experience on the application of finned surfaces for refrigeration machine condensers show that they are extremely effective. This surface technology was very well used in the field of refrigeration, which is why the experiments on refrigerant boiling on these surfaces were initiated. This field attracts interested professionals from other parts of the globe, including a number of Soviet researchers who have made notable contributions to the field [117, 118]; a review is presented by Gogolin et al. [93]. Typical results are given in Figure 4.12. They show the low-finned surfaces boiling heat transfer enhancement in comparison with boiling on smooth tubes. Let us note that heat transfer

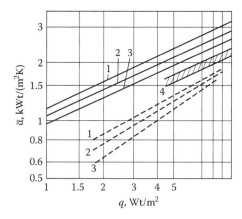

FIGURE 4.12 Comparison of average heat transfer coefficients for low-finned tubes (entire lines) and for smooth tubes (shaded lines) (R-22 boiling); results for six tubes [93].

coefficients have been defined, as they relate to the whole finned surface, and products in the former USSR industry copper low-finned tubes and heat carriers in the boiling temperature range –30°C to +20°C can be calculated by the next empirical correlations [93]:

$$\bar{\alpha} = C_0 q^m P_0^{0.25} \times \varepsilon_n$$

where $C_0 = 18.3$ and $m = 0.5$ for R-12, and $C_0 = 33$ and $m = 0.45$ for R-22; ε_n is a parameter accounting for the finned tubes number in bundle influence on the average heat transfer coefficient.

Only the qualitative explanation was given for the boiling heat transfer on low-finned surfaces enhancement. It could be suggested that by the low-finned surfaces boiling, the slit's phenomenon could play the defined role on heat transfer enhancement. However, the experimentation conditions described in Refs. [117, 118] and other publications, in what was given as the generalization of the corresponding data, did not allow practitioners to really determine the physical reasons for the heat transfer enhancement. However, we note that low-finned tubes boiling have the following peculiarities: existence of a dead-end zone, the slit's gap changeability, and nonisothermal heat input conditions.

The boiling peculiarity conditions were studied experimentally and analytically. The review and analysis of many papers devoted this problem study was done by Tiktin [119]. Generalization reviews on numerous experimental works performed at the High Temperature Institute of the USSR Academy of Science (HTIUAS) are presented in Ref. [120]. It was done as for boiling on separate fins for the slit's system. It was shown in Refs. [119, 120] that boiling on the unisothermal surface has no influence on the total boiling heat transfer characteristics.

The particular experimentation method was developed to include the precise measurements of the local wall temperatures and heat fluxes by boiling on the non-isothermal surface (Refs. [121, 122]), and it allowed the reliable determination of the boiling heat transfer curves at non-isothermal and isothermal conditions for both pool boiling and slits conditions. It was shown that some peculiarities occur only in the transient boiling regime and when pulsating vapor volume was formed at the dead-end zone.

It caused also that the heat transfer regularities on parts washed by the two-phase flow in the slit, and in the pool were the same. These experiments were made for the slits with a gap size not less than 3 mm. Such conditions for water and especially for the refrigerants cannot be not appropriately considered as special condition in comparison with pool boiling, that is, the corresponding regularities have to be identical.

The OSAR experiments, which were conducted by the author's team, showed that the regularities for the short slit's channels did not depend on their position in the gravity field. It was natural to suggest that the slit's boiling inside short vertical channels has to agree with the analogous ones for the horizontal slits. Confirmation for this suggestion was obtained in the experimental results presented by Bezrodny et al. [123, 124] and Sosnovsky [125]. The experiments [125] were conducted on models of the fin's gaps of the long channels in which the heat input was realized by electric direct heating. Parameter ranges were as follows: $20 \leq H \leq 50$ mm (H was the channel's height), the gap $0.3 \leq s \leq 10$ mm, the width $6 \leq b \leq 18$ mm; the observed angle $0 \leq \varphi \leq 150°$; the working pressure $1.0 \leq p \leq 5.0$ bar.

The experimental samples' design variants are given in Figure 4.13. The typical empirical dependencies of the short vertical slit boiling, submerged in the pool, heat transfer, and CHF are given in Figures 4.14 and 4.15. These dependencies showed good qualitative agreement with the experimental data of other authors obtained for the horizontal slits conditions. Experimental data (Table 4.1; [125]) were treated as empirical correlations in the following form for the determined range of every parameter's changing: for R-11 by $1.0 < p < 4.6$ bar, $0.3 < s < 1.2$ mm; $4 \times 10^3 \leq q \leq q_{CR}$ W/m², $\alpha \approx Cq^n s^m$.

For R-113 by $1.0 < p < 2.6$ bar; $0.3 < s < 2.0$ mm; $4 \times 10^3 \leq q \leq q_{CR}$ W/m².

$$\alpha \approx C_i q^{nl} s^{ml}$$

$$q_{CR} = Cq_{CR0}(s/H^{0.5})^k$$

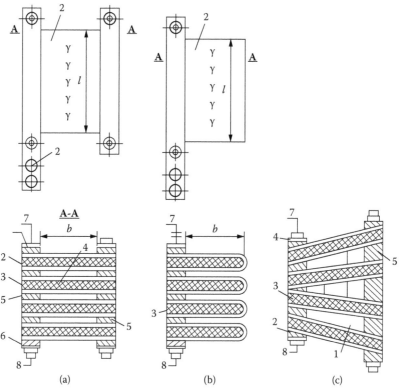

FIGURE 4.13 Working elements of the slit's channels design [123, 125]. (a) Plate parallel channels of the closed type; (b) opened type; (c) trapezoidal type. 1, Slit's channels; 2, heat input elements; 3, glass-textile plates; 4, metallic foil; 5 and 6, cooper and textile plates and wedges, which separated from each other; 7, isolated screws; 8, cooper electric current joints.

$$C = 29.1; \ k = 0.7 \ \text{by} \ 0.001 \le s/\sqrt{H} = 0.01$$
$$C = 1.0; \ k = 0 \ \text{by} \ s/\sqrt{H} = 0.01$$

where q_{CR0} is the CHF by pool boiling.

TABLE 4.1 Empirical Constants Values for the Experimental Data of Sosnovsky [125]

p_s, bar	C	C_1	n	n_1	m	m_1
1.0	0.19	0.65	0.35	0.38	−0.90	−0.67
1.6	—	0.97	—	0.42	—	−0.58
2.5	0.13	—	0.45	—	−0.84	—
2.6	—	0.79	—	0.49	—	0.50
4.6	0.04	—	0.60	—	−0.75	—

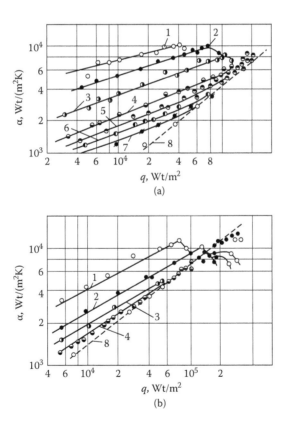

FIGURE 4.14 Refrigerant R-11 boiling inside the vertical slits heat transfer by different saturation pressure ($b = 10$ mm, $l = 50$ mm), based on the data from Ref. [125]. (a) $p_s = 1.0$ bar; (b) $p_s = 4.6$ bar. For parameter s: −1–7, 0.3, 0.5, 0.75, 1.2, 2.0, 3.0, 5.0 mm; 8, calculation for pool boiling conditions based on the Danilova formula base [93].

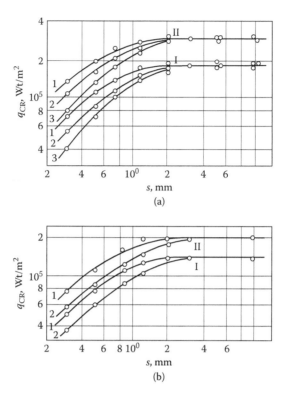

FIGURE 4.15 The CHF dependency on the slit's gap s value ($b = 10$ mm), based on data from Ref. [125]. (a) R-11; I, $p_s = 1.0$ bar; II, $p_s = 4.6$ bar; 1, $l = 50$ mm; 2, $l = 35$ mm; 3, $l = 20$ mm. (b) R-113; I, $p_s = 1.0$ bar; II $p_s = 2.6$ bar; 1, $l = 50$; 2, $l = 20$ mm.

Vaporization by boiling on surfaces with coverings, which did not reliable contact with surfaces, can also be related to the case of the narrow slit boiling. First of all, it could be considered as boiling on the so-called "clothed covering." The positive influence on boiling heat transfer enhancement was discovered at this covering in some of the experiments with the reliable non-heat conductive coverings on the heating surface (sintering or plasma-sprayed). It was initially suggested with respect to well-known physical imaginations that it had to lead to considerable enhancement as soon as the application of such nonwettable covering as Teflon can help in achieving earlier boiling initiation and as a result of a considerable increase in the possible heat transfer enhancement, especially for vacuum conditions. However, it was discovered that when any Teflon net with some holes ("clothed covering") was used, but without reliable contact, it led to a significantly stronger influence on heat transfer than with the reliable contact covering with the same material [126].

The representative experimental data of boiling on "clothed covering" heat transfer surfaces, presented in Refs. [127, 128], were reviewed and generalized by Gogolin et al. [93]. Experimental data of boiling on the "clothed covering" surfaces are presented in Figure 4.16.

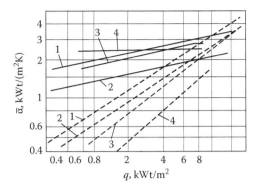

FIGURE 4.16 Boiling on "clothed coverings" surface heat transfer average coefficient correlations (entire lines), and on smooth surfaces (shaded lines) as a function of heat fluxes ($d = 25$ mm). Ammonia (T_s): 1, 20°C; 2, 20°C; 3, 30°C; 4, −30°C.

Obviously, the greatest heat transfer enhancement take place at the first turn, by the relatively small heat fluxes ($q \le 10^3$ W/m²). Enhancement values were decreased when q was growing before 10^4 W/m²; when q reached the value of 10^4 W/m², the enhancement effect "depressed." The dependency $\alpha(q)$ for cases of "clothed covering" from the glass woven screens was close to the analogous one produced from the metallic screens that it is deferred by the weak dependency of α on q (at the correlation $\alpha \approx Cq^m$; $m \approx 0$–0.3). This "clothed covering" type has to be related to the porous coatings. Vaporization by the small perforation density on the "clothed covering" surface will take place by the slit's boiling, that is, $m = 0.6$–0.8.

The cryogenic liquid slit channel and capillary boiling heat transfer and CHF experimentation were developed in recent years, including the liquid helium case.

As a rule, these are experimental studies; results were generalized via empirical correlations. Grigoriev et al. [132] did not consider the heat transfer and crisis phenomenon in his study of flow boiling, including the flow inside the slits and capillaries.

Let us consider some experimentation of heat transfer by cryogenic liquid boiling inside the small size channels. Experimental results of the liquid helium boiling inside the annular channels by the gap $s = 0.5$, 1.0, 1.5, and 2.0 mm, and by the length $l = 100$ mm were presented in two studies [129, 132]. These experiments were conducted by the special initial and outlet parts, without heating, existing. Aside from this, analogous experimentation was conducted by the same authors using tubes with the following parameters: 4×0.26, 7×0.3, and 12×0.3 mm; the length was $l = 300$ mm. Heating was realized via electric current. The working parts were submerged into the liquid pool in such a manner that vaporization initiated the strong circulation.

The circulation intensity depended on the conventional vapor phase velocity

$$w'' = 4 \times (l/d) \times (q/r\rho'')$$

where d and l are the channel's diameter and length. The influence of l/d on heat transfer did not take place by $0 < l/d < 80$. When l/d exceeded 80, the heat transfer essential enhancement was observed. By $q \le 0.25q_{CR}$, the heat transfer law was

$\alpha \approx Cq^{0.7}$; by $q \geq 0.25q_{CR}$, the heat transfer intensity was $\alpha \approx Cq^{0.4}$. This could be explained by the appearance of "dry" spots on the heat transfer surface. Analysis of experimental data show that CHF regularities depend on the channel's width. By $s > 0.9$ mm and $l/d < 15$, the experimental data could be generalized by the follwing empirical correlation:

$$q_{CR}\left[\frac{1}{q_{CR0}}+\left(0.0063\frac{x}{d}-0.04\right)\frac{l}{d}10^{-4}\right]^{-1}+C_1^{\frac{l_{EV}}{d}}$$

Here, l_{EV} is the initial part's length; in this case, $C_1 = -60$ (if l_{EV} is the length of the outlet part, then $C_1 = 30$; by $l/d > 15$, l/d has to insert 15 instead); q_{CR0} is the CHF by this liquid pool boiling. It could be determined on the known hydrodynamic heat transfer crisis theory base using the known constant $C = 0.095$. By any determined size decreasing lower than $s = 0.9$ mm, q_{CR} was decreased and the experimental data were generalized by the following generalized formula:

$$q_{CR} = \left[\frac{1}{q_{CR0}}+0.14\frac{x}{d}10^{-4}\right]^{-1}$$

It was proposed for heat transfer calculation for the discussed condition that the following formula be used:

$$\alpha = 36\times10^{-3}q^{0.7}p^{0.6}\beta^{0.3}$$

where q is used at the next scale p (W/m²) at Pa; $\beta = q_0/q$; $q_0 = 0.25q_{CR}$, at $q \leq q_0\beta = 1$.

The experiments of Petukhov et al. [133] and Shildkret [134] were conducted with the boiled liquid helium inside the tube with the following parameters: $d = 0.8$ mm, and length $L = 181$ mm by the initial part's length $l = 60$ mm, which was heated by the direct electric current. The helium circulation organized by as-forced flow in such a manner that the mass flow velocities were changed from 48 to 313 kg/m² sec, the mass vapor quality of the flow into the channel's inlet was changed from 0 to 0.7, and in the channel's outlet from 0 to 1.0; the heat flux densities ranged from 20 to 7000 W/m². The results showed that it existed at any critical mass vapor quality X_{CR}; when $X < X_{CR}$ the heat transfer intensity by the given q and ρW did not depend on X. But when $X > X_{CR}$, the heat transfer crisis ensued. The heat transfer regularities at the field $X < X_{CR}$ could be prescribed by the following empirical correlation:

$$\alpha = 55.7q^{0.7}(\rho w)^{-0.24}pd^{-0.5}$$

The influence of all other factors agreed very well with the slit and capillary boiling regularities of the low-heat conductivity liquids, that is, $\alpha \approx Cq^{0.7}$. With mass flow rate growth, the heat transfer intensity started to decrease as soon as the boiling partially depressed. The heat transfer coefficient α increased as ρw decreased, etc.

Therefore, we conducted the analysis of the greatest heat transfer and CHF experiments on the slits and capillaries conditions. Other experiments and analytical research have also been conducted on the slit and capillary boiling internal characteristics and the different hydrodynamic phenomenon study by vaporization under these conditions. It is extremely important for the development of fundamental models and working toward the main goal of defining the right form of experimental data generalization.

Some of these works were conducted commonly with the study on the boiling processes integral characteristics [76, 77, 97, 98, 109, 112, 113]. Other studies were mainly devoted to the slit's boiling mechanism. Let us consider these investigations and attempt to define the physical imaginations on the hydrodynamic phenomenon and vaporization regimes by slit and capillary boiling.

4.2 HYDRODYNAMIC PHENOMENA AND VAPORIZATION IN SLIT AND CAPILLARY CHANNELS

When we use visualization of the vaporization in vertical slit channels, it is possible to make the typical imagination of the two-phase flow inside the narrow channels in Figure 4.17.

For some local cases of the annular flow, boiling exists at the channels' entrance, and the economizer part is absent. It can occupy the biggest part of the channel in other cases. This regime can be absent at some flow regimes in short channels. The stuck vapor bubble flow regime is usually absent in the capillaries in the

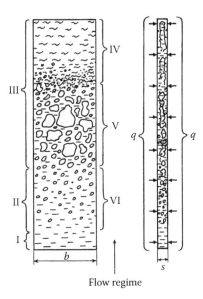

Flow regime

FIGURE 4.17 Boiling flow regimes inside the plate slit channel, based on data from Ref. [93] ($s = 2$ mm; $q = 20$ kW/m^2). The different boiling heat transfer mechanisms: I, single-phase convective flow zone; II, surface boiling zone; III, convective evaporation zone. Flow regimes: IV, slug regime; V, cork stuck vapor bubble zone; VI, bubble zone; b, channel width; s, the channel's gap.

different slits. However, the most general picture of the possible two-phase flow regimes could be presented as shown in Figure 4.17. The most important parameter of the two-phase flow is the flow velocity w_C of the liquid slug. The slug velocity in the adiabatic flow by the gravity could be presented as

$$w_C = 1.2(w_0' + w_0'') + 0.35\sqrt{gL}$$

where L is the channel characteristic size, and the index <0> relates to the inlet parameters. If we consider that by the heat input the slug's growth takes place in the two-phase flow direction, then the liquid slug average velocity can be prescribed as

$$w_s' = \frac{dz}{d\tau} + 1.2(w_0' + w_0'') + 0.35\sqrt{gL}$$

This analysis allows us to make some conclusions relating to the two-phase flow regimes at the slits and capillaries:

- If the following condition take place,

$$\frac{dz}{d\tau} \gg 1.2(w_0' + w_0'') + 0.35\sqrt{gL}$$

  that is, the characteristic velocity of the two-phase flow in the slits is determined by the vapor bubble growth velocity. Then as the hydrodynamic regularities, both the heat transfer ones will be defined by the vapor bubbles dynamics.
- The second possible two-phase flow regime will be associated with the forced convection conditions in the slits and capillaries (the regime takes place, first of all, in the slits and capillaries by the liquid forced supply and small heat fluxes), where the next condition has to be fulfilled:

$$(w_0' + w_0'') \gg \frac{dz}{d\tau} + 0.35\sqrt{gL}$$

- The third possible two-phase flow regime will be associated with the action of Archimedean forces, which are responsible for the vapor slugs' lifting movement. This regime is determined by the following condition:

$$0.35\sqrt{gL} \gg \frac{dz}{d\tau} + 1.2(w_0' + w_0'')$$

This regime will be determined in the vertical channel with the high internal diameter $D_E \cong L$, which has to be much larger than the vapor bubble departure diameter by the pool boiling.

Therefore, the characteristic two-phase flow by the slit's boiling velocity is essentially connected with the value $dz/d\tau = \bar{Z}$ determination.

Let us use the vapor growth energy schematic and propose that, during the vapor slug growth, the condition $q - const$, as a rule, will take place. In addition, we will consider that the entire heat coming to the two-phase border will transfer in the latent heat that it uses on the evaporation process and the corresponding vapor slug volume is increasing.

Then, by the one-dimensional vapor slug growth schematic, when the slug length is considerably larger than its cross size (gap or diameter), we obtain the following formulas:

$$\frac{q\Pi}{r\rho''}d\tau = Fdz; \quad \dot{Z} = \frac{q\Pi z}{r\rho''}$$

$$z = const \, \exp\left(\frac{q\Pi\tau}{r\rho''F}\right)$$

where Π is the parameter and F is the cross section area.

The constant could be determined from the following suggestion that, at the initial moment $Z \equiv S$, we have

$$z = s \, \exp\frac{q\Pi\tau}{r\rho''F}; \quad Z = \frac{q\Pi s}{r\rho''F}\exp\frac{q\Pi\tau}{r\rho''F}$$

The average velocity of prolonged height vapor slug growth will be

$$\bar{Z} = \frac{1}{H}\times\int_{s}^{H}Zdz \text{ by } H \gg s$$

$$\bar{Z} = \frac{1}{H}\frac{s^2}{2}\frac{q\Pi}{r\rho''F}\exp\frac{2q\Pi\tau}{r\rho''F}$$

This leads to the next correlation: $\bar{Z} = q\Pi H/2r\rho''F$.

Using the transit to some undimensional variables (the similarity parameters), it is possible to use existing regime conditions to obtain at the undimensional form:

- The regime by which the two-phase hydrodynamic regularities are determined mainly by the growing vapor bubble's dynamic will be characterized by the next correlation:

$$Re_H = \frac{qL}{r\rho''v'} \gg \sqrt{Ar}, \quad Ar = \frac{qL^3}{v'^2}\frac{\rho'-\rho''}{\rho}$$

  The determined parameter is the Reynolds number (Re).
- The regime of the determined influence on the two-phase hydrodynamics of the external forces are characterized by the next condition:

$$Fr = \frac{w^2_0}{gL} \gg 1$$

- The regime of the determined influence on the two-phase flow gravity forces:

$$\sqrt{Ar} \gg Re_H; \quad Fr \ll 1$$

  The determined parameters is the Archimedes number (Ar).

If the acting forces are approximately equal, then it could be correct to take any combined parameter, for example, $Re + \sqrt{Ar}$ or $Re^2 + Ar$, etc.

The important characteristic of the two-phase flow is the volumetric vapor quality (vapor volumetric concentration), φ. The special meaning has the dependence of δ definition on the main factors. If we have the external forces determine the influence regime (from the aforementioned classification position), this value calculation can be realized using the known correlations, for example, from Refs. [135–137], for the slug's regime:

$$\varphi = 0.833\,\beta\,\frac{1}{1+0.25\sqrt{Fr'}} \tag{4.1}$$

where β is the volumetric flow rate vapor quality.

Therefore, for the calculation of δ, it is necessary to know β or x, the mass flow rate vapor quality. The known uncertainty in these calculations will be associated with the following process peculiarity (it is known that in the real evaporation the entire heat is not transferred into the latent heat). Let us point out the next considerations, which could be useful for such cases. The maximum x, β, and δ will correspond with the total transfer of the whole heat input into the latent heat, that is

$$4(x)_{\max} = \frac{q\Pi H}{rG} \to \beta \to \varphi$$

The minimum aforementioned values correspond to the suggestion that only the heat transferred over the microlayer liquid film will be transient into the latent heat, that is

$$(x)_{\min} = \frac{q\Pi H}{rG}\,\varphi \to \beta_{\min} \to \varphi_{\min}$$

It means that the most probable could be taken as the following estimation:

$$x = \frac{q\Pi H}{rG}\,\sqrt{\varphi} \to \beta \to \varphi \tag{4.2}$$

The common consideration of Equations (4.1) and (4.2) allows us to determine the value $\bar{\varphi}$. The most complex problem appears at the natural convection regimes, when δ becomes an extremely complex process function. The approximated calculations are possible, if all local and length hydraulic resistance coefficients for the two-phase flow circulation the whole way are known as the main two-phase flow parameters w_0, δ (w_0 is the liquid conventional velocity). Unfortunately, these data are not available in most cases, including such representative works as Refs. [92, 96, 97]. Knowledge of these values is necessary not only for the determination of local resistances, but also for the local pressure gradients dp/dz directly caused on the vapor slugs movement, and the determination of local liquid microlayer thickness determination. This is why the single way is to use the well-known two-phase hydrodynamic equation [135–137].

Let us consider what correlations can be used and obtained for the liquid microlayer thickness determination at the growing vapor bubble base. We will use the recommendation for the layer initial thickness definition of the corresponding formulas in Ref. [138]. If we use it, we can see that it requires knowledge of two values: the defined velocity w_0

at $\bar{Z}$ and the pressure gradient dp/dz. Let us imagine in the total view the expression of the pressure determination inside the vapor slug for finding the dp/dz dependency:

$$p'' = p_\infty + \Delta p_\sigma + \Delta p_{in} + \Delta p_v$$

where Δp_σ, Δp_{in}, and Δp_v are the items of the drop pressure between vapor slug and pool, determined by surface tension, inertia, and viscosity forces action: $\Delta p_\sigma = \sigma \times (1/R + 1/s)$ (s is the slit's gap and R is the vapor slug radius at the cross parallel of the heating surface section), that is, in the main stage of the vapor slug growth, we will have

$$\Delta p_\sigma \approx 2\sigma/s$$

Δp_{in} is determined by known pressure losses on the flow acceleration. For two-phase flow, it will be

$$\Delta p_{in} = \rho' \frac{w_0^2}{2}\left(\frac{(1-x)^2}{1-\varphi} + \frac{x^2}{\varphi}\frac{\rho'}{\rho''} - 1 \right)$$

or, for the real values of ρ/ρ'', x, and φ,

$$\Delta p_{in} = \rho' \frac{w_0^2}{2}\left(\frac{(1-x)_{out}}{1-\varphi_{out}} \right)$$

The index "out" is related to the outlet cross section of the channel.

The comparison of the following values Δp_u and Δp_v is equivalent to the following values comparison

$$\left(\sum \xi_{loc} + \frac{C_f}{d_{eq}}H \right)\frac{1}{(1-\varphi)^n} \quad \text{and} \quad \frac{(1-x_{out})^2}{1-\varphi_{out}}$$

For slug regime: $1 - x_{out} \approx 1$; $n \approx 1.5, \dots, 1.7$.

Obviously, for the long channels

$$\left(\sum \xi_{loc} + \frac{C_f}{d_{eq}}H \right) \gg \frac{(1-x_{out})^2}{1-\varphi_{out}}$$

That is, the acceleration losses can be neglected, which is in contrast to the short channels.

$$\frac{C_f}{(d_{eq})}H \ll \frac{(1-x_{out})^2}{1-\varphi_{out}}$$

The main factor will be the inertial forces action. The first case has a place by the forced liquid supply and by the natural flow circulation into the "long channels." The approximated condition to differentiate the "long channels" could be presented by the following equation:

$$\frac{C_f}{(d_{eq})}H \gg 1$$

The second case has to be observed by the boiling into the "short channels." For the first case, we obtain

$$p = p_\infty + \frac{2\sigma}{s} + \frac{C_f}{d_{eq}} \frac{H}{(1-\bar{\varphi})^n} \rho' \frac{w'^2}{2}$$

$$w' = w_0 + 0.35\sqrt{gL} + Z \tag{4.3}$$

$$\frac{\partial w'}{\partial z} = \frac{\partial Z}{\partial z} \approx \frac{q\Pi}{r\rho'' F} \tag{4.4}$$

The prescription of formula (4.4) is that the entire heat in the long channels is transferred into the latent heat. If it is possible to neglect for the slug regime the liquid flow rate decreasing in account of the evaporation ($x \ll 1$), the flow rate constancy condition will have the form $(1 - \varphi)w' = const$, then

$$\frac{1}{1-\varphi} \frac{d\varphi}{dz} = \frac{d\omega'}{dz} \frac{1}{\omega'} \tag{4.5}$$

From (4.3), (4.4), and (4.5), we obtain

$$\frac{1}{\rho'} \frac{dp}{dz} = \frac{C_1}{d_{eq}^{1+m}} v'^m \frac{w'^{2-m}}{(1-\bar{\varphi})^m} \left(1 + \frac{2-m+n}{2} \frac{H}{w'} \frac{qk}{r\rho''s}\right)$$

Here, k is accounting for the heat input from one side or from two sides, and index m is determined by the liquid slugs' flow regimes.

When $w' \approx Z \cong qH\Pi/(2rF\rho'')$ by $m = 1$, then we obtain the following formula for liquid microlayer average thickness definition in the natural circulation:

$$\delta_{0L} = \frac{d_{eq}(1-\bar{\varphi})^{n/2}}{\sqrt{C_1(1+z)}}$$

Then, index $m = 0.25$ for the turbulent flow regime, in the plate channels and by the natural circulation $w' \cong ZF/(2Hs)$.

$$Re = \frac{w'd_{eq}}{v'} = \frac{2qH}{r\rho''v'} = Re_H$$

$$\delta_{0L}/2s = \frac{(1-\bar{\varphi})^{0.75}1.09}{Re_H^{0.37}}$$

The defined points of departure in the present chapter could be added by the "elementary" processes analysis, the associated conditions between their characteristics, and the corresponding boiling integral characteristics at the whole. They could be used for the slit and capillary boiling process models to build.

4.3 EXPERIMENTAL STUDIES OF VAPORIZATION MECHANISM IN PLAIN SLITS AND ANNULAR CHANNELS

The most detailed mechanism and narrow slit channel boiling internal characteristics research were conducted by the author's team and several other scientific groups; results were published in several papers [76, 77, 105, 106–113, 115, 116, 139–142]. Let us consider the methods and the experimental setups. Afanasiev et al. [76, 77] and Koba et al. [105–109] designed experimental setup for the study of the mechanism and internal parameters at boiling in slits. This setup was comprised of a 4 dm^3 volume stainless steel hermetic chamber with a rectangular cross section (see Figure 4.5). All six walls of this chamber had illuminators. There were two special heaters in the chamber's bottom area with a power of 2.5 kW. Besides it, on the top of the chamber were installed the electric power supply elements for the working parts of the high-frequency electric joint to the working part turning on into the measurement schematic and the electric potential wire joints. The working parts used were identical to those used for the boiling heat transfer study and critical flow. Two methods of the volumetric vapor flow quality determination were used: the video observation and the conducting one. Video capture was done by the SKS-1M camera in plain view.

The equivalent vapor slug volume radius was determined in the following manner: it was taken to be equal to the corresponding radius of the circle with an area equal to that deformed by the slit vapor slug. It was projected onto the plane of the chart paper and after that it was measured as the projection area F, occupied by the vapor slug. Then, the vapor slug volume was defined with the increasing scale coefficient K_0. $V = k_0 Fs$; $R \quad \sqrt{F}$, and from here, the equivalent zone radius, R.

The time interval is defined by the notes on the film edge by the special camera time notes. The bubble growth curves obtained from the experimental data are presented in Figure 4.18.

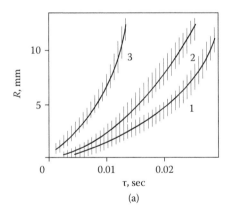

(a)

FIGURE 4.18 Vapor bubbles' growth curves by the boiling inside the horizontal slits ($b = 15$ mm; video data from Refs. [75, 76, 105–109]). (a) $p_s = 1.0$ bar; $s = 1.0$ mm; 1, $q = 175$ kW/m^2; 2, $q = 95$ kW/m^2; 3, $q = 24$ kW/m^2. (b) $p_s = 1.0$ bar, $s = 0.5$ mm; 1, $q = 172$ kW/m^2; 2, $q = 54$ kW/m^2; 3, $q = 24$ kW/m^2.

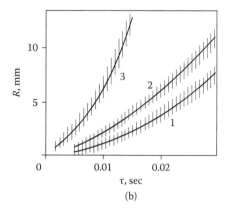

(b)

FIGURE 4.18 *Continued*

The conducting method was used to research the local and average real flow volumetric vapor quality determination by the vaporization into the slits. The usefulness of this method was associated with the real slit's geometry — the plate slit. The conducting method physical base was the significant difference of the liquid and vapor dielectric permeability. The method was consistent with the sensor's (plate electric condenser, formed by the working element external surface and the opposite plate surface) capacity to change while recording. It was directly connected with the internal process characteristics.

The design of the surface average vapor quality determination sensor is presented in Figure 4.19. Its form was determined by the working element design. Electrodes 1 and 2 are forming the electric condenser–sensor. The heating surface was used as electrode 1; the second one fulfilled simultaneously the function of the gap-formed surface, which was defined in such a way by the slit. The third and the forth electrodes are the protected ones. The sensor was produced from foil + textolite; electrodes 2 and 3 were glued by epoxy glue (ID 6) to electrode 4, produced from the plate stainless steel 1CH 18N10T. By the local vapor quality measurement the electrode tip area has to be minimal (not more than 1 mm²).

The results of the volumetric flow vapor quality measurement capacity methods and the possibility to use them for the liquid microlayer thickness definition are shown in Figure 4.20. The horizontal parts existing near the line $\bar{\varphi} = 100\%$ shows that vapor phase sizes became more than the corresponding sensor sizes and the difference between the oscillograms, and the note 100% namely determined the thickness of two liquid microlayers on the heating surface and the opposite one, which formed the necessary gap.

The processed pictures allowed us to obtain the representative information about the main slit boiling internal characteristics. More than 100 objects were studied as separate bubbles for the water boiling into the slits with the gap $s = 0.5$ mm by $0.1 \leq p_s \leq 1.0$ bar.

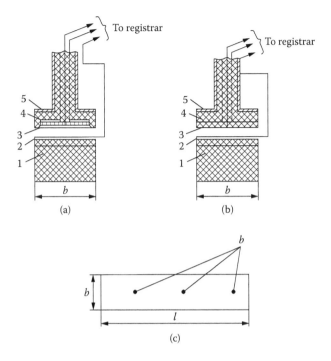

FIGURE 4.19 Design of local and average volumetric vapor flow quality sensors by capacity method [105–109]. (a) Average volumetric vapor quality sensor. 1, Insulator, base; 2, heater + electrode #1; 3, electrode #2; 4 and 5, protective electrodes #3 and #4. (b) Local volumetric flow vapor quality sensor. 1–5, Same data as in panel (a). (c) Local volumetric flow vapor quality sensors location schematic layout.

The important thing in the analysis was the comparison of the suggested physical imaginations relating the vapor bubbles growth law on the energy model base commonly with the suggestion about using constant heat flow with the experimentation results on it. This comparison is given in Figure 4.21.

Photo and video capture were used with the goal of determining some of the peculiarities of the vapor phase removal. The video allowed handlers to imagine some approximate schematics of the vapor phase removal from the slit. This was then used to work out the model. The volumetric flow vapor quality dependence

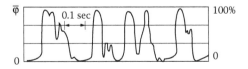

FIGURE 4.20 Average volumetric flow vapor quality measurement results by water boiling into horizontal slits with gaps equal to $s = 0.5$ mm and width $(b) = 15$ mm.

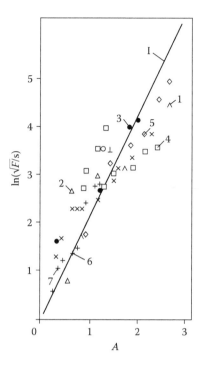

FIGURE 4.21 Vapor bubble growth regularities by water boiling into the horizontal slits with $b = 15$ mm. $p_s = 0.1$ bar, $s = 1$ mm; 1, $q = 160$ kW/m²; 2, $q = 90$ kW/m²; 3, $q = 45$ kW/m². $p_s = 1.0$ bar, $s = 0.5$ mm; 4, $q = 95$ kW/m²; 5, $q = 172$ kW/m²; 6, $q = 54$ kW/m²; 7, $q = 45$ kW/m²; $p_s = 1.0$ bar; $s = 1$ mm; $1 - \sqrt{F} \approx R \approx s \exp q\tau/(r\rho''s)$; $A = q\tau/(r\rho''s)$.

on the heat flux study was conducted by the capacity method using the following sizes: $b \times l = 15 \times 100$ mm; by the slit's gap $s = 0.5$, 1.0 mm, and the saturation pressure $p_s = 1.0$ and 0.1 bar. The heat fluxes range was from $q = 10^3$ W/m² to the CHF. For the different working conditions, the following parameters were defined: local and average volumetric flow vapor quality, the frequency and the two-phase mixture inside the slits pulsation's amplitude, the characteristic times of the boiling process stages.

The oscillogram analysis in Figure 4.22 shows that the slit's boiling character differed essentially on the pool boiling condition. It was confirmed by the video data. The slit's boiling by $p_s = 0.1$ bar and small heat fluxes including $q = 10^5$ W/m² was characterized by the large "delay" times, whose values were dependent on heat flux saturation pressures, and was in the following range: $10^{-2} \le \tau \le 4 \times 10^{-2}$ sec.

The slit's boiling was accompanied by the liquid's unsteady heating when the considerable liquid overheating was appearing. The absolute values of the overheating can be determined from the oscillograms using the known dielectric permeability of the water dependency on the temperature and pressure. The unsteady heating time was equal to the "delay" time.

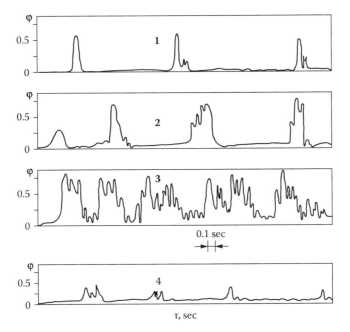

FIGURE 4.22 Average volumetric flow vapor quality by water boiling inside the horizontal plate slit with $b = 15$ mm oscillograms. Water: $p_s = 0.1$ bar; 1, $q = 20$ kW/m²; 2, $q = 48$ kW/m²; 3, $q = 85$ kW/m²; 4, $q = 5.3$ kW/m²; $s = 0.5$ mm.

The vapor slugs' growth inside the slit with the gap $s = 1$ mm took place along the slit with high velocity. As for the slit with gap $s = 0.5$ mm, this result was slightly changed and the vapor volume removal from the slit began before the vapor phase started spreading on the significant length of the slit. The average volumetric flow vapor quality was increasing with the heat flux growth by the slit's gap $s = 1.0$ mm and $p_s = 1.0$ bar, and by $q = 10^5$ W/m² its value became equal to $\bar{\varphi} = 60–65\%$. Then by the next heat flux increasing, the dependency $\varphi = f(q)$ was becoming weaker (Figure 4.26).

The volumetric flow vapor quality pulsation's frequency was increased with the heat flux growth. It was associated with the growth in vaporization centers number. The decrease in the pulsation's amplitudes took place in account of the different vapor slugs interconnection common influence, and appeared simultaneously at different places of the heating surface.

By the heat flux growth, before any "critical" value was reached, the real volumetric flow vapor quality approached its own critical value, $\varphi_{CR} = 0.85$. As seen from the high-speed video picture (Figure 4.25), because of the spontaneous character of the vapor slugs' appearance in the slit at high heat fluxes, their coupling does not take place. A thin liquid film was forming at the interface of growing and vapour slugs emerged.

Based on the local liquid film thickness experimental determination, it was proposed that it is dependent on the slit's position at the gravity field.

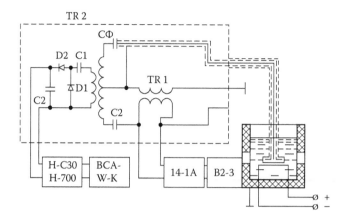

FIGURE 4.23 Average volumetric flow vapor quality and the liquid microlayer thickness measurement principal schematic by capacity method [105–109].

It was discovered that the sensors provided the possibility of direct tare determination for the liquid film thickness. The tare was produced with the mica plates with thickness 25, 50, and 100 μm, which were inserted step by step into the sensor's gap, and the signal was prescribed by the loop oscillograph of the type H080. The loop oscillograph worked at the fast regime and used the high-frequency galvanometer with the strip average speed movement of 10 m/sec.

This device's principal schematic layout for the liquid film thickness registration is presented in Figure 4.23. Liquid film thickness measurements were carried out on the working element with the following size: 15×100 mm and with the gap $s = 0.5$ mm by the pressure $p_s = 0.1$ bar. The specific heat fluxes were varied. The typical oscillograms allowed the definition of the liquid microlayer thickness. They are given in Figure 4.24.

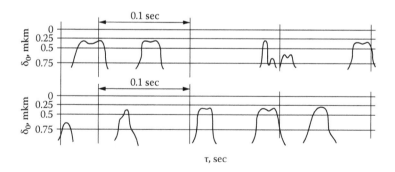

FIGURE 4.24 Oscillograms of the liquid microlayer's thickness under the growing vapor bubble by water boiling inside the horizontal plate slit [105–109]. Water: $p_s = 0.1$ bar, $s = 0.5$ mm, $b = 15$ mm, $q = 83$ kW/m².

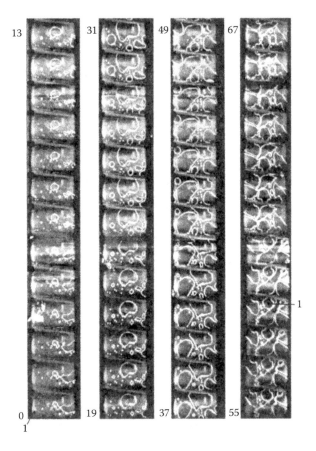

FIGURE 4.25 Horizontal slit channels water boiling high-speed video. Top: video speed, 1000 frames/sec [105–109]; $s = 1$ mm; $b = 30$ mm; $p_s = 1.0$ bar; $q = 116$ kW/m². 1, Vapor slugs (vapor volumes), frame number: 0–13, 19–31, 37–49, 55–67.

The typical experimentation's video results of the boiling into the horizontal slits are presented in Figure 4.25. The video data allowed us to define the whole process in all stages. It is seen that into the vaporization centers, located overall surface, the initial vapor bubbles were initiated; then they grew before the slit's gap size. The vapor bubble's growth speed at this stage was rather high and the corresponding process time was very small. Thus, based on the oscillogram data in the slit $s = 0.5$ mm by $q \cong 10^5$ W/m², $p_s = 1.0$ bar. The vapor bubble growth time from the nucleation size to the slit's gap is $\tau \leq 10^{-3}$. This explains why the time uncertainty, associated with the time measuring choice, had no essential meaning.

The vapor slug's volume deformation started when the vapor bubble reached the external plate. Then it began to stick and to spread along the heating surface. It continued before the phase border reached the slit's edge and the vapor slug began to disappear from the slit. Simultaneous vapor bubble removal from the slit into the pool was also observed. Then the waiting period came; the liquid overheating from the vapor slug whole slit's removal to the new vapor nuclei appearance took place. The waiting time was increasing in the subcooling boiling regimes by the saturation pressure decreasing. It is seen from the boiling processes oscillograms by $p_s \leq 0.1$ bar (Figure 4.22) both from the process characteristics times distribution histograms, obtained from the video.

The deforming bubbles occupied the largest part of the heating surface when the vapor bubble velocity was increased, essentially developing the boiling regimes. The average volumetric flow vapor quality was changed and became more weak. It was taken from the following values: from $\bar{\varphi} = 0.55$ at the undeveloped boiling regimes to $\bar{\varphi} = 0.7$–0.75 (Figure 4.26.). The heat conductivity over the thin liquid microlayer, forming near the vapor bubble base becomes the determined physical mechanism. The evaporating microlayer zone part of the whole process becomes the main direct experimental data, and contains about 30–40% by the average heat fluxes.

The obtained results confirmed the proposition about the existence of the thin liquid microlayer under the growing vapor bubble by slit boiling. This allowed us to define the microlayer average thickness. So, by $p_s = 0.1$ bar, $q = 8.3 \times 10^4$ kW/m², $s = 0.5$ mm, it is calculated as $2\delta = 20$–30 mkµ (Figure 4.24). The decrease in the average microlayer thickness was not discovered as it could have been seen from the oscillograms. The liquid microlayer thickness under the growing vapor bubbles was decreasing by decrease in the slit's height. It may be regarded as one of the major factors that could account for the heat transfer enhancement.

The nonmonotonous changing of the average liquid microlayer thickness in time is seen from these oscillograms. This development requires special attention, as soon as the liquid flow inside the liquid microlayer is proved. It is likely that the boiling crisis on the very thin (20–50 µm) heaters could be the result of some fluctuations in the liquid film and its possible rupture. It could be one of main reasons of the corresponding CHF experimentation data scattering.

Therefore, the boiling internal characteristics study results allowed the authors of Refs. [76, 77, 105–109] to ground their physical imaginations of the slit boiling and build their model. The video was also used for the slit's boiling physical mechanism and for the study of the internal characteristics of the refrigerant R-113 case in Refs. [110–113, 115, 116]. It was made up of the different combinations of the main parameters: the slit's gap (0.5 and 1.0 mm), the saturation pressure (0.1 and 1.0 bar), and the different

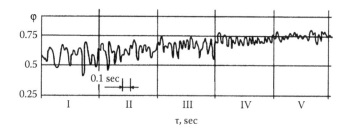

FIGURE 4.26 Average volumetric flow vapor quality changing by boiling into the horizontal plate slits. Water: $p_s = 1.0$ bar; $b = 15$ mm, $s = 1.0$ mm; I, $q = 40$; II, $q = 60$; III, $q = 82$; IV, $q = 110$ kW/m²; V, $q = 135$ kW/m².

heat fluxes. The studied heating surface part comprised 40% of the whole heating surface. The video optimal speed was regarded as the one of the main results, producing
700–1000 frames/sec. The following regime parameters were observed:

- The vapor slug's growth velocities
- The isolated "dry spots," which were formed under the growing vapor slugs
  at the liquid microlayer on the heating surface, "existence times"
- The "vapor slugs throwing away time" from the slit

The isolated "dry spots" development dynamics was studied; the two-phase border velocity defined from the video pictures and beside it was measured for the real
vapor quality and the whole part of the heating surface, occupied by the "dry spots."
The pictures were projected on the plane of a screen, as described in Refs. [77, 109],
and after that corresponding surface areas were defined by direct measurements.
The area was measured by plan-meter. It was used to determine the "dry spots"
and vapor slug's growth velocity too. As was done in the aforementioned studies,
the equivalent radius of the "dry spot" and the vapor slug was determined with an
increasing scale aspect.

From the video, it is seen that by the boiling of the low-heat-conductive dielectric
liquids inside the narrow horizontal slits, two main regimes took place: "slug" and
"emulsion." The "slug" regime is defined by the mutual growth of vapor bubbles,
their integration, and next removal in the form of separate large slugs. The regime
occurs at relatively low heat flux values ($q \approx 1.7 \times 10^4$ W/m^2). While inside the slit
with s = 0.5 mm, both in the slit with s = 1.0 mm, bulky vapor slugs occur surrounded by many small vapor bubbles and nucleation centers (Figure 4.27b.). The
large vapor slug's formation took place as the result of bubble system growth on the
nucleation centers, located in the overall heating surface. When the growing vapor
bubbles attain the upper wall of the slit channel, they start to deform, move along the
heating surface, and form the bulky vapor slugs.

These vapor slugs were blocking a considerable part of the boiling surface at
separate moments. When they reached the slit's edge, the volume is decreased, as the
consequence of the vapor removal channel forming. An output of some vapor slugs
from the slit into the pool was also simultaneously observed. New great vapor slugs
formed on the free surface. There was practically no "waiting time," that is, the liquid
overheating time between the moment of the vapor slugs' output and the appearance
of vapor bubble nuclei. It was associated with the following event: the simultaneous
growth of great vapor slugs took place in a surrounding low-heat-conductive liquid as
the vapor slugs were coming out from the slit. These narrow slit's boiling peculiarities, associated with "waiting" times, were absent, and they were not observed by the
other authors who performed similar experiments. The slit's boiling picture agreed,
as a whole, with known video and physical imaginations of process results reported
by various groups [76, 77, 105, 106–109]. The "emulsion" regime was realized by
more high heat fluxes (Figure 4.27a). The regime was characterized by a stable
existence inside the slit's gap, periodically washed by liquid coming from the pool.
As it could be seen from the presented videos and the volumetric flow local vapor
quality curves, defined based on the treated video data (Figure 4.26), the heating

surface part occupied by the vapor slugs increased from 50% to 95%. The slit's gaps were practically filled by the vapor phase at the heat flux range of $q = (2.5-4) \times 10^4$ W_t/m^2. The liquid that entered the slit via vapor removal was boiled at once. The two-phase mixture movement was observed via vapor removal. This moved the borders of the great vapor slugs. Then it triggered the next processes undergone by the liquid with the washed surface vapor mixture movement: the volume of small vapor bubbles

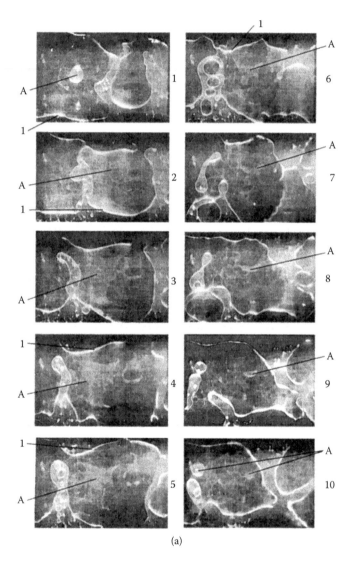

(a)

FIGURE 4.27 Refrigerant R-113 boiling inside the horizontal slit with $s = 1$ mm, $p_s = 1.0$ bar (video data from Refs. [110–113, 115, 116]). "A" — $q = 3.8 \times 10^4$ W/m²; interval between frames was 0.01 sec (growing "dry spots" were seen); "emulsion" regime of boiling inside slit: 1, vapor slugs border; A, "dry spots" (1–10 denote the frame numbers); "B," $q = 1.2 \times 10^4$ W/m²; the interval between frames was 0.02 sec (1–5 denote the frame numbers).

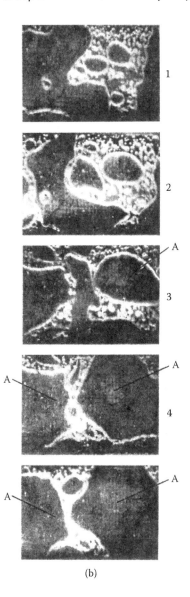

(b)

FIGURE 4.27 *Continued*

increased; they begin sticking together, and with the great vapor slugs, occupied most of the slit; then the next liquid volume began to penetrate into the slit.

Both the time of the vapor-liquid mixture movement from the working element's edge to its central part, when almost the entire liquid volume evaporation and its fusing with the main vapor slugs took place, and size of the heat transfer surface washed by the mixture are depend on the value of heat load. These regimes by the water and alcohols were not fixed for the short slit channels even at heat fluxes near the critical values (see Refs. [76, 77, 105–113]). During the "spheroid boiling" state, it was observed that the difference lay in the microlayer under the vapor slug base. Another difference was in the movement of the great liquid drops on the heating surface, which were being thrown inside the slit's gap. The existence of this regime can be explained by the CHF phenomenon pressing and the development of new boiling forms especially under the electric field action on the massive heater [115, 116].

Video data showed that "dry spots" were formed inside the liquid microlayer under the vapor slug even at very small heat fluxes. They are clearly seen on the images (as slightly lighter spots against a dark background). The "dry spots" formation in the microlayers of the dielectric liquids by low heat fluxes ($q \leq 10^4$ W/m^2) are the extremely high wetting of a typical heating surface, an important new experimental finding (Figure 4.27a,b). Experimental results [77, 106–109] associated with the study of water and alcohol boiling inside horizontal slits show that, even ay very high heat fluxes, including the heat fluxes near CHF, "dry spots" were absent. The detailed explanation of the absence of "dry spots" under water boiling and their existence under refrigerant (R-113) boiling in the mentioned cases is given below. It was shown that this fact was associated with the extreme possibility of the formation of vapor bubbles inside the liquid microlayer films (the so-called the "secondary" vapor bubbles).

The following peculiarities of the low-heat-conductivity liquid slit boiling helped to explain the aforementioned facts: the high overheating of these liquids and the existence of local surface heating parts with their increased temperature as the result of the "dry spots" appearance. It led to the appearance of many small vapor bubbles and boiling centers around the vapor slug near this slug output slit and to the absence of refrigerants' waiting time by boiling. Dry spots were observed by Azarskov [91] and Gogolin et al. [92] in relatively long vertical channels considerably far from the channel's inlet. It was associated with the drying conditions of liquid microlayers in the flow regimes nearby the structure to the

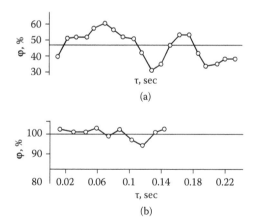

FIGURE 4.28 Changing of the heating surface part, occupied by vapor bubbles when boiling occurs inside the horizontal slit, based on the data from Ref. [116] ($s = 1$ mm). (a) $q = 1.2 \times 10^4$ W/m^2; (b) $q = 3.8 \times 10^4$ W/m^2.

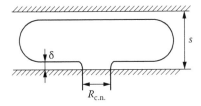

FIGURE 4.29 The microlayer's rupture caused by the single nucleation center action. R_{cn}, dry spot radius; δ, microlayer thickness.

annular or dispersed-annular two-phase flow regime. It was also shown by Klimov et al. [112, 113], on the basis of experiments and analytical modeling, that when the relatively thick liquid film was forming under the vapor slug as a result of the aforementioned process development, and its existence in the slit vapor slug time was small ($\tau = 0.05-0.1$ sec), then the "dry spots" appeared as a result of the microlayer rupture caused by the action of the local nucleation center. A schematic diagram of such rupture is shown in Figure 4.29.

The formation of "dry spots" in the microlayer under the vapor slug was the principal peculiarity of low-heat-conductivity liquid (e.g., refrigerants) boiling at narrow horizontal slits. The appearance of big drying surface parts via the slug regime could be imagined as follows. The secondary nuclei centers were included in the process with any small delay ($\tau \approx 10^{-2}$ sec) after the growth of vapor slugs (Figure 4.30). They continued their action in the liquid microlayer. The combined action of the vaporization center and heat flux led to the microlayer's rupture. The formed rupture — the "dry spot" nucleus (beginning) — and its subsequent growth was defined by the heat flux action. The microlayer's most intensive decrease (for the evaporation account) took place in a relatively small field near the "dry spot" as the result of the most favorable condition of this process. The neighboring dry spots' growth led them to stick together and to the formation of large drying areas. The influence of heat flux q on the slit's filling by the vapor phase is shown in Figure 4.28 [116].

It can be concluded from the aforementioned positions that the density of "dry spots" has to follow the same regularities as the vaporization centers' formation law. The "time delay" of the "dry spot" formation was defined by the liquid microlayer and overheating of the wall. The formation of "dry spots" by emulsion flow regime, which was characterized by the stable existence of the vapor–liquid mixture in the slit, was also caused by the microlayer's drying.

The "dry spots" pressing mainly took place as the vapor slug volume decreased, when it was leaving the slit. The decrease in the number of drying parts on the heating surface was associated with the liquid flow from the external vapor slug border to its center and the decrease in vapor pressure inside the vapor slug. This led to the liquid slug output into the volume filled by the vapor phase. The growth in vapor bubbles, accompanied by the liquid films' intensive movement into the vapor slug volume helped to prolong the liquid spreading on the surface and the pressing of the drying parts.

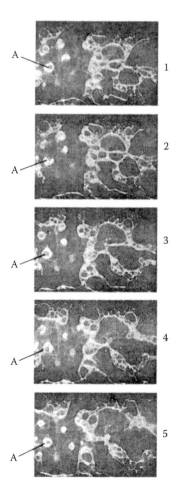

FIGURE 4.30 R-113 boiling inside horizontal slit, based on data from Refs. [110–113]. $s =$ 0.5 mm, $q = 2.57 \times 10^4$ W/m²; the interval between frames is 0.0043 sec; A, the microlayer's incipient boiling places.

Dynamic research shows that growth velocity of isolated "dry spots," when they were not fusing with their neighbors during their growth, by boiling inside the slit with gap $s = 0.5$ mm was 4–5 times higher than in the slit with gap s = 1 mm (Figure 4.31). The obtained results confirmed the assumption about the liquid microlayer's existence under the growing vapor slug by boiling inside the slits. They also showed that when the slit gap decreased, the microlayer's thickness also decreased. It could be considered as one of the main factors accounting for the slit boiling heat transfer enhancement.

The experimental discovery of the existence of "dry spots" even at very low heat fluxes for low-heat-conductive liquids is in excellent agreement with their heat transfer boiling total regularities for the same conditions [110–113, 115, 116]. The heat transfer decrease by heat flux growth (Figure 4.11) also could be explained by the

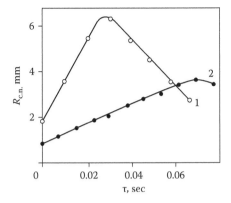

FIGURE 4.31 The "dry spots" dynamic R-113 slit boiling (data from Refs. [112, 116]). $q = 1.2 \times 10^4$ W/m^2; $p_s = 1.0$ bar; 1, $s = 0.5$ mm; 2, $s = 1.0$ mm.

increase in "dry spots" area. The higher heat transfer coefficients by boiling on the working element with indirect electric heating could be explained by the much better heat output at the massive solid wall at the field of the "dry spots" places.

The analogy of considering boiling regimes with the formation of dry spots on heating surfaces were observed by Kozhelupenko et al. [139, 140] via the annular channels water boiling mechanism and CHF study. These experiments were conducted on an experimental setup designed as the separate forced convection loop. Its schematic diagram is given in Figure 4.32. Heat carrier circulation was realized by the three-stage hermetic electropump. The flow rate was controlled by the throttle valves at the working element inlet. The heater was used for the heat carrier to be degassed (2–4 hours) and to be supported by the inlet temperature for a given level. Working element heating was achieved by direct electric current heating.

Video observation was conducted for the entire level of heat fluxes. When CHF values were achieved with the help of the special electronic control block, the video camera was turned on as well as some of the auxiliary devices. This block turning on was carried out with the help of the safety system when any velocity level of wall temperature growth was achieved.

Thus, images were taken of the two-phase flow at the crisis moment. The electronic control block allowed the lighting to switch on to fulfill its function as well as adjustment of the acceleration rate of the video camera (including the time note and recording of an event).

The special peculiarities of this video observation were associated with the task of observing the two-phase flow directly near CHF moment. To attain this, it was organized to first experimentally define the CHF, and only then was the corresponding video capture initiated. It was accompanied by a corresponding decrease in heat flux relating to the CHF value of 20–25%, and then the experiment was repeated.

The video was conducted on every level of the heat flux. At first, video velocity was set at 48 frames/sec. Then it was increased in the dependency of the regime from

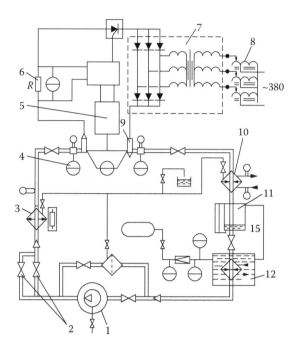

FIGURE 4.32 Experimental setup principal schematic diagram, based on conditions of [139, 140]. 1, Hermetic pump; 2, valve to control; 3, heater; 4, model pressure gauge; 5, CHF recoding device; 6, model electric resistance; 7, electric supply block; 8, electric transformer to control; 9, electric supply input; 10, heat exchanger; 11, volumetric flow rate measuring device; 12; liquid accumulator.

300 frames/sec for "stratified" or "wave" flow to 2600 frames/sec for the "emulsion" regime; for "slug" regime, video velocity was set at 900–1000 frames/sec.

Visual observation, photo, and video capture were conducted with the working element set at a horizontal position and its length, 50 and 100 mm. The annular channel gap sizes were changed, $s = 0.6$–1.5 mm; the inlet channel liquid subcooling was $\Delta T = 70$–5 K; the inlet liquid velocity w_0 was changed from 0.05 to 0.5 m/sec. In the analysis, different two-phase flow regimes were classified and the experimental conditions were divided into the following groups:

- Large velocities and subcooling: $w_0 \geq 0.5$ m/sec; $\Delta T \geq 20$ K
- Small velocities and large subcooling: $0.5 \geq w_0 \geq 0.1$ m/sec; $\Delta T \geq 20$ K
- Small velocities and small subcooling: $0.5 \geq w_0 \geq 0.1$ m/sec; $\Delta T \leq 20$ K
- Very small velocities and large subcooling: $w_0 \leq 0.1$ m/sec; $\Delta T \geq 20$ K

The first regime photograph is given in Figure 4.33. As shown, the bubble structure exists at the main part of the channel and only in the heating channel's outlet is the pulsating slug formation observed. The slug can occupy a small or large portion of the channel's length depending on the channel's gap and the flow velocity.

FIGURE 4.33 Bubbles flow boiling regime into horizontal annular slit. Water: $s = 1.5$ mm; $p_s = 1.0$ bar, subcooling $\Delta T = 35$ K.

By this, the phase border asymmetry is mainly defined as the ratio of inertia and gravity forces, that is, the Froude number.

The two-phase pattern second group (photo) is presented in Figure 4.34.

Beside it, such regimes can be seen in Figure 4.35. Vapor slugs and formation of vapor moving volumes were observed in all cases. The slug's spreading/prolonging channel depth changed, depending on the subcooling value, and was accompanied by flow pulsation. The inlet liquid velocity level determines the vapor slug filling of the channel's cross section area by its movement inside.

The vapor volumes are located by the liquid flow velocity lower than 0.2 m/sec near the heating element top line, (the letter "A" being used to indicate such dry spots) but with increased velocities (more than 0.2 m/sec). The gravity did not exert serious influence on the vapor volume form and its location. It can be seen that the vapor slug front could be spreading at a considerable distance from the inlet.

As shown by video data in all cases in Figure 4.35, vapor slug formation is accompanied by the appearance of some "dry spots" on the heating surface of the working element. The "dry spots" formation on the heating surface was analogous

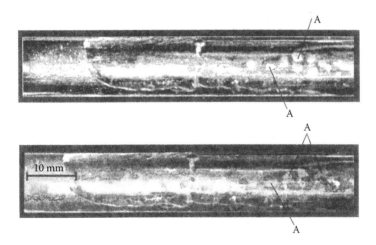

FIGURE 4.34 The slug flow boiling regime inside the horizontal annular slit with $s = 1.9$ mm and with "dry spots" formation. $w_0 = 0.2$ m/sec; $\Delta T = 33$ K, $q = 0.69$ MW/m^2; "A," "dry spots."

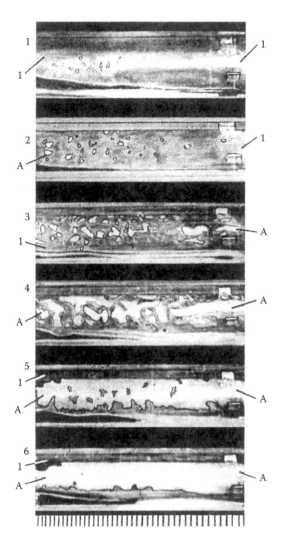

FIGURE 4.35 "Dry spots" dynamic development on heating surface, covered by the liquid microlayer. It was by two-phase flow in the annular horizontal vapor-generating channel [139, 140]. $p = 1$ bar; $s = 1.3$ mm; inlet liquid velocity, $w_0 = 0.22$.

to the appearance of nuclear centers in the liquid film by boiling, as shown in Refs. [139, 140]. Kozhelupenko et al. [139, 140] suggested that the "dry spots" formation in the considered conditions corresponded with the model presented by Klimov et al. [113], and so it was associated with the boiling in the liquid microfilm, its destruction at the appearance of nuclear centers and, consequently, their drying. The "dry spots" growth and their subsequent coalescence led to the entire bulky "dry spot" formation. It was discovered during visual observations that this bulky "dry spot" represents the major reason for the crisis; it occurs because of overheating

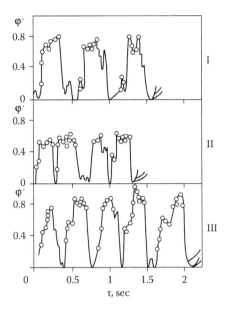

FIGURE 4.36 Slug pulsations inside annular channels by the length 100 mm before CHF moment. The heat-generated wall cooling by liquid slug, growth, vapor slug existence, slug's outlet, crisis subcooling $\Delta T = 34°C$, video velocity. I, $w_0 = 0.22$ m/sec; $\Delta T = 31$ K; $s = 1$ mm, 2300 frames/sec: I, liquid microfilm; A, "dry spots." II, $w_0 = 0.2$ m/sec; $\Delta T = 37$ K; $s = 1.5$ mm; $q_{CR} = 0.57$ MW/m². III, $w_0 = 0.21$ m/sec; ΔT 34 K; $s = 1.3$ mm; $q_{CR} = 0.422$ MW/m². Data from Refs. [139, 140].

the surface to a temperature exceeding the temperature corresponding to the liquid resulting in an irreversible thermodynamic change of state, Thus, liquid becomes unable to wash the whole heating surface occupied by the bulky "dry spot" (Figure 4.35).

The video data treatment was conducted with the assumption about the symmetry in the phenomenon as for the processes in the annular channel's front and opposite sides [139, 140]. The problem led to the dependency of conventional vapor quality determination on time. Results are shown in Figure 4.36 as video data in the view $\varphi = f(\tau)$.

The types of vapor slug movement pulsation are defined at its development by the following factors:

- Slug waiting time (the slug is absent).
- Its development (the slug's growth before it reaches its maximum).
- Its existence time (the period when the slug's front did not move).
- Its outlet time (the slug's "compression" and movement phase border to the channel's outlet).

The treatment results allowed estimation of characteristic periods of the slug regime.

The "emulsion" two-phase flow was realized by subcooling lower than 20 K (Figure 4.37a). The absence of subcooling prolonging the heated part of the heating surface of the annular channel led to the stable existing vapor liquid flow, which was intensively washed off the heating wall.

The "wavy-stratified" flow regime was developed by very small two-phase flow velocities (0.20–0.06 m/sec; Figure 4.37b). By this, the subcooling and heat flux caused a very small influence on the value of velocity. The heat flux, that corresponded to the "wavy-stratified" flow also coincides with the CHF value [140]. It is shown that the "wavy-stratified" flow regime began at the lower velocities defined for CHF lower level.

The local characteristics and physical mechanism of the boiling inside the slits were studied with the help of laser diffraction interferometer; main principal results were presented in Refs. [141, 142, 260]. Boiling inside too narrow slits was also studied. The results could be divided into two main groups: (1) the glass channels with heated walls, when boiling took place on the heated walls; and (2) nonheated channels, in which boiling was initiated by inserting a thin wire (as the heater) into the narrow slit. Boiling inside the heated channels was organized over glass cylindrical sticks wrapped into the wire heaters. These sticks, which were installed near the neighboring flat plate, initiated the boiling on their surfaces. As for the nonheated channels, boiling was instigated, as mentioned, via insertion of wires or plates as heaters into the narrow channel. In this case, these heaters initiated the formation of vapor bubbles between the walls; the transparent walls allowed handlers to obtain good pictures, which illustrated many important details about the liquid microlayer formation on the vapor slug base.

The experiments were conducted under the atmospheric pressure. Water by saturation temperature was used as the heat carrier. The thickness of liquid microlayer, formed on the glass wall, was directly measured in the large vapor slug base during

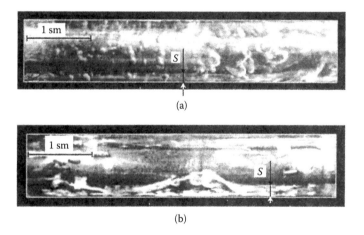

(a)

(b)

FIGURE 4.37 The (a) "emulsion" and (b) "wavy stratified" two-phase flow regime in the annular slit. Data from Refs. [139, 140]. $s = 1.5$ mm $w_0 = 0.217$ m/sec; $\Delta T = 13$ K; $q = 1.2$ MW/m^2.

its growth under different boiling regimes. The dependency of liquid microlayer thickness on the phase border character regime and its movement velocity were determined. The experimental setup and methods are prescribed in more detail in Ref. [142].

The video data illustrating the liquid microlayer structure on the narrow channel's wall (by the bubble's growth and flow) are shown in Figures 4.38–4.42. Both heated and nonheated cases can be discovered in the liquid microlayer's large uncertainties in the forms of "hills" or "waves." The "hills" are shown as lighter points on a dark background, and are designated by arrows (Figure 4.38). The waves are shown as alternating dark-and-white strips, and are also denoted by arrows.

Liquid microlayer wave formations were especially likely when capillary forces were significant. These forces defined the formation of the liquid microlayer wave, as soon as they prevented the phase border uniform movement. The regime showing the formation of liquid microlayer waves was named by Chichkan [142] as "the pulsation evaporation regime." It took place under extremely low heat fluxes (20–30 kW/m²) on well-wetted surfaces and in very narrow slits ($s \le 0.3$ mm). It was noted that, first of all, the formation of "microwaves" and "microhills" was associated with two-phase border movement pulsation.

The transient from the dark line of the interferogram to the light one corresponded to the film thickness changing by 0.5 μm, if it was fixed simultaneously on both sides of the channel. The observed "microhills" height was 4–10 μm and their

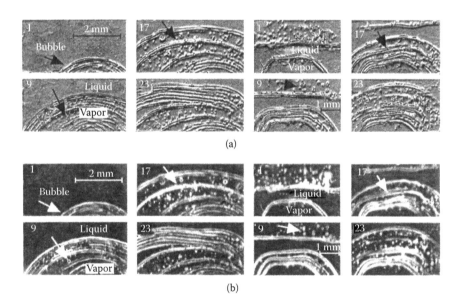

(a)

(b)

FIGURE 4.38 Microhills (lighter points) and "microwaves" (alternating light and dark lines) formation by vapor bubble growth into the glass slit [141, 142, 260]. Liquid boiling by saturation pressure: $p_s = 1$ bar; $q = 25$ kW/m²; a1, a2, video velocity 1300 frames/sec; b1, b2, video velocity 2625 frames/sec; $s = 0.2$ mm; into corner frame's number.

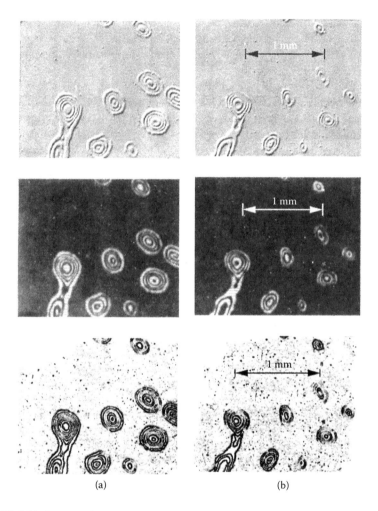

(a)   (b)

FIGURE 4.39 Decrease in "microhills" sizes in the liquid microfilm surface, as the evaporation results in the vapor bubble base [260]. The interval between the frames, $\tau = 5 \times 10^{-3}$; $p_s = 1$ bar; $q = 20$ kW/m^2; $s = 0.45$ mm. Photo treatment by different methods.

base diameter was 10–100 μm. The formation of microhills and microwaves on the liquid microfilm surface did not relate to movement in the surface when the film rupture was not observed. In these cases, their sizes were decreasing only by evaporation (Figure 4.39). The "microhills" movement of the irregular forms (Figure 4.40) was fixed in other cases when the liquid microlayer "curling up" phenomenon took place. This movement is denoted by arrows in the pictures. Microlayer thickness was defined by the values of wave volume and size based σ in any bubble field part.

The coalescence of vapor bubbles had an influence on the change in liquid microlayer thickness, aside from the microlayer "curling up" and the phase border

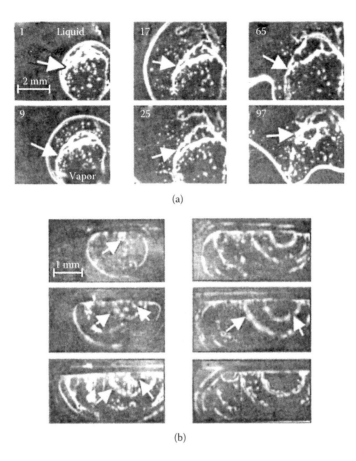

FIGURE 4.40 The liquid microlayer curling up on slit surface video figure by water boiling in the narrow transparent slit [260]. $p_s = 1$ bar, arrows are "following" for the microlayer border movement. (a) $q = 65$ kW/m^2 (heated surface; video velocity, 1515 frames/sec; $s = 0.2$ mm.). (b) Nonheated wall of the slit; vapor bubble is generating by thin wire heater, inserted into the slit with gap $s = 0.86$ mm; video velocity, 1425 frames/sec.

pulsation (Figure 4.41). The mentioned phenomenon is accompanied by liquid movement into the microlayer. The thermal boundary layer was formed during the "bubble's waiting time" near the heater before the bubble appearance (see frame 1, Figure 4.42). Significant heat was accumulated in the liquid during this period; most of it was transformed into latent heat especially in the initial moments of bubble growth. During this stage, the phase border high velocity was fixed (1.5 m/sec). The velocity decreased essentially at once after the movement began (over 2 μsec, velocity decreased to 0.2 m/sec); it is accompanied by the phase border essential deformation and the strong liquid movement into the film. The abrupt increase in meniscus width (frame 4) was the sequence of the returned liquid flow from the periphery to the bubble center. The liquid microfilm profiles are given in

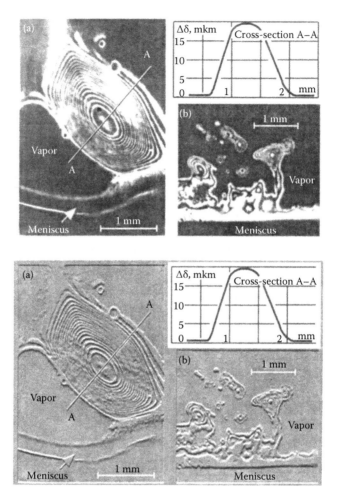

FIGURE 4.41 Liquid microfilm structure near the phase border by the vapor bubbles' coalescence, by slit boiling [260]. The water, $p_s = 1$ bar (photo treatment different methods). (a) $s = 0.43$ mm. View from the top. (b) s = 0.86 mm. View from the side.

the same picture. Microfilm has maximum thickness in the central part (13 μm), and minimum value in the periphery (0.66–0.7 μm).

The connection between liquid microfilm thickness and phase border movement velocity was fixed in the experiments [142]. A field exists in which liquid microfilm thickness was independent of the phase border movement velocity. The formula for calculating the velocity of vapor bubbles' growth for slit was obtained for conditions in which the liquid microfilm whole drying did not take place.

Moreover, it is necessary to note the known information about the single vapor bubble movement into the capillary channel regularities, which was reported by Kutateladze [143]. These regularities are shown in Figure 4.43. As Kutateladze [143]

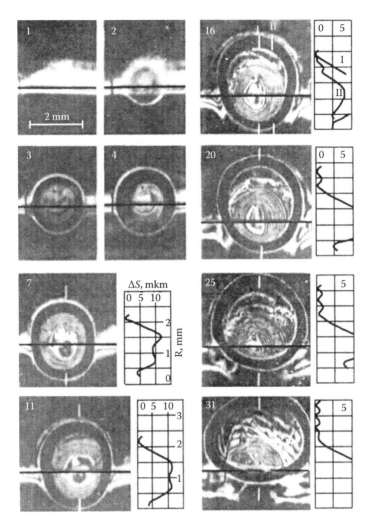

FIGURE 4.42 The plane vapor bubble nucleation, growth, and video movement. The bubble formed into the gap between two plain glass plates on the platinum wire, $d = 0.07$ mm [260]. Slit with the gap: $s = 0.86 = 0.3 \times 10^4$ kW/m²; video velocity, 1425 frames/sec. Dynamics of liquid microlayer formation on the glass surface, in which growth in the slit vapor bubble action is seen.

notes, "by single vapor bubble floating in the cylindrical tube with liquid layer without flow over the tube, the ratio between equivalent bubble radius and channel's radius R_C."

The dependency is presented in Figure 4.43 in the following form:

$$\bar{U}'' = f\left(\bar{R}_C, \bar{R}_{\sigma E}\right), \bar{U}'' = u'' / \sqrt{g(\rho' - \rho'')R_E}$$

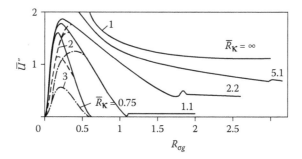

FIGURE 4.43 The vapor (gas) bubbles floating velocity into the vertical channel for different liquids [143]. 1–3, $Ar^* = 2 \times 10^5, 6.7 \times 10^4, 5.5 \times 10^3$.

Here, $\bar{U}''$ is the undimensional vapor (gas) bubble floating velocity, u'' is the whole bubble floating velocity, and $\bar{R}_{\sigma E} = R_E / \delta_{\sigma E}$ is the relative equivalent radius of the bubble by volume V. $R_E = (3V/4\pi)^{1/3}$; $\delta_{\sigma E} = \sqrt{\sigma/f\Delta\rho}$ Laplace size; and $\bar{R}_C = R_C / \delta_{\sigma E}$ where R_C is the capillary channel radius. As the parameter in Figure 4.43 was used, the Ar^* number is

$$Ar^* = \frac{g\delta_{\sigma E}^3 \overline{\Delta\rho}}{v'^2}, \overline{\Delta\rho} = \frac{\rho' - \rho''}{\rho'}$$

The next dependencies that are recommended with respect to Grigoriev and Krochin [97] yields the following for the bubble movement into the plane parallel slit channel with gap δ_C:

$$\text{if } Re^* = u'' \frac{\delta_C}{v'} > 200 \text{ and } \bar{\delta}_C = \frac{\delta_C}{\delta_{\sigma E}} < 0.23, \text{ then } u'' = 6.2g\sqrt{\frac{\Delta\rho\delta_C}{\sigma}}\delta_C$$

$$\text{by } \bar{\delta}_C \geq 0.23, \ u'' = 1.5\sqrt{g\overline{\Delta\rho}\delta_C}$$

It was very difficult to find information about experiments on the internal characteristics of slit and capillary channel boiling between the years 1985–1995. The same thing was shown in the review of Peng and Wang [4]. The last one did not give any reference on such work. The single exception was the report by Diev [144]. Extremely difficult and rarely obtained experimental data were presented about microlayers into the vapor bubble base, which were formed by vaporization in the narrow slit channels (0.3–0.8 mm). However the limited information about the microlayers' thickness fixation in the bubble departure moment without the entire information about process pictures (the "nucleation" places and bubbles "departure" conditions; "neighboring" vapor volumes; the entire "life" times of the bubbles, etc.) did not allow researchers to evaluate the agreement between presented data as well as experimental results from other authors.

4.4 PHYSICAL IMAGINATIONS AND THEORETICAL MODELS FOR HEAT TRANSFER AT VAPORIZATION IN VERTICAL SLITS AND CAPILLARIES

The vapor bubbles' dynamic lower influence on vaporization heat transfer was essential for this process. Visual observations and photo and video data obtained and published in Refs. [92, 93, 97] characterized this process mechanism in the following manner:

- The determined two-phase flow structure was observed at the channel, with dependence on pressure range, heat fluxes, and the slit sizes.
- The economizer's zone was formed at the channel's inlet part with the height H_{EC}. The liquid overheating power transferred into this zone with the overheating level before temperatures, corresponding with the initial boiling T_{IB}.
- The separate and coalesced bubbles forming the vapor slug's zone appeared behind the economizer zone. At this zone, the heat was transferred by the heat conductivity over the liquid microlayer and convective/conductive mechanism from the wall to the internal liquid flow. The domination of convective or conductive mechanism can be estimated with Fo number, $Fo = \alpha' \tau / L^2$, where L is the determined size (the channel diameter, the slit's gap, etc.).

Estimations of the Fo number for different cases of boiling into vertical slits are presented below. It is seen that the average $Fo \geq 1$ [91], and the "elementary" heat transfer from the wall to the liquid process intensity can be calculated based on the known convection heat transfer formulas.

Reference	Circulation Schematic, Calculation Base	Fo Number, Range
[96, 97]	Natural circulation: $G = 1.2 G_{EV}$, $w_0 = G/(\rho' bs)$	$0 \leq Fo \leq 400$
[99, 100]	The forced circulation: G, w_0 take part in the simple conditions, $w_0 = 0.001$–0.01 m/sec	$0.4 \leq Fo \leq 3.0$
[92]	The natural circulation $G = G_{EV}$ in the absence of circulation; $G > G_{EV}$ in the presence of recirculation	$0.5 \leq Fo \leq 2.0$
[91]	$w_0 = 0.1$–0.5 m/sec	$0.01 \leq Fo \leq 0.50$

The vapor dynamic influence has no determined influence on the boiling heat transfer by the forced circulation case [100]. It was very important to define in the entire thermal resistance the main items of the theoretical model of complex capillary heat transfer at boiling. For the forced circulation model, this item was the economizer part thermal resistance, which was equal to $(H_{EC}/H)(1/\alpha_{EC})$. Using the following formula, Grigoriev et al. [96] calculated the economizer part thermal resistance as

$$\frac{H_{EC}}{H} \frac{1}{\alpha_{EC}} = \frac{G c_p}{H q \Pi} \left(\Delta T_{IB} - \frac{qs}{Nu_{ES} \lambda'} \right) \left(\frac{\Delta T_{EC}}{q} + \frac{s}{Nu_{EC} \lambda'} \right) \tag{4.6}$$

Here, Π denotes the channel perimeter. Nu_{EC} was defined by the convection heat transfer formula with respect to the forced flow case:

$$\frac{Gc_p}{Hq\Pi}\left(\Delta T_{IB} - \frac{qs}{Nu_{EC}\lambda'}\right) = \frac{H_{EC}}{H}; \quad \frac{\Delta T_{IB}}{q} + \frac{s}{Nu_{EC}\lambda'} = \frac{1}{\alpha_{EC}}$$

As soon as the liquid velocity inside the economizer zone reached the minimum value, it led to the flow laminar regime and $Nu_{EC} = Nu_{min}$. The structure of Equation (4.6) had such form that the specific conventional thermal resistance on the economizer part was caused to be essentially higher than the heat conductivity maximum thermal resistance (by small heat fluxes). It was shown by Grigoriev and Krokhin [97] that it was associated with the temperature calculation on the next schematic base: $\overline{T}_W - T_s$ instead of $\overline{T}_W - T_L$. This peculiarity called for this thermal resistance input even for the very short economizer part, $H_{EC}/H \leq 0.1$, to be determined. It can be experimentally determined even if the single thermocouple is located on the economizer part. That is why, if $Gc_p/Hq\Pi\,(\Delta T_{IB} - qs/Nu_{EC}\,\lambda') \geq 0.1$, the whole thermal resistance can be considered equal to the economizer part thermal resistance and the liquid microlayer thickness; the heat transfer from the wall to liquid slug and from the wall over microlayer intensity made no sense.

Heat transfer can be calculated based on the following equation:

$$\frac{1}{\alpha} \cong \frac{GC_p}{\Pi\overline{q}\,H}\left(\Delta T_{IB} - \frac{qs}{Nu_{EC}\lambda'}\right)\left(\frac{\Delta T_{IB}}{q} + \frac{s}{Nu_{EC}\lambda'}\right) \tag{4.7}$$

The initial boiling temperature pressure in Equation (4.7) is an unknown value. If it could be considered the identical boiling conditions (the surface roughness, wall material, the heat carrier, the channel's geometry, the flow regime, etc.), and with the uncertainty of the aforementioned value, then we could come to the idea that Equation (4.7), accompanied by one empirical constant allowed to generalize the capillary channel boiling heat transfer by the forced circulation experimental data, when the next in-equality will be right: $Gc_p/H\overline{q}\Pi\,(\Delta T_{IB} - qs/Nu_{EC}\lambda') \geq 0.1$. As the first approximation for the average heat transfer coefficient $\overline{\alpha}$ determination, it is possible to take ΔT_{IB} on pool boiling by the same pressure data recommendations. Comparison of the results in Equation (4.7) based on the experimentation data of Labuntsov et al. [100] confirmed the process model correspondence.

The calculation of the liquid flow rate for boiling in slits by natural circulation had a particular complexity. If liquid forced circulation is absent (not full flooding) or insufficient, then $G = G_{EV}$ (where G_{EV} is the two-phase mass flow rate at kg/sec). In Ref. [97], $G = 1.2G_{EV}$. In Ref. [92], for experimentation part, $G = G_{EV}$, then

$$\frac{h_{EC}}{H} \approx \frac{Gc_p}{\Pi\overline{q}\,H}\left(\Delta T_{IB} - \frac{qs}{Nu\lambda'}\right) = \frac{c_p}{r}\left(\Delta T_{IB} - \frac{qs}{Nu\lambda'}\right) \tag{4.8}$$

Apparently, for the considered processes, as soon as

$$\Delta T_{IB} \leq 10 \text{ K}; \quad \frac{c_p \Delta T_{IB}}{r} \leq 0.1 \quad \text{and} \quad H_{EC}/H \cong 0,$$

it means that the economizer part is absent. In this case,

$$\frac{1}{\alpha} = \frac{1}{\alpha_L}(1-\bar{\varphi}) + \frac{1}{\alpha_{EV}}\bar{\varphi}, \tag{4.9}$$

In this expression, α_L is the liquid convection heat transfer coefficient at the economizer part, and α_{EV} is the heat transfer coefficient at the evaporation part.

The economizer zone pressing took place for low-temperature liquids (refrigerants) by $G \gg G_{EV}$. Substituting for real meaning of ΔT_{IB}, s, and λ' even for very small average q gave $(\Delta T_{IB} - qs/Nu_{EC}\lambda') \leq 0$. The velocity of liquid slugs was determined by the action of vapor bubble dynamic and gravity forces for the conditions $G = G_{EV}$ (where circulation was absent), that is,

$$w' = \bar{Z} + 0.35\sqrt{gd_E}$$

Based from both factors having equal meaning ($\bar{Z}$ and $\sqrt{gd_E}$) to the complex determined criteria at the following equation:

$$Re^* = Re_L + \sqrt{Ar}$$

The correlation (4.9), with respect the calculation formulas for α_{EV} and δ, if it could not account for the decrease in the liquid microlayer thickness as evaporations result, could be presented in the following dimensionless form:

$$\frac{1}{Nu(1-\bar{\varphi})} = \frac{1}{Nu_L} + \frac{1.1\bar{\varphi}}{(1-\bar{\varphi})^{0.25} Re_L^{0.37}} \tag{4.10}$$

There are difficulties with calculation of values $\bar{\varphi}$ and $(1-\bar{\varphi})$ definition for the natural circulation condition. The experimentation conducted and reported by Grigoriev et al. [96] was added in the two-phase flow hydrodynamic at the vertical plane slits. The approximated empirical correlation was obtained, which was defined as the connection between $\bar{\varphi}$, regime, geometry, and thermophysical parameters in the following form:

$$\bar{\varphi} = 0.8\left(1 - \frac{\ln(1+\bar{w})}{\bar{w}}\right); \quad \bar{w} = 0.57\frac{qH}{r\rho''s}\frac{1}{\sqrt{gs}} \tag{4.11}$$

where $\bar{w}$ is the vaporization relative velocity in the slit (ratio of the vaporization velocity to the gas bubble floating velocity). The direct comparison of results in formula (4.10) with the experimental data of Grigoriev et al. [96] was complex as soon as flow regime of the main liquid slugs was transient by the experiments' [96] conditions:

$$10^3 < Re_L \equiv \frac{2qH}{r\rho''v'} < 10^4$$

The experimental data [97] treatment based on formula (4.10) showed that the model and the experimental results had qualitative agreement. The transient flow regime peculiarities did not allow obtaining the reliable formulas for calculation. The liquid slug flow peculiarity could have serious influence on the transient from laminar to turbulent regime conditions. It is likely that the increasing Nu_L was associated in the comparison with known limitations at the field range $10^3 \leq Re^* \leq 2 \times 10^3$ and also the dependency character in the form $Nu \approx C(Re^*)^{0.8}$. Reliable checking in the correspondence, and, if necessary, to correct presented correlation, could be obtained after the special experimentation on the local heat transfer for the corresponding flow regimes.

It would be especially interesting to compare the calculation on formula (4.10) base results with the data on long vertical slit boiling heat transfer for natural circulation conditions [92]. The reliable determination of $\bar{\varphi}$ and $\overline{1-\varphi}$ was not possible as soon as information about the two-phase flow at the experimental conditions [92] both for this one and for the circulation loop geometry was absent.

The direct use of correlation (4.11) was not correct as for the difference at the heat carrier's thermophysical properties for both serious differences at the channel geometry: $H/b = 90/150 = 0.6$ in Ref. [96] and $H/b = 400/140 = 2.8$ in Ref. [92]. It was used in the following considerations for the approximate correction of empiric correlation $\bar{\varphi} = f(\bar{w})$ with possible influence of ratio H/b. Let us assume the following: if $H/b \to \infty$, the vapor quality $\bar{\varphi} \to 1$, and if $H/b \to 0.6$, $\bar{\varphi} = 2.86\varphi_0/(1 + 1.86\varphi_0)$, and

$$\overline{1-\varphi} = \frac{(1-\varphi)_0}{1+1.86\varphi_0} \qquad (4.12)$$

The considerable part of the experiments [92] was conducted for the long vertical slits with the following gaps: $s = 2$ and 4 mm, and by pressures $p_s \gg 1$ bar.

It was possible to see by the determination of the vapor bubble departure diameters D_0 for refrigerants R-12 and R-22 with pointed slit's gaps S that $D_0 < S$ and even $D_0 \ll S$. So, with Gogolin et al.'s [93] recommendation with respect to R-12, $D_0 = D_{00}(p/p_0)^{-0.46}$ by $t_s = -20°C$, it gave $D_0 = 0.54$ mm by $t_s = 20°C - D_0 = 0.3$ mm (D_{00} is the bubble departure by any $p_0 = 0.1$ MPa).

Naturally, it was not possible to relate the slits with gaps $s = 2$ or 4 mm to some special "narrow" boiling conditions. The convective heat transfer between the wall and two-phase flow had to be so for pool boiling or to take part in both of these mechanisms. It meant that principally the theoretical model has to account for the action of both of these mechanisms. It could be realized with respect to the simultaneous action of these mechanisms analogous for the known principal approach to boiling by strong convection such as the schematic: $\bar{\alpha} = \sqrt{\alpha_L^2 + \alpha_{EV}^2}$. Then, formula (4.10) will take the next form:

$$\frac{1}{Nu(1-\varphi)} = \frac{1}{\sqrt{Nu_L^2 + Nu_{EV}^2}} + \frac{1.1\bar{\varphi}}{(1-\varphi)^{0.25} Re_H^{0.37}} \qquad (4.13)$$

where $\bar{\varphi}$ and $\overline{1-\varphi}$ have to be calculated based on formula (4.12).

It is not difficult to see that the heat transfer combined mechanism at these conditions was near the vapor generation channels working state and it was possible to assume that generalization of the experimental data [92], first of all for the slit channels with the gaps $s > 2$–4 mm, will agree with correlation (4.13). However, checking this assumption was impossible as soon as the information about the circulation velocities $\bar{w}$ was absent. This is why the qualitative estimation of Equation (4.13) correspondence to the real regularities was made by the comparison of the separate local dependencies, following from the calculation of Equations (4.12) and (4.13) with experimental dependencies obtained in experiments [92].

Determination of Nu^* was done on the next correlation base, which generalized the experimental data [96]:

$$Nu^* = \frac{\alpha H}{\lambda'} = 0.013 Re_H^*(Pr')^{0.43}$$

where $Re^* = Re_H + \sqrt{Ar}$, $Re_H = qH/r\rho''v'$.

The complex character of the Re_H^* number has an essential positive influence on the experimental boiling heat transfer results, generalization of the "wide" slit channels $s \ll D_0$. The comparison shows that correlation (4.13) gives the correct account of the main interconnections between the average heat transfer coefficients and the main influenced parameters. Unfortunately, direct comparison of formula (4.13) transformation with respect of formula (4.12) and coming to the empiric formulas [92] was not possible, as soon as the last ones gave the empiric treatment for the whole experimentation volume with joining different thermal regimes:

$$Re \gg \sqrt{Ar}; \quad \sqrt{Ar} \gg Re_H; \quad Nu_L \gg Nu_{EV}; \quad Nu_{EV} \gg Nu_L$$

$$\varphi \sim \bar{w}^{0.4}; \quad \varphi = const, \text{ etc.}$$

Local comparison can be done with respect to the next approximated degree correlations using

$$\bar{\varphi}(\bar{w}) \text{ so } 0.1 \leq \bar{w} \leq 1; \quad (\bar{\varphi}) \approx C(\bar{\varphi})^{0.5}; \quad (\overline{1-\varphi}) \approx C(\bar{\varphi})^{0.2}$$

If $1 \leq \bar{w} \leq 10$ and $\varphi \approx C(\bar{w})$, accounting for the average correlations,

$$\bar{\varphi} \approx C\overline{(w)}^{0.35} \quad (\overline{1-\varphi}) \approx C\overline{(w)}^{0.3}$$

for most often spreading regimes $Nu_{EV} \gg Nu_L$ and $Re_H \gg \sqrt{Ar}$, we can obtain the following from (4.13):

$$\alpha \approx q^m H^p s^{-n} (\rho'')[f(p)]^k \qquad (4.14)$$

where

0.2 < m < 1.05, that is, $\bar{m} \sim 0.62$

0.2 < p < 0.3, $\bar{p} \sim 0.25$

0.7 > n > 0.45, that is, $\bar{n} \approx 0.55$; $\bar{k} \approx 0.5$

At the generalization obtained by Azarskov [92], values of the following degree items were taken: $\bar{m} = 0.6$; $\bar{n} = 0.45$; $\bar{k} = 0.45$; an approximate agreement with calculation results based on formula (4.14) was observed.

From the correlation (4.13), other main heat transfer regularities were noted, discovered during the experimentation [92]. The heat transfer coefficients by the whole channel flooding coincided between these values for vertical slits $s \geq 2$ mm and the vertical wall, that is, $Nu_{EV} \gg Nu_L$, $(1 - \varphi) \gg \bar{\varphi}$. By this, the average heat transfer intensity at the small heat fluxes field ($10^3 \leq q \leq 3 \times 10^3$ W/m²) and low pressure of R-12 ($t_s = -20°C$) will be $\bar{\alpha} \approx q$.

Actually, $\overline{(1 - \varphi)} \gg \bar{\varphi}$ and

$$\frac{1}{Nu} \sim \frac{1 - \varphi}{\sqrt{Nu_L^2 + Nu_{EV}^2}} \sim \frac{1}{q^n} \quad 1.0 < n < 1.1$$

It agreed with the results obtained by Azarskov [92]. The ratio between $\bar{\varphi}$ and $(1 - \varphi)$ changed with q growth, that is, $\overline{(1 - \varphi)}$. At the limit by the high φ

$$\frac{1}{Nu} \sim \frac{\overline{\varphi(1 - \varphi)}}{Re_H^{0.37}} \sim \frac{1}{q^{0.25}}$$

The average value $\alpha \approx Cq^{0.6}$ agrees well with the average value n between the two aforementioned limit cases. The decrease in the slit's gap led to increase in α. The influence of gap value took place at high pressure values $10 \leq t_S \leq 20°C$ and by $s < 1$ mm. $Nu_{EV} \gg Nu_L$; $(1 - \varphi) \to 1$; α did not depend on q.

If the slit flooding height decreases, then it is possible to assume that the average volumetric liquid concentration $(1 - \varphi)$ will be proportional to the flooding relative level h_H/H. Then, the regime with the flooding level $h_H/H < 1$ will correspond with the conditions $\bar{\varphi} \gg 1 - \varphi$ and $1/Nu \sim \overline{(1 - \varphi)}^{0.75} \bar{\varphi}/Re_H^{0.37}$.

It seen that when heat flux influence on heat transfer intensity was becoming lower, $\alpha \approx Cq^{0.4}$; this agrees agreed with Azarskov's [92] generalization, etc.

Therefore, the proposed model is qualitatively correct to account for boiling at long vertical slit heat transfer by natural circulation. It would be necessary to develop some experimentation devoted to the study of the following parameters: (1) real vapor quality definition; (2) the local heat transfer coefficients determination by the vapor and liquid slug's movement, etc. But, as a whole, if we consider that comparison of the model results with the generalized correlation by the forced flow boiling (experimentation [100]) was done satisfactorily, both with empiric correlation for the natural convection cases, including the data from two studies [92, 97], then it could be possible to consider the suggested model's satisfied requirements for such matter and to recognize their possible application as the basis for the following analysis and development.

It is necessary to know that, for the different problems of the small heat generated elements' cooling (including the electronic equipment elements), it can be attractive to use the boiling process in the horizontal and short vertical slits, submerged into the liquid pool.

4.5 PHYSICAL IMAGINATIONS AND THEORETICAL MODELS FOR HEAT TRANSFER AT VAPORIZATION IN HORIZONTAL AND SHORT SLITS AND CAPILLARIES

The vapor bubble dynamics exerted the main influence on the process mechanism of boiling inside the horizontal, the short inclined, and vertical slit channels. The same boiling conditions appeared at the gravity evaporative cooling systems of heat-generated electronic elements and equipment such as: (1) electromagnetic wave guide elements, (2) interfinned channels of high-frequency electronic devices collectors and electronic lamp anodes, etc. The most appropriate factors for the complex experimentation statement, including the internal characteristics determination, were the horizontal slits with the one-single side heat input. Because of this it was possible to study the vapor bubbles nucleation, and their growth and outlet from the slit, to develop the systematic experimentation of the local and average vapor quality, and to conduct measurements of the representative liquid microlayer thickness. This experimentation was done in various studies [76, 77, 105–109]. The photo and video data and the real volumetric vapor quality data allowed the researchers to propose the key assumptions for vaporization at this condition model and to generalize experimental results (typical video data of the boiling inside the horizontal slits are given in Figure 4.25). It was done by water boiling into the plane horizontal slits with gaps $s = 0.5$ and 1.0 mm and pressures $p_s = 0.1$ and 1.0 bar.

Let's consider the process physical model, based on cited observation results (Figure 4.44).

1. The vapor bubble nucleation and growth took place simultaneously in different heat transfer surface points. In the first approximation, the appearance of vapor nucleation centers density has to be the same as for pool boiling. It means that the nucleation centers density n_F (using recommendations of Labuntsov [145]) can be defined as

$$n_F \sim \left(\frac{r\rho'' \Delta T}{\sigma T_s} \right)^2 \qquad (4.15)$$

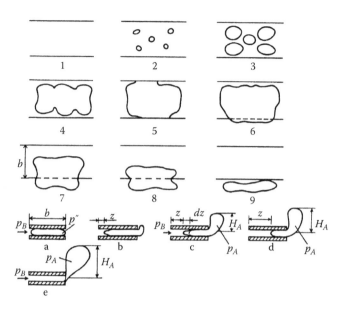

FIGURE 4.44 Schematic diagram of a conventional model for boiling inside horizontal slit (nucleation, growth, movement, and outlet from the slit of the vapor phase). 1–9, View from the top; a–e, view from the side; 1, "waiting" time; 2, 3, "growth" time; 4, vapor bubble coalescence in the whole vapor "slug"; 5, "existence" time of vapor slug into slit; 6–9, vapor slug outlet.

2. The average velocity of vapor bubble growth can be defined from the suggestions about the cylindrical form of the vapor slug and from the vapor slug growth energy model. The initial differential equation has the following form:

$$2\pi R r \rho'' s dR = k\left(\int_0^R \frac{\lambda'}{\delta}\Delta T 2\pi R dR\right)d\tau \qquad (4.16)$$

The liquid microlayer thickness δ and temperature drop $\Delta T = T_W - T_s$ at this equation are the unknown functions of the radius R (any vapor slug average radius at the slit) and τ; $k = 1$ by the one side heat input at the slit; $k = 2$ for two-side heat input. Difficulties in using this method, which appeared during the model building attempt and were associated with the complex character of the dependencies (R, τ) and $\Delta T(R,\tau)$ can be overcome by some assumptions. The simplest assumption contained what could not be accounted for in the local instant heat source distribution:

$$q(R,\tau) = \lambda'[\Delta T(R.,\tau)]/\delta(R.,\tau)$$

and to take $\left(\int_0^R \frac{\lambda'}{\delta}\Delta T 2\pi R dR\right) = \pi..R^2 \times q$

Then, Equation (4.16) can be transformed to

$$R = \frac{qkR}{r\rho''s} \tag{4.17}$$

The integral of Equation (4.17) by the initial condition, $\tau = 0$, $R = C_R^S$, have the following form:

$$R = C_R S \exp\left(\frac{qk\tau}{r\rho''s}\right); \quad \frac{dR}{d\tau} = C_R \frac{qk}{r\rho''} \exp\frac{qk\tau}{r\rho''s} \tag{4.18}$$

The simplest law (4.18) did not give the vapor dynamic growth guarantees at boiling in the slits' reliable generalization at the wide range of pressure values obtained from experiments. However, in the limited range it was right to prescribe vapor bubble growth (for the atmospheric pressure $p_s = 1$ bar [Figure 4.21]). Knowledge on the velocity of dependency bubble growth allowed researchers to come to the next step in the modeling stage: calculation of the average thickness of the liquid microlayer, formed during bubble growth.

3. Formation of the average microlayer thickness took place during the process of the vapor bubble system's simultaneous growth. By it, the average liquid outlet flow from slit w in the main stage of the growth (in the "plane disks forms") could be defined from the correlation of the material balance:

$$kb1n_F R dR = wd\tau; \quad w = \pi R R b n_F k$$

It was proposed that the equal probability of the appearance of vapor bubble nuclei on both surfaces formed the slit, by $k = 2$.

4. The pressure drop between the vapor bubble volume p'' and the liquid volume p_∞ was defined mainly by the inertia forces:

$$p'' - p_\infty = C\rho'(\pi R R b n_F k)^2 \tag{4.19}$$

5. The formed initial microlayer thickness δ_0 value could be defined with respect to the recommendations in Ref. [138] on the next formula base:

$$\delta_0 = \sqrt{\frac{v'R}{\left|\frac{1}{\rho'}\frac{dp''}{dR}\right|}} \tag{4.20}$$

If we differentiate (4.19) on the vapor bubble's radius, we will obtain

$$\frac{1}{\rho'}\frac{dp''}{dR} = C_1(\pi b n_F k)^2 (R^2 \times R + R^2 \times R) \tag{4.21}$$

Substituting (4.21) in (4.20), we obtain

$$\delta_0 \sim \frac{1}{\pi b n_F R} \sqrt{\frac{v'R}{R^2 R + R^2 R}} \tag{4.22}$$

The average the microlayer thickness $\overline{\delta_0}$ over the surface determined the thermal resistance of the heat transfer over the evaporative liquid microlayer

$$\overline{\delta_0} = \frac{1}{\pi R_{max}^2} \int_s^{R_{max}} 2\pi\delta_0 R dR \tag{4.23}$$

$$n_F \pi R_{max}^2 \approx 1$$

where R_{max} is the vapor bubble maximum radius inside the slit. It could be defined from the geometry considerations: $n_F \pi R_{max}^2 \approx 1$.

Substituting (4.15) into (4.22) and in (4.23), and accounting that $n_F \pi R_{max}^2 \approx 1$ and supposing that $R_{max} \gg s$ (Figure 4.27), we obtain

$$\delta_0 = \frac{1}{bk} \sqrt{\frac{r\rho'' v's}{qkn_F}}$$

6. The microlayer thickness changing under the evaporation action by the condition $q = const$ has the following form:

$$\delta = \overline{\delta}_0 - \frac{q\overline{\tau}}{2r\rho'}$$

The average value of microlayer thickness over time can be defined as

$$\overline{\delta} = \delta_0 - \frac{q\overline{\tau}}{2r\rho'}$$

where $\overline{\tau}$ is the "microlayer existence" average time.

The statistical treatment of the video data of the main boiling in the horizontal slits show that the decrease in microlayer thickness by the determined conditions could have an important influence on the process regularities. This influence allowed researchers to explain the observed data on boiling at vacuum with some stable heat transfer regimes with the following correlations: by $\alpha \approx Cq^m$, where $m = 1$–2. The determination of the character of evaporation time dependency on the main factors required building the model of the elementary process "vapor phase output" from the slit.

7. As soon as a considerable amount of gravity forces is absent in the horizontal slit, it means that the vapor phase outlet from this channel was possible only after the vapor volume has been removed, when at least a part of the vapor volume came out of the slit into the surrounding liquid pool and leading to the appearance of any gravity force (Figure 4.44). The average pressure inside the output part of the vapor volume (zone A) became lower than the pressure into the vapor slug in the slit as the value $\Delta p''$ at the quasi-stable state.

Therefore, the vapor phase outlet from the horizontal slit internal volume can be conventionally divided into the following two stages:

- The vapor phase border movement inside the horizontal slit under the action of internal pressure drop $\Delta p''$ by the vapor volume growth inside the slit (the vapor phase border velocity and the vapor "existence time" there can be defined from the balance of the external pressure, friction, inertia, and other possible forces).
- The growth in vapor volume at the liquid pool after the slit's borders. This led to the decrease in the vapor slug's internal pressure; maximum pressure drop into the vapor slug relating the pressure at the liquid pool at the level of the slit without accounting for the inertial effects can be defined in the following equation:

$$\Delta p'' \sim g(\rho' - \rho'')H_A$$

The vapor slug part (zone A) departure under the action of gravity forces from the slit could take place earlier, and then the internal part of this slug will likewise depart. Such process development will be most favorable for the initiation of "vaporization of the surface." Let us consider the one-dimensional vapor slug growth into the horizontal slit. We will disregard the possibility of their destruction near the outlet.

The vapor slug border displacement into the horizontal slit dz for the time interval $d\tau$ will correspond to the vapor slug surface area of zone A, thereby increasing dF. The entire zone A cross section changing dF comprised value dF' and the vapor volume that appeared as evaporation from the surface part $(b - z)lk$ for time interval $d\tau$. The corresponding differential equation takes the following form

$$s\frac{dz}{d\tau} + \frac{qk}{r\rho''}(b - z) = \frac{dF}{d\tau} \qquad (4.24)$$

The flow inside the slit is caused by the action of pressure drop $\Delta p''$. If for $\Delta p''$ we will neglect the inertia effects, accompanying the zone A growth, and consider that its height $H_A \approx C\sqrt{F}$, then

$$\Delta p'' \sim g(\rho' - \rho'')k_1\sqrt{F}$$

Let us consider that, by liquid movement inside the slit, the resistance $\Delta p'$

$$\Delta p' \sim C\rho'\left(\frac{dz}{d\tau}\right)^2$$

The liquid movement momentum equation can be written as

$$\Delta p'' \approx \Delta p''; \quad g(\rho' - \rho'')k\sqrt{F} \sim C\rho'\left(\frac{dz}{d\tau}\right)^2$$

$$\frac{dF}{d\tau} = \left(\frac{C\rho'}{g(\rho' - \rho'')k_1}\right)^2\left(\frac{dz}{d\tau}\right)^3\frac{d^2z}{d\tau^2} \qquad (4.25)$$

The common consideration (4.24) and (4.25) led to the one-dimensional phase border movement inside the horizontal slit by the vapor volume outlet.

It is right to consider two limited states corresponding to the initial and final stages of the vapor volume output from the slit: $b \gg z$; $(b - z) \to 0$.

By $b \gg z$, we can obtain for the approximated estimation

$$\tau_1 = 1.25 b^{0.8} \left(\frac{s r \rho''}{4 q b} \right)^{0.2}$$

For the next limiting case $(b - z) \to 0$ and characterized time τ_2, it was proposed that the first part of the movement had an essential meaning for the calculation of $\bar{\delta}$. It could be possible to take $\tau = \tau_1$.

Therefore, the thermal resistance of the microlayer evaporation process is determined by the next correlation:

$$\frac{1}{\alpha_{EV}} = \bar{\varphi} \frac{\delta}{\lambda'} = \bar{\varphi} \left[\frac{C}{b k \lambda'} \sqrt{\frac{r \rho'' v' s}{q k n_F}} - \frac{q}{2 r \rho' \lambda'} 1.25 b^{0.8} \left(\frac{s r \rho''}{4 q b} \right)^{0.2} \right]$$

The average "waiting time" from the moment of vapor phase removal to the first appearance of vaporization centers with respect of the video data (Figure 4.25) was $0.01 \leq \tau \leq 0.03$ sec. These values corresponded to Fo numbers from 10^{-3} to 10^{-2}. The "contact" heat transfer mechanism between the wall and the liquid was suggested as the predominating one.

Supposing that the microlayer mechanism has the determined input into the whole thermal resistance of the process by such small Fo, we obtain the following:

$$\frac{1}{\bar{\alpha}} = \frac{1}{\alpha_{EV}} = \bar{\varphi} \frac{\bar{\delta}}{\lambda'}$$

As noted, the determination of vapor quality was an essential complex problem as the dependency on main regime, geometry, and other factors for vaporization into the vertical slits. By vaporization inside the horizontal slits, this value's dependence on heat flow density was considerably weak (Figure 4.26). If we prescribe this dependence by the formula $\varphi \sim q^m$ (as follows from the experiments, $m \approx 0.1$), consequently, at the first approximation it could be possible for the generalized experimentation data of heat transfer into horizontal slit boiling to use $\bar{\varphi} = const$. Then,

$$\frac{1}{\bar{\alpha}} = const \frac{1}{\alpha_{EV}} = \frac{const_1}{k b \lambda'} \sqrt{\frac{r \rho' s v'}{q k n_F}} - const_2 \frac{q b^{0.2}}{r \rho' \lambda'} \left(\frac{r \rho'' \alpha_1}{q b k} \right)^{0.2}$$

Taking into the consideration the correlation for n_F from the formula, the following was obtained for $\bar{\alpha}$:

$$\bar{\alpha}^{-2} \frac{const_1}{k b \lambda'} \sqrt{\frac{s v' (\sigma T_s)^2}{q^3 k r \rho''}} - const_2 \frac{q b^{0.8}}{r \rho' \lambda'} \left(\frac{r \rho'' s}{q b k} \right)^{0.2} \bar{\alpha} - 1 = 0$$

Including the generalized variables, we have

$$Nu = \frac{\bar{\alpha}}{\lambda'} \frac{\sigma T_s \rho' c_p}{(r\rho'')^2}; \quad Re = \frac{qb}{r\rho'' v'}; \quad Ar = \frac{qb^3}{v'^2} \frac{\rho' - \rho''}{\rho'}$$

We obtain the equation that generalized heat transfer by boiling inside horizontal slits from experimental data, in the form

$$Nu\, Re^{-2.3} \left(\frac{s}{b}\right)^{0.5} Ar^{0.4} \frac{\rho'}{\rho''} \frac{1}{Pr'k^{1.5}} = C_1 \left[1 + \sqrt{1 + C_2 \frac{Ar^{0.8}}{Re^{3.1}} \left(\frac{\rho'}{\rho''}\right)^2 \frac{1}{k^{1.1}} \left(\frac{s}{b}\right)^{0.5} \frac{1}{Pr'} \frac{l^*}{b}} \right]$$

$$(4.26)$$

This equation on the thermodynamic similarity form base

$$\frac{\bar{\alpha} s^{0.5}}{q^{2.3} b^{1.6} k^{1.3}} f_1(\pi) = C_1 \left[1 + \sqrt{1 + C_2 \frac{s^{0.5} f_2(\pi)}{q^{3.1} b^{2.2} k^{1.1}}} \right];$$

$$f_1(\pi) = A_1 \pi^{0.83}; \quad f_2(\pi) = A_2 \pi; \quad \pi = \frac{p_1}{p}; \quad p_1 = 1 \text{ bar}$$

$$A_1 = 0.73 \times 10^{-3} \text{ for } H_2O; \quad A_1 = 0.58 \times 10^{-2} \text{ for } C_2H_5OH$$

$$A_2 = 1.2 \times 10^{11} \text{ for } H_2O; \quad A_2 = 1.3 \times 10^9 \text{ for } C_2H_5OH$$

The experimentation data generalization, based on formula (4.26) positions, is presented in Figure 4.45. The satisfied generalization quality and good agreement with analytical solution, which was reached by the next empirical constants using $C_1 = 1.43 \times 10^{-3}$; $C_2 = 154$ confirmed this approach to be right and, based on it, the model to be perspective.

It expected that heat transfer by boiling into the short vertical channels have to be the same for horizontal slits. From this point of view, it would be especially interesting to use these considerations and to check it based on the corresponding experimentation data, using the same model positions as for horizontal slits generalization. Such a possibility appeared with respect to its use for the known experimental facts, discovered by the experts from Kiev National Polytechnic University, who published their results in several papers [123–125] (Figure 4.14), based on their experiments on the boiling of refrigerants R-11 and R-113 inside the short vertical slits (with height ranging from 20 to 50 mm), in the wide range of values of other factors, $0.3 \le s \le 2$ mm; $1 \le p_s \le 4$ bar; $20 \le H \le 50$ mm. The treatment of experimental data (Figure 4.14) on the proposed model base confirmed the possibility of using the aforementioned model's main positions for some cases of horizontal slits to spread them on the short vertical slits by the determined influence of the vapor bubble dynamics on heat transfer regularities.

It is possible to recognize to be satisfied by the generalized correlation (4.27) presented in graphic form at the same coordinates as for horizontal slits (Figure 4.45)

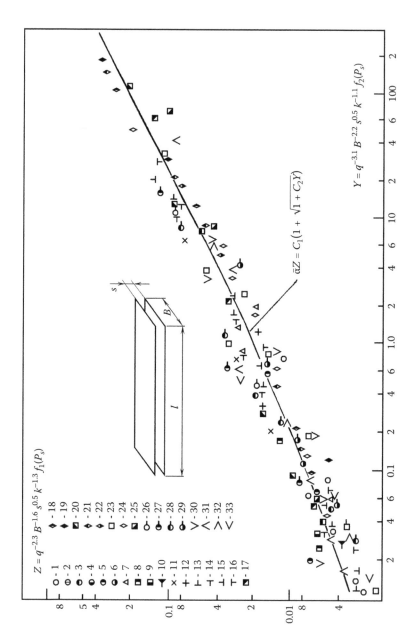

FIGURE 4.45 Generalization of experimental data on heat transfer boiling inside the horizontal slits [105–109]. Heat carriers: water, refrigerant R-11; ethanol: $0.05 \leq p_s \leq 1$ bar; $b = 15$; 30 mm; $0.2 < s < 1$ mm. $Z_1 = Nu^* Re^{-2.3} s^{0.5}/b \; Ar^{0.4} \rho'/\rho'' (Pr'k^{1.5})^{-1}$; $Y_1 = Ar^{0.8} / Re^{3.1} (\rho'/\rho'')^2 \; 1/k^{1.5} \sqrt{s/b} \; 1/Pr' \; \sigma T_s \rho' C_p /(rp'')^2 b$. 1, Calculation on the model base). $Z_1 = 1.43 \times 10^{-3} (1 + \sqrt{1 + 154 Y_1})$.

and compare it with the correlation presented in Figure 4.46. The equation structure generalized the experimental data for the short vertical slits, adjusted to be rather near the correlation (4.26) and has the following form:

$$Nu^*Re^{2,3}\left(\frac{s}{b}\right)^{0,5}Ar\frac{\rho'}{\rho''}\frac{1}{Pr'k^{1,3}}=C_1\left[1.02\sqrt{1+C_2\frac{Ar^{0.8}}{Re^{3.1}}\left(\frac{\rho'}{\rho''}\right)^2\frac{1}{k^{1,1}}\left(\frac{s}{b}\right)^{0,5}\frac{l^*}{b}\frac{1}{Pr''}}-1\right]$$

(4.27)

Correlation (4.27) is spreading to the next complex range:

$$Y_1=\frac{Ar^{0.8}}{Re^{3.1}}\left(\frac{\rho'}{\rho''}\right)^2\frac{1}{k^{1.5}}\sqrt{\frac{s}{b}}\frac{1}{Pr'}\frac{\sigma T_s\rho'C_p}{(r\rho'')^2b}$$

from 10^{-7} to 10^{-2}. As for (4.26) ranges, it will be valid for range $10^{-2}<Y_1<10^4$.

The two-phase flow inside the short vertical slits can be applied over height H, which is why in the dependency (4.27) it is used as the determined size instead of $b-H$.

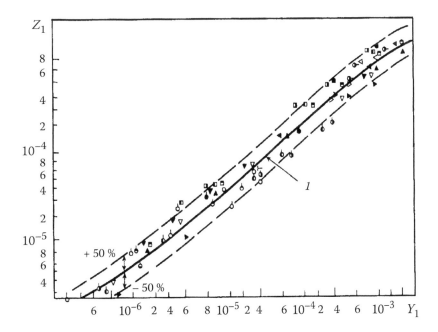

FIGURE 4.46 Generalization of experimental data on heat transfer on boiling inside the short vertical slits [123–125]. Heat carriers: water; R-11; ethanol: $1\leq p_s\leq 4.6$ bar; $0.02\leq H\leq 0.05$; $0.3\leq s\leq 1.2$ mm (Figure 4.45). Z_1 and Y_1, see Figure 4.45; 1, calculation on the next formula: $Z_1=1.57\times 10^{-4}(1.02\sqrt{1+1.2\times 10^{-4}Y_1}-1)$.

Therefore, the generalized correlations (4.26) and (4.27) can be recommended to the heat transfer boiling calculations inside the horizontal slits or short vertical ones, if the slit's gaps are smaller than vapor bubble departure size by the pool boiling conditions $s/D_0 \leq 1$, and the liquid circulation determined mainly by the vapor bubbles dynamic. The last condition can be approximately prescribed by the following manner:

$$\frac{\rho' \bar{w}^2}{2} \gg (\rho' - \rho'')gH \tag{4.28}$$

where H denotes the conventional circulation loop height (for horizontal slit, $H = s$); $\bar{w}$ is the liquid average velocity by action of the growing vapor bubbles. As for the slit, approximately, we have $\bar{w} = qkb^*/r\rho''s$, where b^* is the cell radius; $\pi b^{*2} = 1/n_F$. Then, condition (4.28) at the dimensionless form will be

$$Re_H \gg \sqrt{Ar_H \frac{1}{\pi n_F s^2}}$$

Therefore, heat transfer intensity becomes more noticeable when the vertical slit height H is lower, the heat flux density q is higher, vapor phase density ρ'' is lower, and the slit's gap s is lower than the influence of the growth in vapor bubbles.

5 Heat and Mass Transfer at Vaporization on Surfaces with Capillary-Porous Coverings

5.1 EXPERIMENTAL INVESTIGATIONS OF VAPORIZATION HEAT TRANSFER ON SURFACES WITH POROUS COATINGS

5.1.1 EXPERIMENTAL INVESTIGATIONS OF BOILING HEAT TRANSFER ON COATED SURFACES IN SUBCRITICAL THERMAL REGIMES

Hundreds of papers are devoted to experimental investigations of vaporization inside capillary-porous structures (CPS) attached to a heat transfer surface. The number of publications and research groups in this field is increasing continuously. However, a majority of these investigations is done in such a manner that presented results had a fragmented character. Sometimes, results reveal a contradictory nature and often there is not enough information for analysis. Therefore, we are going to consider only the experimental research having a wide range of main factors and containing representative experimental data and sufficient information. We will not analyze in detail the experiments representing local episodic nature. The publications dedicated to physical explanations and theoretical models of vaporization inside porous structures will be examined more carefully.

The first extensive experiments on vaporization heat transfer in thin-layer porous structures were performed for the operating conditions of low-temperature heat pipes. They were performed with the goal of obtaining the essential data about heat pipe thermal resistance.

Experimental data on vaporization heat transfer on surfaces covered by detachable porous structures were presented by Ferell et al. [146]. The working chamber design schematic layout is given in Figure 5.1. The working element was the stainless steel flat plate with square cross section of 2.5×2.5 inches. The plate was soldered to the copper block with an inserted electrical heater. The maximum heat flux was 45×10^4 W/m^2. The system of thermocouples was installed in stainless steel block.

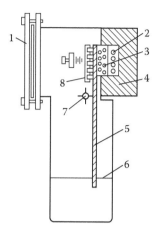

FIGURE 5.1 Working zone [146]. 1, Cover; 2, heater; 3, thermocouples; 4, thermal insulation; 5, wick; 6, liquid level; 7, rotation axis; 8, the wick compressing device.

Heat flux value and wall surface temperature were determined by temperature drop value. Porous samples (width, 2.5 inches; length, 12 inches) were pressed by a special device to the plate surface.

Several experiments were preformed to study heat transfer regularities. Available capillary potential (by capillary lifting height) and permeability (by Darcy method) were determined for each tested porous sample. The following porous structures were studied:

- Monel ball layers with average particle diameter ranging from 0.7 to 0.09 mm; nine variants of layer.
- Glass ball layers with average particle diameter ranging from 0.7 to 0.16 mm; six variants of layer.
- Metallic felt with average pore size ranging from 0.38 to 0.06 mm; layer thickness, 2–3 mm; skeleton materials were stainless steel, nickel, copper; six variants of structure.
- Sintered powder layers from copper and stainless steel particles with average pore size ranging from 0.06 to 0.016 mm; thickness, 0.7–3 mm; six variants of structure.
- Metallic screen layers from stainless steel with mesh sizes 40×49, 100×100, 150×150, and 200×200 μm.

Working element position (inclination angle), capillary lifting height, saturation pressure, structure types, and samples were varied in the experiments. Typical results are given in Figure 5.2.

The critical heat flux was selected as the main objective for the analysis. Crisis appearance was fixed in the curvature changing point on temperature dependence, $Q = f(T_W - T_s)$ (see Figure 5.2). Experimental data treatment for Q_{max} was based on the known hydrodynamic equation associating the moment of crisis appearance

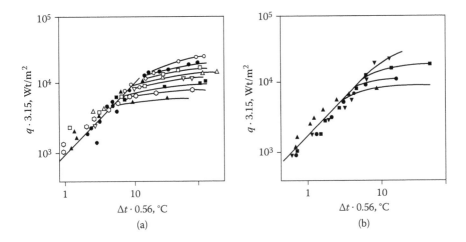

FIGURE 5.2 Typical experimental results [146]. (a) Fiber–metal nickel: vertical and horizontal wicks; (b) fiber–metal copper: inclined and vertical wicks.

$(Q = Q_{max})$ with equality between available capillary potential and sum of pressure drops in liquid filtration through the porous structure along the heating surface and the vapor–liquid mixture filtration perpendicular to the heating surface, from the deepened interface to the exit from the capillary structure.

It was supposed that the dimensionless coordinate of the interface deepening $\bar{X}$ is independent of the heat flux. Actually, experimental data revealed that $\Delta Q \approx \Delta T$, that is, $\alpha \cong const$. This result could be explained by the constancy of liquid layer thickness between the surface and interface. Calculations based on this assumption resulted to $\bar{X}$ values of about 0.2 for nickel felt samples, demonstrating acceptable concurrence between experimental data on heat transfer coefficient and calculation results determined as $\alpha = \lambda_E / \delta \bar{X}$, where λ_E is the effective thermal conductivity and δ is the thickness of the capillary structure.

However, for other types of CPS, for example, when sintered copper samples were considered, essential (several times) nonconformity between experimental and calculation results was observed. It was explained by the appearance of the vapor layer between the heat transfer surface and the liquid layer inside the structure. To explain such a phenomenon in such conditions, when special measures were done to ensure reliable contact between the capillary structure and the wall, it was supposed that dimensions of the gaps remaining between the structure and the wall considerably exceed small pore sizes. However, this assumption was not verified.

As discovered in analogous experiments [147–149], the order of magnitude for values of α corresponds to the assumption that major thermal resistance is associated with the wetted skeleton of the capillary structure at the interface deepening up to the heating surface.

Experiments [150] were conducted on the experimental setup with glass chamber filled by the working fluid. Working element was positioned in the central part of the chamber. It was a tube having an internal diameter of 17 mm with inserted heater placed in the copper block. The porous structures (screen wicks) made from stainless steel, phosphorous bronze, and nickel felt were wrapped on the tube surface. The wicks compressing to heating surface was maintained by special hinges. The number of layers was changed from 1 to 11. Structure thickness was varied from 0.28 to 0.89 mm. Screens with two main mesh sizes (0.042 and 0.095 mm) were used.

Experiments were conducted under pool boiling conditions, when the capillary structure was submerged into the liquid at capillary feeding conditions. Typical results are given in Figure 5.3. As seen from Figure 5.3b, function $q = f(T)$ has two main zones, which differ by heat transfer regimes. In zone I (small heat fluxes), $q = const\Delta T$. It is important to note that vapor bubble generation was not observed in this zone. However, experimental values of heat transfer coefficients were so high that it was impossible to explain by convection and by effective heat conduction through the wetted structure alone. Only the assumption of the existence of stable vapor regions inside the porous structure and heat transfer over these regions by loop evaporation from heating surface and condensation on external surface, justifies such high values of heat transfer coefficients. Transition from zone I to zone II (Figure 5.3b) is accompanied by changing of heat transfer dependency to $q = const\Delta T^m$, where $m \approx 3-4$, similar to pool boiling heat transfer on smooth surface. Conformity has qualitative as well as quantitative character. In addition, intensive vapor removal from the capillary structure external surface was observed.

Heat transfer regularities for capillary feeding conditions were the same as for pool boiling. This coincidence for acetone and ethanol was observed only at small

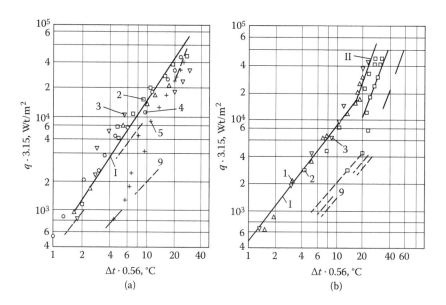

FIGURE 5.3 Typical experimental data [150]. Experimental data: $p = 0.1$ MPa. (a) Water boiling on tube wrapped by wick; (b) Same conditions for ethanol, acetone boiling; wicks: stainless steel and brass screens, fiber–nickel; mesh size $a = 0.042, 0.095$ mm; layers number was 1–11. Calculations: 9, boiling on smooth tube.

heat fluxes (approximately 10^4 W/m^2). As soon as heat fluxes exceeded this value, the significant superheat of the heat transfer surface was detected (approximately 1 order of magnitude), that is, the porous structure was completely filled with a vapor. The same results were obtained in another study [152].

Experiments [152] were conducted with the goal of defining boiling heat transfer curves for coated surfaces. The experimental setup consisted of copper block coated by nickel. Powerful electric heaters were placed in the block. Temperature measurements were performed by 11 thermocouples located 1 mm (16 × 125 mm strips) from the heat transfer surface. Porous structures were sintered copper and nickel plates and nickel screens with a porosity of 0.7–0.85 and thickness of 0.46–3.3 mm. Water was used as the working fluid at saturation pressure, from 0.007 to 0.25 MPa.

Tests were performed for the horizontal position of the plate under conditions of boiling on submerged surface and at capillary feeding. It was discovered that the shape of the boiling curves was independent of liquid supply conditions and the structure skeleton material. It was noted that although the evaporation general mechanism from coated surfaces was not clear, many specialists assumed that heat is transferred to the liquid–vapor interface by heat conductance. This assumption contradicted the test results. As shown experimentally, heat transfer intensity was not dependent on structure thickness. Such results can be explained only by assuming that the interface is deepening and located near the heating surface.

In addition, the influence of saturation pressure on boiling heat transfer intensity was determined. If heat transfer dependency is treated as $q = const\Delta T^m$, m decreases from 4 to 1 as ΔT increased. Similar heat transfer regularities for the boiling on coated surfaces were presented by Marto and Mosteller [153].

Extensive experimental data on vaporization inside porous structures were obtained during investigations on heat pipe thermal modes. The following are specific features of these data:

- In heat pipe experiments, variations of main independent parameters are interconnected, for example, change in heat flux considerably influences saturation pressure, liquid supply, etc.
- Maintaining similar thermal and hydrodynamic conditions of vaporization along the heat pipe length and diameter is complicated and is even not possible in many cases.
- Manufacturing technological factors — such as sustaining constant thickness of porous structure and uniformity of contact between the wall and porous structure; degassing of container, structure, and heat carrier; heat pipe filling — significantly influence vaporization conditions in heat pipes.

Therefore, analysis and discussion of experimental data on evaporation heat transfer in heat supply zones of heat pipes have a qualitative nature. Because of the above-mentioned reasons, experimental generalizations for heat transfer coefficients within the heat supply zones of heat pipes are essentially nonmonotonous and nonreproducible in different heat pipes [155].

5.1.2 EXPERIMENTAL INVESTIGATIONS OF BOILING HEAT TRANSFER ON SURFACES COVERED BY SCREEN WICKS

Experimental investigations [154, 155] on low-temperature heat pipes were performed at Moscow Power Institute (MPI). Tested heat pipes had two- or three-layer screen wicks and heterogeneity in thickness, that is, increase of screen mesh size was directed from the heating surface to the evaporating surface. Typical dependencies between heat fluxes and wall superheats obtained in MPI tests are presented in Figure 5.4. Considerably less heat transfer intensity than at pool boiling conditions was observed due to wick partial filling with a vapor in places of screen detachment from the wall.

Direct experimental confirmation of the essential dependence of vaporization mechanism and heat transfer regularities inside heat pipes on the wick compression to the wall was obtained in experimental investigations [156–158]. Experiments [156] were conducted on heat pipes with multilayer screen wicks and with a special wick holder. It was a bent perforated plate fixed in a most decompressed state. The temperature distribution along the heat pipe length and perimeter was measured. Considerable temperature inequality at high heat fluxes was observed. It was evidence of thermal regime stability for the case of the porous structure partial drying.

Experimental research on heat pipe with glass container is presented by Asakyavichus et al. [156]. Wick consisted of three or four copper screen layers with mesh size of 0.1 mm or the same number of brass screen layers with mesh size of 0.287 mm. The quality of the contact between wick and wall was poor because

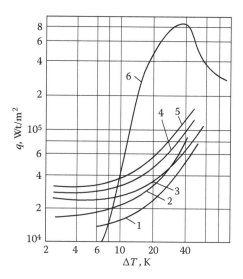

FIGURE 5.4 Dependencies between heat flux and wall superheat at vaporization in nonuniform heat pipe screen structures [155]. 1, Three-layer screen wick with mesh sizes 0.112, 0.2, and 0.23 mm; 2, two-layer uniform wick 0.2, 0.2 mm; 3, single-layer, $a = 0.28$ mm; 4, three-layer glass wool wick with mesh sizes 0.112, 0.2, and 0.28 mm; 5, two-layer metal wick 0.2, 0.28 mm; 6, ethanol pool boiling on smooth surface.

the glass tube internal surface had significant microroughness. The wick inside the glass tube covered about 90% of the internal surface [156].

Visual observations of processes occurring in the glass heat pipe demonstrated that in certain areas of the wick (in gaps between the wall and the wick), blocked vapor volumes appeared, even when heating was absent. After turning on the wall heating, the area of these vapor volumes increased abruptly and new vapor volumes were developed in the wick. Asakyavichus et al. [156] ascribed this phenomenon to poor contact quality in the corresponding locations. It was also discovered that with the increase in wick compressing size, vapor volumes decreased, and eventually disappeared completely. In locations of reliable contact, deepening of the interface down to the first screen layer (near the wall) was observed at small heat fluxes (approximately $0.5–1.0 \times 10^4$ W/m²).

In the depth, the liquid meniscuses pulsated with a frequency of 10–100 Hz [156], which increased with heat flux growth. The visual method used for measurements of such high pulsation frequencies is not described in the paper. The characteristics of the working element used by Asakyavichus et al. [157, 158] for refrigerant R-113, ethanol, and water heat transfer coefficient determination were similar to those presented by Seban and Abhat [151]. The wick consisted of 2–12 screen copper or stainless steel layers with mesh sizes 0.1×0.1 and 0.07×0.07 mm. The screen layers were compressed to the heating wall by means of eight tightened wire rings or by the perforated screen holder. At low heat fluxes (up to $q \leq 10^5$ W/m²), heat transfer intensity between the surface and water at $p_s = 1$ bar was higher in the presence of screen wick [156–158], because the entire heat transfer surface and the number of nucleation centers increased. Further heating caused worsening of wick heat transfer intensity due to the escalating impact of the vapor removal process, which caused a decrease in heat transfer intensity.

Moreover, it was discovered that heat transfer coefficients for the 12-layer wick was 2.5 times higher than for the two-layer wick fixed by the rings because the thermal contact for the 12-layer wick was better, and effective thermal conductivity λ_E of the wetted porous structure was significantly higher [157]. Comparison of the eight-layer wick and thread-type finned surface demonstrated that starting from $q = 10^5$ W/m², intensities of heat removal were approximately the same, in spite of the fact that during heat transfer, area of the finned surface was 2.5 times larger (Figure 5.5). Thus, these examples reveal significant contradictions to physical models [156–158].

The authors obtained empirical generalization of experimental data in the following form:

$$Nu_L = 0.241 \, Re_L^{0.24} \, (Pr')^{4.2}; \, Nu_L = \frac{\alpha l_L}{\lambda_E};$$

$$\lambda_E = k \lambda_L^n; k = 2.3C; n = 0.885 \times (0.07 \leq \lambda_L \leq 0.17 Wt/mK);$$

$$k = 0.91C; n = 0.37(0.17 \leq \lambda_L \leq 0.68 Wt/mK); C = C_C C_V C_{SC}$$

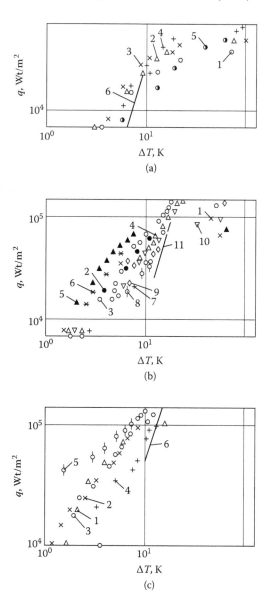

FIGURE 5.5 Water, ethanol, and refrigerant R-113 vaporization heat transfer at boiling on cylindrical surfaces with uniform screen structures [156–158]. (a) Experimental data for R-113: 1 and 2, submerged and nonsubmerged wick holder, eight layers, $a = 0.1$ mm; 3 and 4, the same with rings; 5, with rings at q decreasing; calculation: 6, without wick. (b) Experimental data for ethanol: 1 and 2, submerged and nonsubmerged wick holder, eight layers, $a = 0.1$ mm; 3 and 4, the same with rings; 5 and 6, nonsubmerged wick holder, $a = 0.007$ mm; 7–10, rings, $a = 0.01$ and 0.07 mm; calculation: 11, without wick. (c) Experimental data for water: 1, nonsubmerged wick holder, eight layers, $a = 0.1$ mm; 2 and 3, submerged and nonsubmerged rings; 4, submerged wick holder without wick; 5, submerged microfinned surface; calculation: 6, without wick.

Constant C accounts for the influence of contact surface, capillary structure geometry, and thermal conductivity:

$$C_C = 0.709F_C/F_W + 0,291; C_V = 2,8 - 1,89a; C_{SC} = 3,3\lambda_{SC}/10^4 + 0,872$$

Representative character of the experimental data and reliability of experimental methods allowed handlers to determine the influence of the main factors independently. Therefore, these experimental correlations have a general nature for vaporization in screen wicks attached to the heat transfer surface by compressing. This was also confirmed by experimental investigations on heat transfer at vaporization inside screen wicks [76–79].

Experiments were conducted in the hermetic rectangular chamber. The working element based on the thermal wedge principle with the main and auxiliary heaters allowed heat flux values to reach up to 0.8×10^6 W/m^2. Experimental conditions are presented in Table 5.1.

In the table, screen structures were compressed to the heating surface by the special clips grill. In experiments, the margin compressing force value was determined, at which the heat transfer regularities remained unchanged with the increase in applied force. The mesh size variations of the clips grill simulated different options of contact point's regularity between the screen wick and the heat transfer surface. Special tests were performed for the investigation of vaporization conditions in flat artery heat pipes with a one-side heat supply zone. For certain experimental conditions, when unheated surface was transparent, the visual observation of vaporization phenomenon under natural conditions corresponding to heat supply zones of heat pipes was completed (Figure 5.6).

The artery wicks consisted of screen layers with mesh sizes $a = 60$ and 125 μm. The wicks were arranged on the ledges and compressed by the glass or metal plate. Artery wicks were combined with different screens and consisted of different thin-layer structures in the near-wall zone (one or two layers with mesh sizes 40, 60,

TABLE 5.1
Experimental Data on Heat Transfer inside Screen Wicks

Experimental Conditions	Water	Ethanol	Refrigerant R-113
Pressure, MPa	0.005–0.1	0.003–0.1	0.1
Screen structures (material)		mesh size, μm:	
Stainless steel	40, 60, 125, 200, 450, 1600	40, 60, 125, 200, 450	40, 60, 125, 200, 450
Brass	80, 160	80, 160	160
Copper	45	45	45
Layers number	1–24	1–9	1–3
Liquid level, mm	50	—	−140
Heat fluxes range, $\times 10^4$ W/m^2	1–97	1–50	0.3–25

and 130 μm). Experiments with artery wicks revealed that heat transfer regularities in the structures were the same as for the plain uniform single-layer and multilayer wicks having similar near-wall layers.

As shown in Table 5.1, liquid supply was created by capillary feeding from the liquid lowered level (−140 mm) and at submerged conditions (50 mm). It was observed that all main heat transfer regularities for two-, three-, and multilayer uniform structures determined in several experiments [75–79] concur both qualitatively and quantitatively with the experimental data of several studies [156–158], which investigated the correlation between heat transfer and working fluid level, layer thickness, working fluid supply conditions, working fluid and wick structure properties, mesh size, etc. (see Figure 5.7).

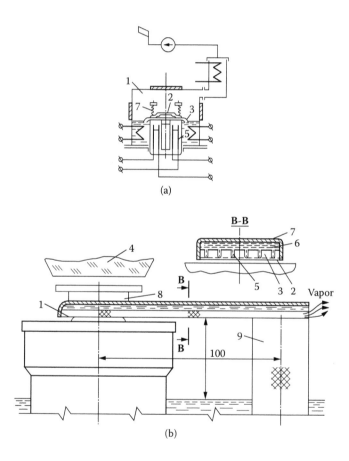

FIGURE 5.6 Experimental setups used in experiments [75–79]. (a) Experimental setup for the study of boiling heat transfer at vaporization on the surface of thermal wedge covered by single-layer and multilayer screen wicks: 1, chamber; 2, thermal wedge; 3, capillary structure; 4–6, heaters; 7, compressing grill. (b) Test sample of artery heat pipe with separated heat input, transport, and heat removal zones; 1, heating surface; 2, structure; 3, vapor channel; 4, compressing glass; 5, artery partition; 6, arteries; 7, container; 8, press; 9, feeding wick.

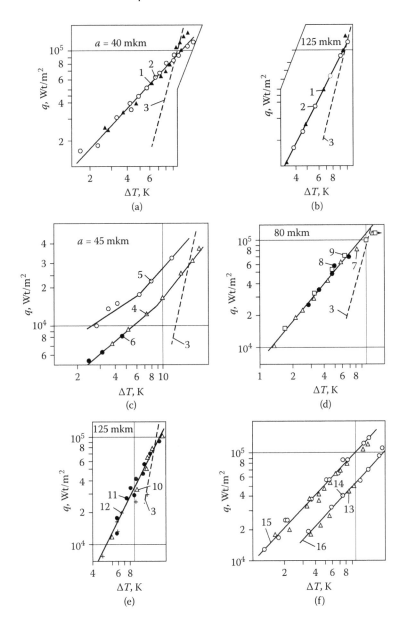

FIGURE 5.7 Heat transfer at vaporization inside uniform multilayer screen structures [75–79]. (a, b) Liquid level influence on boiling heat transfer; water, 0.1 MPa (three stainless steel screen layers); 1, submerged structure; 2, capillary feeding; 3, pool boiling on smooth surface. (c) R-113: 4, submerged copper two-layer structure; 5, the same, single-layer; 6, the same, capillary feeding. (d, e) Layer thickness influence on boiling heat transfer (water and ethanol, respectively; 0.1 MPa); 7–9, brass, two, three, and five layers; 10–12, stainless steel, two, three, and eight layers. (f) Boiling heat transfer inside screen structures with different skeleton thermal conductivities: 13, stainless steel; $a = 40\ \mu m$, two layers; 14, copper, $a = 45\ \mu m$; 15 and 16, water and ethanol, 0.1 MPa, approximations of experimental data.

Largely new heat transfer regularities were obtained for vaporization inside nonuniform and single-layer mesh screen structures, including nonuniform screen wick structures containing screen layers with different geometrical parameters. Experimental data on refrigerant R-113, ethanol, and water vaporization heat transfer in nonuniform two- and three-layer structures with three-stage increasing mesh sizes from the wall to the external surface are shown in Figure 5.8.

The existence of two main thermal regimes is typical for all tests. These regimes differed by heat transfer correlation, $q = C\Delta T^m$ form. In the region of low heat fluxes, heat transfer at vaporization in nonuniform structures concurred quantitatively and qualitatively with similar regularities for uniform structures, when their mesh size is equal to the mesh size of the near-wall layer of nonuniform structure. These experimental results consistently demonstrated that basic thermal and hydrodynamic

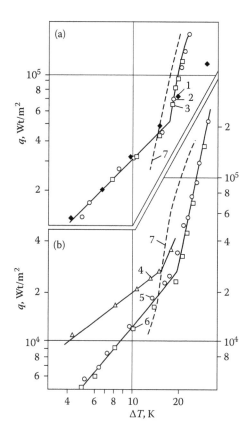

FIGURE 5.8 Boiling heat transfer at vaporization of (a) ethanol and (b) R-113 inside nonuniform screen structures [75–79]. 1–3, Stainless steel screen from three layers: $a_1 = 60 \, \mu m$, $a_2 = 200 \, \mu m$, and $a_3 = 125 \, \mu m$ at 0.01, 0.03, and 0.1 MPa, respectively; 4, the same: $a_1 = 40 \, \mu m$, $a_2 = a_3 = 125 \, \mu m$; 5, screen from two layers: copper $a_1 = 45 \, \mu m$ and brass $a_2 = 160 \, \mu m$; 6, screen, three layers: copper $a_1 = 45 \, \mu m$ and brass $a_2 = a_3 = 160 \, \mu m$; 7, pool boiling on smooth surface.

phenomena associated with vaporization heat transfer regularities are determined by structure parameters of the first screen layer located near the wall, and such phenomena develop mainly inside this first layer. It is also confirmed by the fact that when heat flux exceeds certain value, q_1^*, the thermal regime in the nonuniform structures appears similar to that in pool boiling (see Figure 5.8).

When $q < q_1^*$, the boiling heat transfer intensity inside uniform two-, three-, and multilayer structures are considerably higher (3–5 times) than under pool boiling conditions. When heat flux attains the value of q_1^*, heat transfer intensity at vaporization on surfaces covered by screen wicks appears the same as in the case of pool boiling. Furthermore, increase in wall superheat ($\Delta T > \Delta T^*$ and $q > q_1^*$) maintains favorable conditions for nucleation. It leads to vapor bubble formation and increase in the number of bubbles.

This leads to vapor blanket formation in uniform structure due to the significant increase in hydraulic resistance. Upper layers of nonuniform structures (with mesh size increasing from bottom to top) exhibit higher permeability and lower hydraulic resistance, maintaining stable boiling conditions in the thin liquid film found on the heat transfer surface and hold by capillary forces. This vaporization mechanism was confirmed by photo and video observations of hydrodynamic phenomena in screen structures (see Section 3.3).

Similar regularities of boiling heat transfer were observed in uniform structures (Figure 5.9). This also confirms the decisive influence of phenomena occurring in the near-wall zone on vaporization processes inside porous structures.

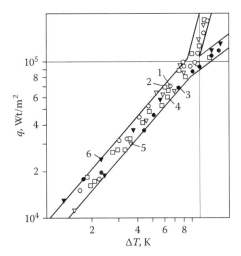

FIGURE 5.9 Comparison between experimental data on heat transfer intensities at vaporization inside uniform and nonuniform screen structures (water at 0.1 MPa) [75–79]. 1, Three-layer stainless steel screen: $a_1 = 40 \ \mu m$, $a_2 = a_3 = 125 \ \mu m$; 2, the same: $a_1 = 60 \ \mu m$, $a_2 = a_3 = 125 \ \mu m$; 3, two-layer stainless steel screen: $a_1 = a_2 = 40 \ \mu m$; 4, the same: $a_1 = a_2 = 60 \ \mu m$; 5, one layer of copper screen: $a_1 = 45 \ \mu m$ and two layers of stainless steel screen: $a_2 = a_3 = 125 \ \mu m$; 6, two-layer copper screen: $a_1 = a_2 = 45 \ \mu m$.

The presented experimental data were obtained for conditions of reliable contact between CPS and the wall. The experimental data [75–79] demonstrated that, for the majority of stainless steel screen wicks having high values of flexibility and rigidity, reliable contact with the wall was maintained by the largest grill used in the experiments with step equal to 7 mm.

When copper screens and single- or two-layer structures of brass and stainless steel with $a < 90$ μm were used, consistency of experimental data was obtained by grills with step 4 and 6 mm were used. Pressing force was also varied from 2 to 20 N; however, for higher compressing force values, there was no noticeable change in boiling heat transfer coefficients.

Experimental data on compressing force influence on boiling heat transfer regularities are presented in Figure 5.10. As shown, the presence of thermal regimes similar to the MPI's data for heat pipes (see Figure 5.4) is feasible in case of uncontrolled thermal contact between the mesh screen and the wall.

When a gap between wall and porous structure appears, the heat transfer intensity could be considerably higher compared to that under pool boiling conditions; however, vapor layer formation essentially happens earlier. Hence, appearance of the gap between wall and porous structure defines the conditions of two-phase flow filtration, vaporization center activation, vapor phase appearance, and development similar to pool boiling in slit channels. This similarity will be proximate when vapor filtration resistance is perpendicular to the wall, that is, through screen structure, will be higher.

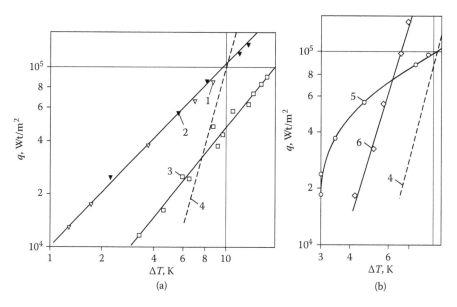

FIGURE 5.10 Influence of contact between structure and wall and gap formation conditions on heat transfer intensity at vaporization of (a) water and (b) ethanol inside screen structures at $p = 0.1$ MPa [75–79]. (a) Influence of contact between structure and wall (copper, $a = 45$ μm): 1, compressing grill with step of 4 mm (five layers); 2, step 2.6 mm (two layers); 3, step 7 mm (two layers); 4, pool boiling on the smooth surface. (b) Boiling at presence of the gap (stainless steel, two layers): 5, gap size is 450 μm; 6, gap size is 0.7 mm.

Therefore, it is obvious that, for screen structures disjoined from the wall, mesh size is smaller, the number of screen layers is higher, and heat transfer regularities will be proximate to conditions typical for boiling in slit channels. Hence, heat transfer regularities of slit channel boiling should disappear for single-layer screen structures with large mesh sizes, and heat transfer correlations should be similar to those found under pool boiling conditions. The nature of the vaporization process in screen structures disjoined from the wall should be accounted for in the analysis of experimental data obtained from the Institute of Thermophysics of the Ukrainian Academy of Sciences. The most typical results were presented in several papers [159–164].

The experiments were performed by using the two following working elements:

- The first one was a thermal wedge covered by single-layer screen structure top [159–162]. The structure fixing was mounted outside of the working element surface, allowing reliable contact between the screen and the heat transfer surface. Such measure ensured reliable contact along the working element perimeter only, because the diameter of the thermal wedge was equal to 28 mm. Thus, a gap between the wall and the screen was presented in the central part of the working element.
- The second working element represented a model of the screen structure single cell [163, 164]. The working surface was produced from the 0.07-mm-thick permalloy plate. The round copal bar was welded inside the plate center. Bar diameters were 0.5, 1.15, and 2.5 mm, and its length was about 2 mm.

A local heat transfer intensity study for surfaces covered by CPS was planned. Tolubinsky et al. [163] and Kudritsky [164] did not indicate how they carried out reliable thermal contact between the structure and the heat transfer surface. As noted by Tolubinsky et al. [163], the greater part of the experimental program was performed for the single cell covered by CPS. The model obtained from the single cell experiments was verified by tests on the extended surface with CPS. However, the essence of such experimental verification disagrees with Kudritsky's [164] statement that boiling on the modeling surface was fundamentally different from boiling under actual conditions. It means that using these experimental data should be considered doubtful. It was also confirmed in the authors' empirical correlation, which generalized the experimental data on heat transfer at boiling on surfaces covered by screen wicks: $Nu \sim (L)^{0.7}(Pr')^{-0.2}$, where $L = d_s/D_0$ is the ratio of heating $L = d_s/D_0$ surface dimension d_s to the bubble diameter D_0.

Here, D_0 is defined by the correlation for the dynamic regime of bubble growth and removal [164]: $D_0 = const \times Ja \times a'^{2/3} \times g^{-1/3}$, where $Ja = (\rho'C_p'\Delta T)/r\rho''$ is the Jacobs number and a' is the liquid thermal diffusivity.

The obtained generalization revealed the influence of microsurface (MS) size d_s. Therefore, it excludes the possibility of using the data obtained for MS in models of extended surface heat transfer. It was also confirmed by the data illustrating the known peculiarities of the MS boiling curve in wide range of heat flux values (Figure 5.11).

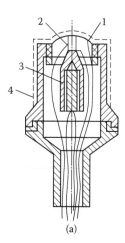
(a)

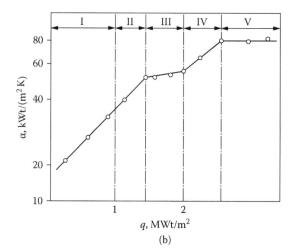
(b)

FIGURE 5.11 (a) Working element and (b) heat transfer intensity at vaporization of water on external surface covered with single-layer screen at $p = 0.1$ MPa [163, 164]. 1, Permalloy membrane; 2, copal bar with diameter of 1.15 mm; 3, heating element; 4, screen covering; I–V, heat flux measurement ranges.

Heat transfer regularities presented [159] for the thermal wedge conditions demonstrated concurrence with pool boiling correlations. It means that several thermal regimes observed in one study [164] are not appropriate for boiling heat transfer on extended surfaces and especially for vaporization in low-temperature heat pipes.

In several papers [165–168], the physical models were not considerably changed. The presented experimental results were based on the same methods and similar experimental conditions were used. However, the restricted character of the investigations [159–168] was acknowledged by Kudritsky [169], who illustrated this by introducing the concept of small size surface. Hence, the small size surfaces should fit the following condition: $0,1 < d_s / \sqrt{\sigma / g(\rho' - \rho'')}$, where d_s is the characteristic size of the heating surface.

Unfortunately, comparative analysis with the data of corresponding studies completed by other researchers has not been carried out in several papers [159–169].

High-speed video capture was used in a number of experimental investigations [159, 164]. Nevertheless, none of these images was presented in the papers, thereby making any discussion on visualizations almost useless. Because of the lack of information on the change in the working element design associated with visualization, one can suppose that the high-speed video capture was done from the top. In such a case, boiling in the disjoined structure could be confused with boiling in slit [105–113].

In addition, single-layer structures used in the experiments had a large mesh size (about 1×1 mm). This means that such structures are very specific and do not represent a common type of screen CPS, including the case of low-temperature heat pipes.

To conclude the critical analysis, it is important to consider vaporization thermal regimes in low-temperature heat pipe wicks that appeared from the data of Kudritsky [164], but introduced by the author as a general statement. The first regime was defined as the heat conduction through the structure skeleton and liquid film to the interface [164]. The working fluid evaporates from the meniscus surface.

The author made no effort to differentiate between single-, two-, three-, and multilayer structures, although the characteristic of vaporization appears significantly dissimilar for these screen structure types. In multilayer structures, the deepening of the evaporation front occurs down to the near-wall layer even at very small wall superheats (about 1–3°C or less). Then, in a wide range of heat flux values, heat is transferred through the wetted capillary structure skeleton to the interface. Therefore, in multilayer structures, the regime of heat transfer by heat conduction through the wetted porous structure to its external surface is really presented. It should be the first regime, "I" (Figure 5.11).

The heat transfer through the wetted skeleton to the deepened meniscus surface represents the second thermal regime, which is essentially different from the first one by thermal and hydrodynamic parameters. In this regime, the generated vapor can flow out of the capillary structure and then on its external surface some vapor bubbles or cavities can be periodically developed. Vapor condensation near the external structure surface (at some superheats range) is feasible. It is the so-called "microheat pipes" regime.

In single-layer screen structure, there are no major differences between the considered regimes, that is, in several investigations [159–164] the second regime, including "the micro heat pipes" regime, is unachievable. Therefore, observation of corresponding changes in temperature profiles was unattainable in these studies.

Regime "II," in which bubble boiling occurs in the meshes of the porous structure, is the third thermal regime; note that in multilayer screen structures, this regime is not feasible for some cases [163]. In the uniform structure with a significant thickness, the beginning of this regime can lead to vapor blanket formation in the wick. Actually, this regime — accompanied by the boiling in liquid film with deepened meniscus — will be developed at significantly higher wall superheat values than in pool boiling case.

Tolubinsky et al. [163] noticed significant contradiction between their own data on initial boiling superheats and the experimental results of a study [171] using very small values of superheats, $T_W - T_s$, sufficient for boiling initiation in metal–fiber wicks. Actually, there is no contradiction, because the metal–fiber wicks should be treated as multilayer wicks, realizing the second vaporization thermal regime accompanied by deepening of interface down to the first layer level. This regime could not be visually fixed, because vapor flowing out of the structure's internal volume is not necessitated. It will start when only the wall superheat will be sufficient for the activation of large pores. The appearance of this regime confirms the boiling initiation in the submerged porous structures, and from this point, there are no peculiarities in boiling initiation between submerged and capillary feeding structures. However, for the single-layer structures, such peculiarities can be observed. Accordingly, if Tolubinsky et al. [163] accounted for the fact that their experiments covered only very limited local cases of boiling on surfaces covered by porous structures, their conclusions could have been more valuable.

The following regimes, "III," "IV," and "V" (see Figure 5.11), were associated with "dry spots" formation, and existence in the cell center, followed by disappearance, and finally, total dry-out [164]. Because of doubts relating to the frequency, stability, and reproducibility of these regimes in real conditions, a reliable evaluation of these results is impossible.

Representative experimental investigations [75–79] demonstrate that thermal regimes accompanied by strong throwing of liquid were observed mainly in nonuniform thin-layer and single-layer screen structures.

Experimental results [172] devoted to single-layer screen structures on the surface of thermal wedge and obtained for a limited range of parameters concurred with data of several studies [75–79].

Conditions, when several heating elements are located on the mutual evaporation surface of heat pipe with arbitrary heat fluxes distribution, are prospective for the practical application of screen coverings. It is the case of discrete heat supply. From the physical viewpoint, it is clear that a strong interaction between different thermal regimes is feasible under these conditions. How advantageous would such interactions prove to be? Which factors will be significant? How strongly would regime interactions depend on geometry, heat fluxes, thermo-physical properties, and other parameters? Answers to these questions were not found in the majority of theoretical and experimental studies.

These circumstances prompted corresponding experimental investigations performed by the author's scientific team at the Odessa Academy of Refrigeration; results were published in several papers [170, 342, 343].

Specific results were obtained for two copper and stainless steel horizontal heat pipes at different arrangements of heat-generating sources and heat supply regimes. Heat pipe design schematic diagram, capillary structures, and model of heat generating elements manufactured in the shape of copper concentrators with quadratic cross section of 1×1 sm and soldered to heat pipe surface are shown in Figure 5.12a. Experimental results demonstrated that during the operation of separate heat sources, heat transfer regularities remain unchanged.

Temperature curves of mutually operating heat-generating elements for different arrangements were approximately identical. Some distinctions connected with wall superheat decreasing with heat flux growth were discovered at heat flux values preceding the crisis of heat transfer. The principal peculiarities were observed for critical heat flux regularities.

Obtained experimental data revealed that in case of discrete heat supply, worsening thermal regimes were caused by the interaction of two main heat pipe limits, such as hydrodynamic and boiling.

Further experiments were performed for heat pipes with uniform capillary structures in the near-wall zone and in the artery (Figure 5.12b). Specific arrangement of heat sources excluded the option in which the working fluid will flow out of heating zones. Plain arterial heat pipe (length, 220 mm; width, 2.5 mm; height, 14 mm; stainless steel container) was used as a test sample. The gap in capillary structure between heat sources "A" and "B" was made. It offered better reliability in determination of hydrodynamic parameters and corresponding calculations of hydrodynamic path length for the most remote heat source.

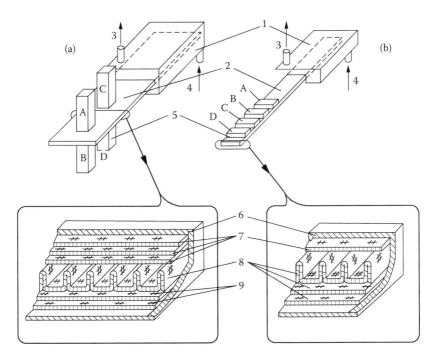

FIGURE 5.12 Design schematics of heat pipes [170, 342, 343]. (a) Two-side heat removal. (b) Single-side heat removal. 1, Heat exchanger; 2, heat pipe; 3, water outlet; 4, water inlet; 5 (A–D), heat generating elements; 6, heat pipe container; 7 and 9, capillary structures; 8, arteries.

Good concurrence of experimental data was observed for each heat pipe position and heat source location (heat load was changed simultaneously for four operating heat sources), except for heat source "A" because it was at the maximum distance from the condenser (Figure 5.13a). The heat transfer characteristics of mutually operating three sources, "B", "C," and "D," at simultaneous changing of their heat load are presented in Figure 5.13b. The same peculiarities in thermal regime change for the most remote from the condensation zone source "B" were observed with respect to heat load and position in gravity field.

Therefore, in the case of mutual operation of discrete heat sources, deterioration of heat transfer regime could occur for the heat-generating element that is most remote from the condenser, while the heat pipe continues to operate steadily.

An important drawback of screen structures in industrial elements and devices, including heat pipes, is the strong dependence of their thermal and hydrodynamic characteristics on the quality of thermal contact. Another disadvantage is connected with the high and fixed porosity values ($\varepsilon = 0.65$–0.70) restricting heat transfer enhancement opportunities. These disadvantages of screen structures could be significantly eliminated by transition to fiber–metal capillary structures sintered with a heat transfer surface.

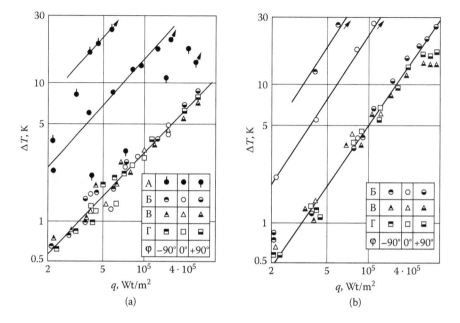

FIGURE 5.13 Dependency between superheat of capillary structure in the heat pipe evaporator and heat flux at water boiling. (a, b) Simultaneous turning on four and three heat-generating elements, respectively. A–D, Superheat measured under the heat source (see Figure 5.12); $\varphi = -90°$, vertical position of heat pipe; heat input zone at the top; $\varphi = -90°$, the same, heat input zone at the bottom; $\varphi = 0°$, the same, horizontal position of heat pipe.

5.1.3 Experimental Investigations of Vaporization Heat Transfer on Fiber–Metal Surfaces

Results of experimental investigations of vaporization heat transfer on heat pipe heat input zones with fiber–metal wicks were published in several papers [19, 173, 175, 180]. Experiments were performed mainly on six samples of cylindrical heat pipes with such working fluids as water, ethanol, methanol, acetone, R-113, R-11, and liquid nitrogen. Copper, stainless steel, and nickel were chosen as skeleton materials having structural porosity in the range 0.6–0.93, characteristic pore size ranging from 10 to 300 μm, and a capillary thickness from 0.5 up to 2.8 mm. Heat transfer experimental investigations on vaporization inside fiber–metal porous structures (FMPS) were performed on the special experimental setup presented in Ref. [174]. Experimental data on temperature drops corresponding to boiling initiation in the structure were obtained (Figure 5.14).

The principal advantage of several experimental investigations [19, 171, 173–180] lies in the extensive information about structural and thermophysical parameters of fiber–metal wicks: ε, D_{max}, D_{min}, λ_E, K, etc. D_{max} and D_{min} are the maximum and minimum pore diameters, respectively. It offered an accurate correlation between boiling initiation condition, ΔT_{IB}, and structural and thermophysical properties of structures as follows:

$$\Delta T_{IB} = \frac{4\sigma T_S}{r\rho''D_{MAX}}$$

Experimental validation of this correlation is given in Figure 5.14 [19]. Based on the analysis, in case $\Delta T < \Delta T_{IB}$, vaporization inside fiber–metal wicks is absent.

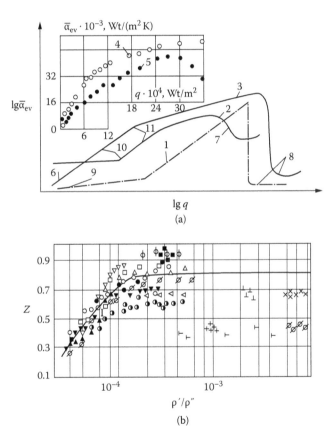

FIGURE 5.14 Heat transfer at vaporization inside fiber–metal structures [19, 173, 174]. (a) Dependency between vaporization heat transfer and heat flux: 1, smooth surface pool boiling; 2 and 3, surfaces with low-porosity and high-porosity MFCPS; 4, heat pipe #4; $p_s = 50$ kPa; 5, heat pipe #6; $p_s = 20$ kPa; 6, heat conduction; 7, transient regime; 8, film boiling; 9, evaporative regime in heat pipe and natural convection on smooth surface; 10, vaporization inside porous structure (boiling zone I and boiling zone II) (b) Comparison between experimental data and calculation; curve, calculation results: $Z = \dfrac{\alpha_{EV}}{q^{2/3}(\lambda'^2 \,/\, v'\sigma T)^{1/3}(D_{max}/\delta_{CS})^{1/3}}$. (c) boiling zones I, II, and III, calculation results at $D_{max} = 50, 100, 300 \ \mu m$.

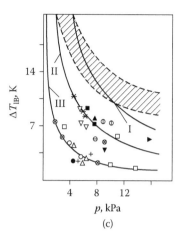

FIGURE 5.14 *Continued*

It means that heat is transferred by effective heat conduction through the FMPS. In thermal regime I, effective thermal conductivity is independent of heat flux, and can be calculated as

$$\alpha_{EV} = \left(\frac{\delta_{CS}}{\lambda_E} + R \right)^{-1}$$

where λ_E is the effective thermal conductivity of wetted FMPS, δ_{CS} is the structure's thickness, and R is the specific thermal resistance of the contact between the structure and the wall, equal to $(3–7) \times 10^{-5}$ m^2 K/W for fiber–metal wicks soldered to the wall and $(20–100) \times 10^{-5}$ m^2 K/W for structures compressed to the wall [19].

It is probable that such low estimations of specific thermal resistances required considerable experimental validation, because they have poor concurrence with other data for compressed porous structures.

Changing of heat pipe position in a gravity field causes structure flooding or partial/complete dry-out with deepening of the interface, etc. In such a case, corresponding change in the calculation correlations of heat transfer coefficients should be accounted for.

When high wall superheat values ($\Delta T > \Delta T_{IB}$) were attained due to heat flux growth, separate vapor bubbles were observed on the external surface of the structure. The number of vapor-generating pores increases with heat flux growth, but a greater part of the surface remains unoccupied by bubbles. This thermal regime can be considered as a transient regime from evaporation to a developed boiling. Such a transition appears more extended in the range of heat flux values for fiber–metal structure use than for the case of pool boiling.

Further heat flux increase causes development of vaporization inside the structure and appearance of the thermal regime, when dependence $q(\Delta T)$ is similar to that in the pool boiling case. This range of q and ΔT values determines a boiling zone I (see Figure 5.14). When a superheat value close to $\Delta T = 4\sigma T_s/r\rho''$ is reached, the dependency between q and ΔT deteriorates, and approximately corresponds to the correlation $\alpha_B \approx q^m$, where $0.1 \le m \le 0.3$ [19]. Finally, heat transfer surface dry-out and a flooding of capillary structure by vapor occur at some heat flux value q. It is accompanied by a significant decrease of heat transfer coefficients and increase of wall superheat.

Typical dependencies for two basic FMPS variants (high- and low-porosity) are presented in Figure 5.14a. It should be noted that in low-porosity structures, the heat conduction regime range is considerably more extensive than that for high-porosity because λ_E values of low-porosity structures are higher and consequently, equal values of ΔT_{IB} correspond to larger heat fluxes.

Semena et al. [19], in their explanations on vaporization heat transfer inside porous structures, used the model analogous to known pool boiling model [181]: $\alpha_{EV} = \lambda'/\delta_E$, where δ_{CS} is the effective thickness of the liquid layer on the wall. It was supposed that higher boiling heat transfer intensity in porous structures in comparison to pool boiling is determined by higher nucleation centers density due to the large number of unlocked pores of different diameters and higher vapor bubble formation frequency caused by boiling in thin layer of capillary structure having a high thermal conductivity.

Direct confirmation of the increasing number of nucleation centers and their formation frequency is not presented. However, it was assumed that change in the number of vaporization centers in boiling zone I can be estimated by the ratio between maximum FMPS pore diameter and vapor bubble nucleation (critical) diameter: $n = D_{max}/D_{CR}$; $1 < D_{max}/D_{CR} < D_{max}/D_{min}$. The ratio $D_{max}/D_{CR} = 1$ corresponds to boiling initiation, and $D_{max}/D_{CR} = D_{max}/D_{min}$ is a transient regime from boiling zone I to zone II (Figure 5.14a).

Then vapor bubbles formation frequency f_V was supposed to be proportional to the ratio between vaporization velocity $w_V \approx q/r\rho''$ and capillary structure thickness [19]: $f_V \sim q/(r\rho''\delta_{CS})$.

Using the approach introduced by Labuntsov [181], Semena et al. [19] obtained $\delta_E \sim \sqrt{v'\sigma T_s \delta_{CS} / \Delta T q D_{max}}$ and then the correlation for "developed boiling" zone $\alpha_{EV}{}^I \sim C_0 q^{2/3}(\lambda'^2/v'\sigma T_s)^{1/3}(D_{max}/\delta_{CS})^{1/3}$.

This correlation is valid in the range:

$$\frac{4\sigma T_S}{r\rho'' D\,\text{max}} < \Delta T < \frac{4\sigma T_S}{r\rho'' D\,\text{min}}$$

Because the obtained correlation for small ratios ρ''/ρ' (vacuum conditions) generalized poorly with presented experimental results (Figure 5.14b), the following empirical correlation was suggested for estimation of constant $C_0 = C_1 th(1.5 \times 10^4 \rho''/\rho')$, where C_1 is dependent on skeleton material. Considerably less influence of heat flux on heat transfer intensity in the high heat flux region is caused by the low surplus in the number of vaporization centers [19]. The following empirical correlation was supposed for this region:

$$\alpha_{EV}^{II} = \alpha_{EV}^{BR} \left(\frac{q_{EV}}{q_{BR}} \right)^{0,25} \left(\frac{A}{A^{BR}} \right)^{0,1} ; A = \frac{\lambda'^2}{v'\sigma T_s}$$

where q_{BR}, α_{EV}^{BR}, and A^{BR} correspond to the transient border from zone 1 to zone 2 and to the following superheat: $\Delta T_{BR} = 4\sigma T_s / r\rho'' D_{min}$.

Therefore, the authors did not succeed in developing a model that accounted for the influence of all factors in the frame of a common approach.

There are several major distinctions in the heat transfer mechanism at vaporization inside porous structures in comparison to pool boiling:

- Heat is transferred from the heating surface to the interface both from wetted pores basis and through porous structure skeleton elements. The input of the second term increases with the decrease in porosity and pore size.
- Superheat inside pores determining evaporation heat flux value is $\Delta T = 4\sigma T_s / r\rho'' D$. Accounting for nucleation superheat is especially essential at low saturation pressures and for small pores.
- Liquid flow to the heat transfer surface is defined not by Archimedean forces, as in the case of pool boiling, but by capillary forces overcoming the effect of friction forces at filtration inside porous structures. The simple estimation of the ratio between capillary and gravity forces for FMPS via the Weber number, $We = (\sigma / \rho' g d_0 \delta_{CS})$, yields We values of 10–40, that is, $We \gg 1$. Thus, generalized correlations [19] are mostly empirical.

Experimental investigations [175, 176] of vaporization heat transfer on surfaces covered by FMPS in capillary feeding and submerged conditions were performed in an experimental setup analogous to those used in Refs. [75–79]. Typical experimental results for water and acetone boiling heat transfer on surfaces covered by FMPS [176] are presented in Figures 5.15, 5.16, and 5.17.

These results were generalized by empirical correlations. Hence, for calculation of heat transfer coefficient in submerged FMPS, the following correlation is recommended:

$$\alpha = Cq^n \delta_{CS} \lambda_{SC}^{0.6} \left(\frac{1 - \varepsilon_{max}}{1 - \varepsilon} \right)^{0.5} D_E^{0.5} \left(\frac{\lambda'^2}{v'\sigma T_s} \right)^{1/3}$$

where $C = 2 \times 10^4$, $n = 0.15 \times \delta_{CS}^{-0.14}$ at $0.1 \times 10^{-3} \le \delta_{CS} \le 0.8 \times 10^{-3}$ m, and $n = 0.0535 \delta_{CS}^{-0.28}$ at $0.8 \times 10^{-3} \le \delta_{CS} \le 10 \times 10^{-3}$ m; D_E is the effective pore diameter defined by the capillary lifting height.

Experimental data presented in Figure 5.16 reproduce results obtained in experiments conducted at Kiev National Technical University: dependence between heat transfer intensity and FMPS thickness is represented by a curve with a maximum.

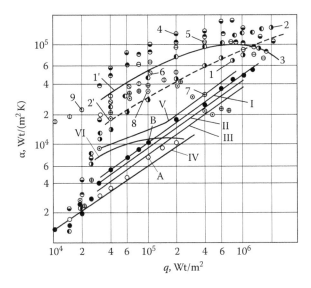

FIGURE 5.15 Heat transfer at water and acetone boiling on surfaces coated by fiber–metal structures [176]. Boiling on smooth surface: A, acetone; B, water; calculation: I, water, by the Tolubinsky formula; II, water, by the Kutateladze formula; III, water, by the Labuntsov formula; IV, acetone, by the Tolubinsky formula. Heat transfer at boiling on surfaces covered by capillary-porous structures (CPS): $\varepsilon = 40\%$, water; experiment: FMPS thickness: 1–8, $\delta_{CS} = 0.1, 0.2, 0.4, 0.8, 1.0, 2.0, 4.0, 10.0$ mm; calculation: acetone, $\delta_{CS} = 0.66$ mm; 1′, $\varepsilon = 70\%$; $\delta_{CS} = 0.8$ mm; 2′, $\varepsilon = 84\%$; $\delta_{CS} = 0.8$ mm; screen capillary structures: V, data from Refs. [75–79] for water; VI, data from Refs. [150, 151].

When FMPS thickness was equal to 0.8 mm, the maximum value of heat transfer intensity was obtained.

Shapoval et al. [175, 176] demonstrated that establishment of different correlations in the generalized experimental data [19] was based on an empirical

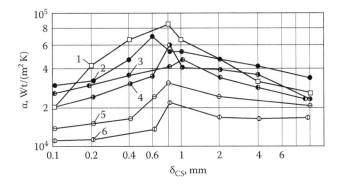

FIGURE 5.16 Influence of FMPS thickness on water boiling heat transfer [176]. $p = 0.1$ MPa, $q = 5 \times 10^5$ W/m², copper FMPS: 1–6, $\varepsilon = 40\%, 50\%, 60\%, 70\%, 80\%, 84\%$.

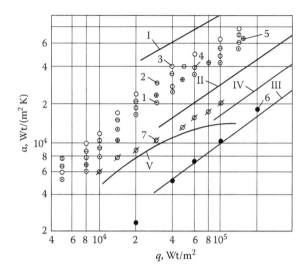

FIGURE 5.17 Heat transfer at water boiling on surfaces covered by FMPS at capillary feeding, $p_s = 0.1$ MPa [176]. Experiment: copper, $\varepsilon = 40\%$, $\lambda_{SC} = 69$ W/m K; coating thickness: 1–5, $\delta_{CS} = 0.4, 0.6, 0.8, 1.0, 2.0$ mm; 6, Water pool boiling on smooth copper surface; 7, $\delta_{CS} = 1.0$ mm (MFPS is compressed to surface); I, maximum heat transfer intensity at the same conditions; II, minimum heat transfer intensity at the same conditions; III, calculation by the Tolubinsky formula, water, smooth surface; IV, screen capillary structures, data from Refs. [160–163]; V, nickel and stainless steel MFPS data [152].

approach only (and not on the physical concept). They also confirmed the fact that concurrence between the presented correlations and those previously reported is completely absent.

Unfortunately, comparison (Figure 5.15) of experimental data [19, 176] with data on screen capillary structures [75–79, 150–151] was not analyzed.

Moreover, experimental investigations on capillary feeding conditions were performed by Shapoval [176]. Typical results of these experiments are presented in Figure 5.17 and generalized by the following correlation:

$$\alpha = Cq^{0.6}\delta_{CS}^{m}\lambda_{SC}^{0.25}[(1-\varepsilon_{max})(1-\varepsilon)]^{0.15}D_E^{0.1}\left(\frac{\lambda'^2}{v'\sigma T_S}\right)^{1/3}$$

where $C = 200$, $m = 0.65$ at $0.4 \times 10^{-3} \le \delta_{CS} \le 1.12 \times 10^{-3}$ m; and $C = 0.5$, $m = -0.2$ at $1.12 \times 10^{-3} \le \delta_{CS} \le 9 \times 10^{-3}$ m.

Later, Y. Fridrikhson, representing the same research group, investigated heat transfer at boiling on surfaces covered by FMPS under submerged conditions. Equipment and methods were the same as those used earlier in experiments performed by A. Shapoval. Main results were published in several papers [177–180]. Qualitative concurrence between experimental data [180, 176] was demonstrated.

Physical concepts [180] were developed by using the model [71] and statement on interaction between thermal boundary layer and nucleation centers activation conditions. Experimental dependencies between initial boiling superheats on structure thickness and pressure allowed obtaining generalizations [180] for conditional thermal boundary layer thickness δ_T and nucleation centers density n as follows:

$$n = C_1 \left(\frac{r\rho''\Delta T}{\sigma T_S} \right)^m \delta^l \left(\frac{1 - \varepsilon_{max}}{1 - \varepsilon} D_E \right)^p$$

where $C_1 = 8 \times 10^{-4}$, $m = 4.2$, $l = 1.4$, $p = 1$ at $\Delta T < \Delta T_{BR}$ and $C_1 = 1.4$; $m = 1.55$; $l = 0$; $p = -0.07$ at $\Delta T > \Delta T_{BR}$.

The value of the border superheat ΔT_{BR} was also determined empirically

$$\Delta T_{BR} = C_2 \left(\frac{\sigma T_s}{\rho''} \right)^m \left(\frac{1 - \varepsilon_{max}}{1 - \varepsilon} D_E \right)^k \delta_{CS}^s$$

where $C_2 = 246$, $s = -0.15$, $k = -0.44$ at $\delta_{CS} < \delta_{BR}$, and $C_2 = 282$, $s = 0$; $k = -0.53$ at $\delta_{CS} > \delta_{BR}$.

Two estimations were suggested for the calculation of capillary structure thickness δ, appearing as equal to δ_{CS}; this excludes cases when $\delta_{CS} > \delta_{BR}$, where it is assumed as equal to δ_{BR}

$$\delta_{BR} = 92.7 \frac{1 - \varepsilon_{max}}{1 - \varepsilon} D_E$$

All auxiliary parameters n, δ_{CS}, δ_{BR} were substituted into correlations for boiling heat transfer on surfaces covered by FMPS, as follows: at $n < n_F$, where $n_F \geq 4\varepsilon/\pi D_E$ is specific pore number [180]:

$$Nu = 0.03(nD_E) \frac{\lambda_{SC}}{\lambda'} \frac{\delta}{D_E}$$

where $Nu = \alpha D_E/\lambda'$; $\delta = \delta_{CS}$ at $\delta_{CS}/D_E \leq 10$.

When $n = n_F$, the correlations have the following form

$$Nu = 1.25\, Re\, Pr\, \frac{\rho''}{\rho' - \rho''} \frac{\delta_{CS}}{D_E} \quad \text{at} \quad \delta_{CS}/D_E \leq 10 \text{ and}$$

$$Nu = 15.3\, Re\, \frac{\rho''}{\rho' - \rho''} \frac{\lambda_{CS}}{\lambda'} \frac{\delta_{CS}}{D_E} \quad \text{at} \quad \delta_{CS}/D_E > 10$$

As noted by Fridrikhson [180], the presented correlations generalize about 90% of experimental data with scattering of less than ±35%. Physical explanations of the changing dependencies of the Nusselt numbers, Nu, on different factors and their interrelation with suggested physical concepts were not analyzed by the author. Moreover, correspondence between generalizations [180] and correlations [176] is ambiguous. Therefore, the presented correlations should be treated as empirical recommendations only.

5.1.4 EXPERIMENTAL INVESTIGATIONS OF VAPORIZATION HEAT TRANSFER ON SURFACES COVERED BY CORRUGATED STRUCTURES

During the search for simplified heat pipe technology, applications for corrugated structures were discovered. Aside from the simplicity of their technology, their principal peculiarities were associated with the nonexistence of the wall–structure contact problem. In common cases, the necessity of reaching reliable thermal contact, as a rule, causes considerable technological complications. Thermal contact in corrugated structures occurs in specific places, that is, in corrugation sockets maintaining capillary potential simultaneously.

Application of corrugated structures significantly simplifies manufacturing technology due to the nonexistence of the problem of fixing corrugated structures to the wall by welding or by other methods. Their contact is achieved by pressing corrugation sections to the internal surface of the heat pipe by the elastic deformation forces. Two basic variants of the corrugated structure design are known: from metal screen [212, 213] and from metal foil [214–218]. So far, there have only been a few studies devoted to experiments on corrugated heat pipes.

Excluding two papers [217, 218], extensive original experimental studies devoted to heat transfer on surfaces covered by corrugated wicks were unknown in the literature. Zigalov et al. [214], Afanasiev et al. [215], and Silinsky [216] focused their attention on studies of critical heat fluxes, when data on average heat transfer coefficients were obtained from the measurements of heat pipe wall temperatures in heating and cooling zones, and saturation temperature was assumed as equal to the average of wall temperatures in evaporation and condensation zones.

Therefore, only experimental data from Refs. [217, 218] can be treated as reliable. Experiments [217, 218] were completed based on the experimental setup [79] cited and described in detail earlier in Section 5.1.4 of this book. Some distinctions were connected with working elements of the capillary structures, that is, corrugated metal foil, and small changes in the design of the compressing device.

Corrugation types are presented in Figures 5.18 and 5.19, and their basic dimensions are given in the following table.

Module, m	Corrugation Height (h_c), mm	Corrugation Wall Thickness (δ_c), mm	Ratio of External and Internal Pitches Along Tube Perimeter (t_o/t_i)	Corrugation Number Along the Tube Perimeter, N_c
0.5	1.0	0.1	1.68/0.98	18
0.3	0.56	0.1	0.85/0.64	30
0.2	0.21	0.1	0.64/0.6	49

Experiments [218] were performed with samples of corrugated structures produced from the copper and stainless steel. Influence of thermal conductivity of the structure material on heat transfer was not observed.

Experimental results revealed heat transfer enhancement due to covering heating surface by corrugated structures was 1.3 ($m = 0.5$) and 3.4 ($m = 0.3$) times higher in comparison with the case of pool boiling.

The nonmonotonous dependency of heat transfer intensity on corrugation module, that is, characteristic size, should be noted. When the module was decreased from 0.5 to 0.3, heat transfer intensity was increased. However, when the module was changed from 0.3 to 0.2, heat transfer intensity was slightly decreased.

The influence of corrugated strip width on heat transfer is shown in Figure 5.20. When strip width was increased from 5.5 to 40 mm at the same module value, heat transfer intensity likewise increased. It is probable that the vapor slug appears more extended along the capillary length, thereby decreasing liquid microfilm thickness.

The experiments were performed at the horizontal position of the heating surface covered by corrugated strips submerged in the liquid layer with thickness h (Figure 5.21). In addition, heat transfer regularities were studied in the case of capillary feeding, when the liquid layer was gone and the liquid level was lower than the heating surface on depth h.

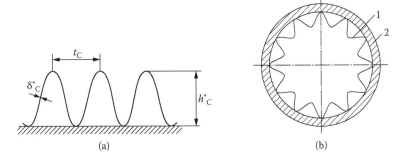

FIGURE 5.18 Corrugated capillary structures from the metallic foil. (a) Corrugated profile: t_c, pitch; h_c^*, height; δ_c^*, thickness of metallic foil. (b) Corrugated capillary structure (1) inside cylindrical ($d = 10$ mm) heat pipe (2).

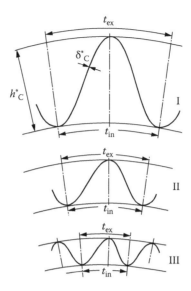

FIGURE 5.19 Corrugated structure profiles inside the heat pipe container: I–III, module value, $m = 0.5, 0.3, 0.2$, respectively.

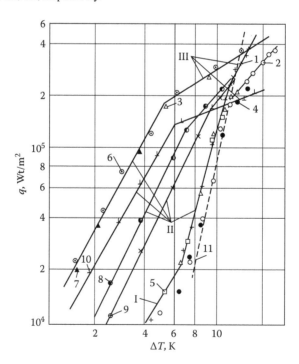

FIGURE 5.20 Water boiling on surfaces covered by capillary corrugated structures, $p_s = 0.1$ MPa, module $m = 0.5$: 1–4, stainless steel, $b = 5.5, 10.2, 20.7, 40$ mm; 5, copper, $b = 40$ mm; $m = 0.3$: 6 and 8, stainless steel, $b = 20.7, 10.5$ mm; 7 and 9, copper, $b = 20.7, 10.5$ mm; $m = 0.2$: 10, copper, $b = 19$ mm; 11, pool boiling on the smooth surface; b is corrugated strip width.

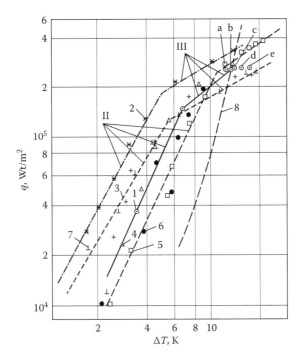

FIGURE 5.21 Heat transfer at water boiling on corrugated structures with modules $m = 0.2$ and $m = 0.3$. 1, Stainless steel, $b = 20$ mm, $m = 0.3$; 2, copper, $b = 20.7$ mm, m = 0.3; 3, the same, $m = 0.2$; 4, copper, $b = 20.7$ mm, $m = 0.2$; 5, stainless steel, $b = 10.6$ mm, $m = 0.2$; 6, copper, $b = 10.5$ mm, $m = 0.2$; 7, copper, $b = 19$ mm, $m = 0.2$; 8, pool boiling on the smooth surface. a–e are levels of $h = 20, 1, 0, -5, -7$ mm, respectively.

Moreover, the influence of thermal conductivity of corrugation material on vaporization heat transfer intensity was not established.

It could be supposed that the dynamic vapor flow does not support thermal contact reliability, thus eliminating the influence of corrugation material thermal conductivity.

Moreover, it is reasonable to consider that dynamic vapor flow causes larger elastic corrugation deformations when the module was decreased, that is, corrugated structure rigidity was reduced. These considerations explain the decrease in heat transfer intensity when the corrugation module was changed from $m = 0.3$ to $m = 0.2$.

Comparison of heat transfer regularities at boiling in slits and capillaries (see Chapter 4) with the data obtained for boiling on surfaces covered by corrugated structures demonstrated their similarity.

Thus, it was supposed that physical mechanisms are also general under these conditions. The physical nature of vaporization at boiling on surfaces covered by the corrugated structures could be presented in the following manner.

At small heat fluxes ($q \leq 2 \times 10^4$ W/m² for water at $p = 0.1$ MPa), evaporation regime from the internal corrugated surface becomes evident. Internal volume of the structure is filled with heat carrier and the heat is transferred from the wall to the interface via heat conduction.

Increase of thermal load ($10^5 \geq q \geq 2 \times 10^4$ W/m^2) causes the appearance of vapor slugs in the corrugation internal channels. Heat is mainly transferred in the latent form and the average thermal resistance is determined by the joint effect of the following mechanisms: convective (contact at liquid slugs flow) and heat conduction through the liquid microlayer at the vapor slug movement. This regime corresponds to zone II of the curves presented in Figures 5.20 and 5.21.

Experiments performed for the capillary feeding conditions revealed that stable vaporization regimes were sustained only at small heights of the capillary lifting (about 10 mm for water) and heat transfer regularities coincided with the existence of a large liquid layer. Lowering of the liquid layer level below the heating surface increased surface superheat in thermal regime III.

Experiments devoted to thermal regimes of heat pipes with corrugated structures were also considered by Vinogradova [218].

Temperatures of both heating and cooling zones' surfaces, and vapor temperatures inside the heat pipes were measured in these experiments. It allowed the author to obtain reliable experimental information on the average heat transfer in evaporation zones of the heat pipe covered by corrugated structures.

Direct comparison of these data with the preceding experimental results revealed good qualitative concurrence.

A generalized model for the experimental data treatment was obtained for boiling heat transfer on corrugated surfaces [217, 218]. It was accounted for in the above-mentioned similarity of boiling processes in slits and capillary channels and on corrugated surfaces. Experimental results obtained by means of both thermal wedge and heat pipes were generalized by the model, as follows:

$$Nu_L = C_0 Re_L^n K_L$$

where $Nu_L = \alpha d_E / \lambda'$, $Re_L = q d_E / r \rho'' \nu'$; d_E is the effective corrugation diameter determined by the height of liquid column keeping inside the corrugation by capillary forces

$$K_L = \left(\sqrt{\frac{\sigma}{\rho' - \rho''}} \right)^{-0.23} \left[(r\rho'')^2 \frac{\nu'}{\sigma T_S \lambda'} \right]^{1.3} (bt_C)^{0.765}$$

$C_0 = 3.19 \times 10^{-3}$, $n = 0.65$; b is the corrugated strip width.

Maximum scattering of the experimental data relating to the generalized equation was ±30%.

5.1.5   EXPERIMENTAL INVESTIGATIONS OF HEAT TRANSFER INSIDE EVAPORATORS OF LOOP HEAT PIPES

Loop heat pipes (LHP) and capillary pumped loops were recognized as perspective devices for heat transfer on significantly extended distances, primarily, for thermal control systems of space and flying apparatus. Intensive research on these devices has been performed in the past 30 years by many scientific groups from the Ural Branch of Russian Academy of Sciences, the Lavochkin Association (Russia), several U.S. firms and companies, and later in other countries, including Germany, China, France, etc. It was recognized that the teams led by Prof. Yu. Maidanik, Dr. K. Goncharov, and Prof. V. Kiseev presented outstanding results on the parameters and characteristics of LHP. Evaporators represented the key part of these devices, maintaining two basic functions: (1) intensive heat removal from the heating surface and (2) reliable capillary pumping, preventing entire hydraulic losses in liquid, vapor, and two-phase flow pipelines.

The experimental and theoretical research on vaporization in LHP porous structures were performed mostly with samples corresponding to real process conditions, that is, processes were studied in their interrelation, such as liquid and vapor two-phase flow in pipelines and condensation heat transfer, etc. [182–184, 187]. However, several studies [81, 185, 186] are particularly attractive for further detailed analysis.

The necessity for comprehensive research on LHP was prompted by significant interconnections between processes. A number of publications devoted to LHP problems are very extensive. However, the most representative and typical experimental data are obtainable only in a few papers, including those by Maidanik et al. [182–184, 187].

Design and technology of LHP evaporators were improved and significantly modified during the research process.

Design of first-generation LHP evaporators is shown in Figure 5.22. In these devices, heat input to the heating surface was realized through the low-porosity structures, when the liquid filled all pores. However, the liquid could also occupy all vapor channels' internal volumes. At LHP start-up, depending on the superheat level, vaporization occurred inside the vapor removal channels (VRC).

At small heat fluxes, vaporization in the VRC is not extensive and some liquid could remain, representing additional thermal resistance for heat transferred through the wetted porous structure to the interface in the VRC.

The vapor flow dynamic action increases with the increase in thermal load, and when a specific value of heat flux is reached, partial or complete liquid removal from the VRC is achieved. The total thermal resistance over the vaporization process was defined by the effective heat conduction through the porous structure filled by the liquid to the interface located inside the VRC. At given values of thermal conductivity λ_E, such thermal resistance can be calculated by solving the corresponding two-dimensional heat conduction problem (three-dimensional in some cases). Approximation can also be obtained from the one-dimensional problem [183].

Further heat load growth causes increase in the interface curvature, $K = 2/R$ (R is the main curvature radius), inside the VRC depending on the thermophysical properties of the working fluid and total sum of hydraulic resistances of LHP pipelines. It causes an increase in equilibrium temperature on the value $\Delta T^* = \sigma T_s K/r\rho''$, equivalent to thermal resistance increase in the vaporization process.

Hence, when vaporization occurs inside the capillary structures of first-generation LHP evaporators, the minimum thermal resistance has to exist at the specific heat flux value. Depending on the cooling conditions, such minimum value could determine the nonmonotonous change in the saturation and wall temperatures in the LHP. It is confirmed by typical experimental results obtained from research on

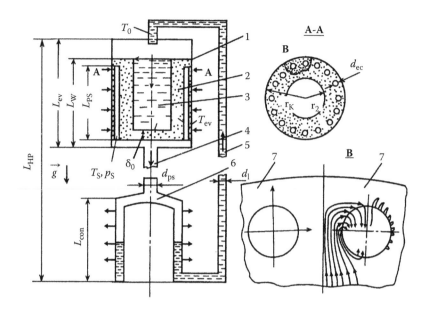

FIGURE 5.22 Schematics and evaporation zones of LHP with separated vapor and liquid channels. 1, Evaporator; 2, wick; 3, compensation chamber; 4 and 5, vapor and liquid channels; 6, condenser; 7, element of evaporator; r_k, r_1- container and compensation chamber radii; d_{PS} and d_L, vapor and liquid channel diameters; L_{CON}, L_{EV}, L_W, condenser, evaporator, porous structure lengths; L_{PS} vapor channel length in the evaporator; T_{EV}, container temperature at heat input; L_{HP}, whole heat transfer length of LHP; δ_0, thickness of porous locked liquid of the wall; T_0, LHP evaporator inlet temperature of liquid.

first-generation LHP at different inclination angles, φ, and ambient temperatures, t_{AM} ([182], Figure 5.23).

Different shapes of the LHP temperature curves depend on the ratio between the pumping capabilities of the LHP and the total hydraulic losses. Some of them are shown in Figure 5.24. As seen, under some conditions the range of LHP thermal loads, at which increase in thermal resistance R_T causes growth of the thermal load under small values of $\partial R_T/\partial Q$, could be extended sufficiently to enable the effective use of LHP in various thermal stabilization systems.

In first-generation LHP evaporators, separation of the liquid supply zone from the compensation chamber (which was positioned in the center of vapor-generating porous structure, situated partly between the heating surface and the VRC) from the heat input was not adequately achieved due to the combined countercurrent flows of liquid and heat inside the near-wall porous structure. It caused improvement of vaporization inside the pores and significant decrease of heat transfer intensity in some thermal regimes. Therefore, in second-generation LHP evaporators, the VRC represented a system of grooves on the contact surface between the wall and the porous structure, that is, there are evaporator designs with vapor-removal grooves formed on the heating wall or on the external surface of the porous structure.

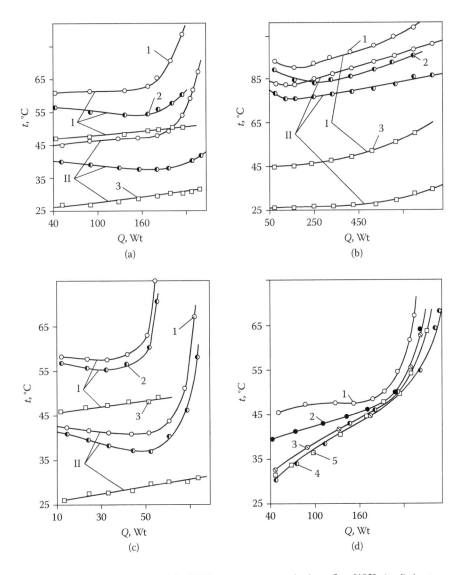

FIGURE 5.23 Dependencies of the LHP temperature on the heat flux [182]. (a, d) Acetone. (b) Water. (c) R-11: 1, evaporator wall; 2, vapor channel; 3, condenser wall; I, $t_{AM} = 45°C$; II, $t_{AM} = 25°VC$. (d) Evaporator wall temperature: 1–5, $\varphi = 90, 30, 0, -90, -30$; $t_{AM} = 25°C$.

The mentioned variants of LHP evaporators are presented in Figure 5.25. These types of vaporization surfaces were studied experimentally with the intention of determining the fundamental heat transfer mechanisms at vaporization inside these structures.

A scientific team under the supervision of Prof. Yu. Maidanik [81, 187] conducted experiments devoted to vaporization internal characteristics and thermal contact influence.

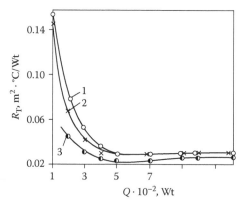

FIGURE 5.24 Dependency of the LHP thermal resistance on heat flux [184]. LHP #3-11, ammonia, nickel: 1, $\varphi = 60°$; 2, $\varphi = 30°$; 3, $\varphi = 0°$.

Autonomous research on heat transfer regularities inside LHP evaporators was performed by a team led by Prof. V. Kiseev [185, 186].

Similar to LHP, the experimental setup representing closed loop (Figure 5.26) consisted of evaporator 2 and condenser 7 interrelated by the vapor 5 and condensate 6 pipelines. The porous structure 3 was located in the heat input zone, and the wick was placed directly on the polished surface of heater 4.

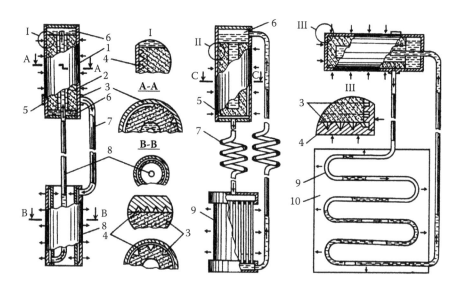

FIGURE 5.25 Basic design schematics of second-generation LHP (I–III) [187]. 1, Evaporator; 2, capillary structure; 3 and 4, extend and azimuthal vapor removal channels (VRC), respectively; 5, vapor removal collector; 6, compensation chamber; 7, vapor pipeline; 8, liquid pipeline; 9, condenser; 10, radiator.

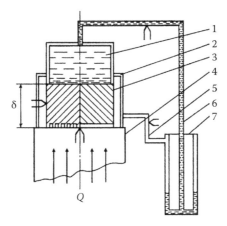

FIGURE 5.26 Experimental setup [185, 186]. 1, Compensation chamber; 2, evaporator; 3, wick sample; 4, heater; 5, vapor pipeline; 6, liquid pipeline; 7, condenser.

The VRC were arranged inside the wick. The cooling water heating in the condenser measured the heat output performed by this device. Temperatures of the heating and cooling surfaces, the outlet vapor flow, and the inlet liquid flow were measured by the copper-constantan thermocouples.

The operating regime was created and maintained by varying the power of the electrical heater. When the heater is turned on, a temperature field is formed in the evaporator. Temperature difference on the wick's backside appears and, when an interface exists, it is accompanied by the formation of excess vapor pressure. This pressure drop causes the liquid to leave the vapor channels and condenser until the moment when the hydraulic resistance of the external loop can compensate for the excess pressure. Under the evaporator's excess pressure, the working fluid enters the compensation chamber maintaining the feeding of the vaporization zone.

Typical results of experiments [185, 186] for dependencies of the average heat transfer coefficient on the geometrical parameters of the porous structure and heat fluxes are given in Figures 5.27 and 5.28.

As shown, dependencies of the heat transfer coefficient on the heat flux have their own maximum rate for every type of considered porous wick geometries. Explanations of the enhancement of heat transfer intensity at small heat flux values remain similar to the case of first-generation LHP.

Decrease of heat transfer coefficients in the high heat fluxes region [185, 186] was mainly associated with the appearance and development of the stable vapor layer inside the porous structure separating the moving interface from the heating wall. The appearance of stable vapor regions in the near-wall zone of porous structure represents interlayer crisis of vaporization heat transfer on surfaces covered by porous structures. It is clear that the vapor layer appearance causes the growth of thermal resistance, increasing with the reduction in the effective thermal conductivity of the wetted layer and with the growth of its porosity.

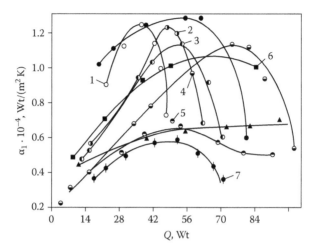

FIGURE 5.27 Dependencies of average heat transfer coefficient on the heat flux for LHP evaporators with various values of wick's locking wall thickness h [185, 186]. 1–7, h = 11, 9, 7, 5, 3, 4.5, 3.5 mm.

The interlayer crisis regimes were also realized inside first-generation LHP evaporators. However, at reliable contact between the wall and the structure, the nucleation barrier for the interlayer crisis development could be lower and the heat transfer characteristics could be better.

The design of the contact between the wall and the porous structure of second-generation LHP evaporators completely corresponds to the inverted meniscus concept.

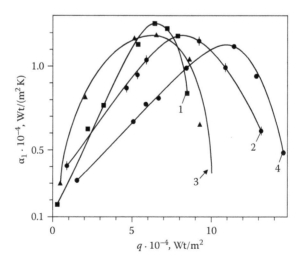

FIGURE 5.28 Dependencies of average heat transfer coefficient on the heat flux for LHP evaporators with various vapor removal pitch [185, 186]. Eight radial VRC; 1–4, 4, 1, 3, 2 mm.

Direct experimental results [185, 186] contradict the information about the absolute efficiency of this heat-removing principle.

Therefore, the rational design of LHP evaporators requires making the locking part of the porous structure as thin as possible. On the other hand, it is necessary to provide long-term reliability and good thermal contact for the locking part. Indeed, it was hard to simultaneously realize these features in first-generation LHP; in contrast, these problems appear to be easily solved for second-generation LHP operating by the inverted meniscus principle.

Theoretical models of second-generation LHP evaporators are based on physical imaginations about the heat transfer and hydrodynamic mechanisms presented by Kovalev and Soloviev [36]. Unfortunately, for comparison of calculations and experimental data, the authors only used the values of absolute temperatures of heating surfaces and saturation temperatures [187].

Experimental data [185, 186] were not compared with calculations. Hence, the conclusion about good conformity between experimental and calculation results meant concurrence of trends in experimental values of the maximum heat transfer coefficients and the authors' hydrodynamic model. The continuous curves presented in the corresponding figures are related to approximations.

Therefore, the presented physical imaginations and the corresponding calculation models of LHP evaporators are incomplete and require further research and development in the field.

5.1.6 EXPERIMENTAL INVESTIGATIONS OF BOILING HEAT TRANSFER ON SURFACES COVERED BY SINTERED AND GAS-SPRAYED COATINGS

It is probable that the first extended experiments on heat transfer at vaporization on surfaces covered by porous structures were performed following the experimental methods developed in the Leningrad Technologic Institute of Refrigeration Industry (LTIRI) with the goal of using the refrigerator's effective evaporating surfaces. The main results were generalized by Gogolin et al. [93]. In this work, we shall analyze only the principal results of these studies.

The LTIRI experimental setup and prescription of methods devoted to the study of refrigerants boiling on coated surfaces were presented by Dyundin et al. [188, 191]. The most important results of the experiments were published in several reports [189, 190, 192]. Heat transfer enhancement on the base of the porous structures for application in improvement of refrigerating machines' evaporators was investigated. The experiments were performed by using refrigerants R-11, R-12, R-22, and ammonia in pool boiling conditions with the different coverings' technology: electrochemical, gas and plasma spraying, sintering, and metal and glass screen wicks wrapped around the tubes. The experiments were completed in the following range of thermal parameters: $243 < T_s < 303$ K and 5×10^2 W/m^2.

Stainless steel tubes (diameter, 5.5×0.2 mm; length, 90 mm) were covered via the electrochemical method. The porous layer was precipitated from salt–water solutions via the electrochemical method, and was in the composition of Fe–Ni, Fe–Ni–Mo. The layer was sintered with the base surface and its thickness was in the range 10–140 μm. Porous structure analysis revealed the existence of pores with diameters ranging from 1 to 100 μm on the structure external surface. It was discovered that the structure parameters of the porous layer are dependent on the electrochemical precipitation technology regimes. Formed by the raising metal dust method, coverings were produced from the copper M3 type and were brought on the steel and copper tubes with a diameter of 20–25 mm. Layer thickness was changed in the range from 75 to 580 μm; porosity was varied from 0.14 to 0.65. The unlocked coupled pores were discovered inside the structures obtained via the raising metal dust method. During the first stage of the experiments, steel tube samples with metallized coverings, which were brought by gas spraying and electrocoating methods on tubes with 25 mm diameter and 400 mm length, were tested [190]. The covering layer thickness was changed from 65 to 300 μm. It was discovered that the best coverings were obtained by the electrocoating method. However, even for this case, structure formation significantly depends on such technological factors as thickness, layer number, pore width and length, their quantity and size distribution, porosity, etc.

Different types of metallized coverings were considered in [190]:

1. Dense coverings without layers separation
2. Porous coverings with large pores opening to the surface
3. Combined coverings consisting of elements from (1) and (2)

Experimental results on heat transfer at refrigerants boiling on surfaces with metallized coverings are given in Figure 5.29 simultaneously with the corresponding data for sintered coverings. Structures from steel powder particles with diameter $(d) = 63$–280 μm were brought on the surfaces of stainless steel tubes with diameters ranging from 20 to 25 mm. Covering thickness ranged from 0.3 to 1 mm, and porosity was varied from 0.45 to 0.55. Visual observations revealed the existence of stable bubble boiling at small temperature drops ΔT corresponding to ineffective natural convection thermal regime in case of application of finned and smooth tubes.

Influence of pressure and heat load on heat transfer at boiling on the porous structure was less compared to that on a smooth surface. In the LTIRI experiments, index m from the dependency $\alpha \approx Cq^m$ was equal to 0.1–0.2, whereas in case of the boiling on smooth surface, $m = 0.6$–0.8. Strong dependency of heat transfer intensity on the porous layer structure parameters was observed. Hence, for surfaces with metallized coverings, the highest heat transfer intensity is achieved when using coverings with developed opened porosity of the type 2 (see preceding list).

The heat transfer intensity on surfaces with sintered coverings depends on particle dispersion and their packaging. Some optimal thickness exists for each structure type, that is, further thickness increase does not cause additional increase in heat transfer intensity. This value for the present coverings is $\delta \approx 0.1$–0.3 mm, that is, it is similar to the thickness of some near-wall layers.

Boiling on surfaces covered by nonmetallic coverings has approximately the same character as boiling on metal porous coverings. Quality of the contact between the wall and the covering has an essential value for such coatings. It substantially influences the effective thermal conductivity of the layer [190].

Comparison of data on heat transfer intensity at boiling on surfaces with different coatings presented in Figure 5.30 demonstrated that sintered coverings were most effective. The application of sintered coverings for heat transfer enhancement at R-12

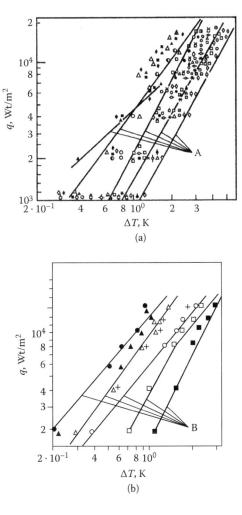

FIGURE 5.29 Heat transfer at R-12 boiling on surfaces with porous coverings ($T_s = 263$ K [190]). (a) Boiling on surface with metallized coverings. (b) Boiling on surface with sintered coverings; A, B, approximating lines.

boiling with saturation temperature of 20°C, for example, allowed increasing heat transfer coefficient by 6–10 times in comparison with boiling on the finned surface.

Experimental results on different methods of heat transfer enhancement at refrigerants boiling on tube surfaces were presented and discussed by Danilova et al. [189], including the metallized and glass-weave coverings. The level of heat transfer enhancement on surfaces with different coverings was similar to the case of slit channels. It also relates to the local heat transfer coefficients changing along the tube length in comparison with the corresponding regularities for the slits. Comparison for boiling on irrigated tubes, representing the most perspective method of heat transfer enhancement, with results for covered surfaces was performed by Bukin et al. [192]. Information about the main parameters of working elements is given in Table 5.2.

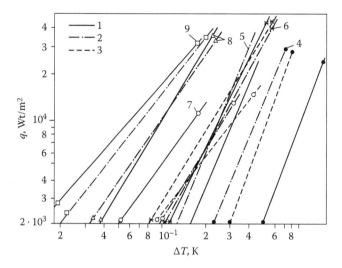

FIGURE 5.30 Comparison of data on heat transfer intensity at boiling on the surfaces with different coatings [190]. 1, R-12; 2, R-22; 3, NH$_3$; 4, smooth tube; 5, finned tube ($s = 0.76$ mm; $h = 1.5$ mm); 6, electrochemical coverings; 7, porous nonmetallic coverings (glass-weave material, $\delta = 0.33$ mm); 8, metallized coverings; 9, sintered coverings.

Comparison of heat transfer regularities of different enhancement methods at R-12 boiling on covered surfaces is shown in Figure 5.31.

A significant influence of structure parameters on heat transfer intensity was observed [192]. Particles' shapes and sizes, specific surface dimension, and porous

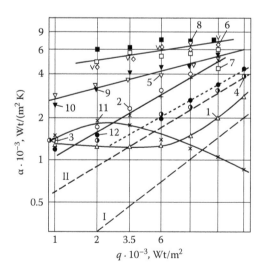

FIGURE 5.31 Influence of choice of heat transfer surface on heat transfer intensity at refrigerant R-12 boiling in the irrigation conditions [192]. Irrigation density $V = 1.2 \times 10^{-4}$ m^3/(m s); $T_s = -10°$C: I, pool boiling on the smooth tube surface; II, pool boiling on finned tube surface; 1–12, tubes #12, 13.

TABLE 5.2

Covering's Parameters of Experimental Samples [192]

Covering's Characteristics

Number of Tube	Tube Material	Material	Thickness, mm	Porosity	Particle Diameter, μm	Manufacturing Method
3	copper M3	copper Ml	100	0.37	–	electro-arc spraying
4	copper M3	copper Ml	150	0.31	–	electro-arc spraying
5	copper M3	copper Ml	170	0.24	–	electro-arc spraying
6	stainless steel	steel	350	0.46	63–100	sintering
7	stainless steel	steel	300	0.46	63–100	sintering
8	stainless steel	steel	1000	0.50	100–200	sintering
9	stainless steel	steel	500	0.52	160–250	sintering

Covering's Characteristics

Number of Tube	Tube Material	Material	Thickness, mm	Threads Number on 1 sm Warp	Woof	Fiber Diameter, μm	Fibers Number in the Thread
10	copper M3	glass fiber	0.1	22	18.4	4.8	200
11	copper M3	glass fiber	2×0.1	22	18.4	4.8	200
12	copper M3	glass fiber	0.25	26	21	5.5	200
13	copper M3	steel	0.08	120	120	40	1

layer thickness had influence on the heat transfer. Visual observations at boiling on covered tubes demonstrated that vapor was generated mainly under the coverings. The premature vaporization and considerable decrease in heat transfer intensity were possible in case of small permeability of the coverings (see Figure 5.31, results for tube #12). Use of a weaker material opposes premature vaporization. The research on covered tube bundles revealed that all tubes in the bundle had the same characteristics.

Similar results were obtained in the study of heat transfer at vaporization of cryogenic liquids. Experimental results were reported [193, 194] on heat transfer at nitrogen boiling on aluminum tube surfaces with porous coverings obtained via the gas thermal raising dust method from the aluminum alloy. The coverings' thickness was changed from 0.05 to 0.7 mm, the unlocked porosity varied from 0.2 to 0.4, and the effective pore diameter was in the range 40–60 μm. The saturation pressure was equal to 0.0196, 0.012, 0.047, and 0.098 MPa. An internal electric heater was used for heat supply. Results revealed that the initial boiling temperature drop decreased significantly from 2 to 0.3 K with the heat transfer coefficient increasing by 5–10 times.

Heat transfer at liquid nitrogen and R-113 boiling on porous coated surfaces was studied at the considerably wide range of heat fluxes and temperature drops for two metallized porous coverings obtained by the gas thermal raising dust method and by sintering. Copper and aluminum coverings were brought by the gas thermal raising dust method on tube samples with a diameter of 22 mm. The unlocked porosity value was 0.2–0.36. The thickness of copper coverings was not higher than 0.15–0.2 mm, and for the aluminum ones thickness was in the range 0.4–0.8 mm. Copper powders from the dendrite particles with grains averaging 50–250 μm in size were used to obtain sintered coatings. The obtained porosity was in the range 0.58–0.74. The basic characteristics of these coverings are presented in Table 5.3.

Experimental data on heat transfer at liquid nitrogen and refrigerant R-113 reported in several papers [193–196] as well as specific data on R-113 boiling on coated surfaces with sintered coverings presented by Tunik et al. [197] and Takakhiro et al. [198] are shown in Figures 5.32 and 5.33. The gas thermal raising dust technology maintains the surface porosity and the sintering supports the volumetric porosity [196]. The above-mentioned studies yielded the following conclusions about the heat transfer peculiarities for boiling on porous coated surfaces:

- Vaporization on these surfaces was initiated at significantly less superheats in comparison with smooth surfaces due to considerably easier bubbles nucleation inside the superheated liquid located in the pores.
- Transition from natural convection to vaporization is accompanied by heat flux growth and decrease in wall superheat.
- Heat transfer intensity at boiling on surfaces with porous coatings is significantly higher than in the case of pool boiling on the smooth surfaces.
- At small heat fluxes, heat transfer intensity increases with growth in thickness of coverings ($q \leq 0.6$ W/sm^2 for nitrogen and $q \leq 2$ W/sm^2 for R-113). However, further increase of heat flux causes the heat transfer to worsen as covering thickness increases.
- If heat transfer regularities at vaporization on coated surfaces is represented by the correlation $q = C\Delta T^m$, then regime influence on values C and m will be specific for different samples as follows:
  - For some coverings, C and m have constant values.
  - For some coverings, increase in heat flux causes growth in the value of m.
- For thick sintered coverings with thickness exceeding 10 particle sizes, this dependence approaches the extremum, that is, reaching the maximum heat transfer intensity.

Comparison of experimental data [196] with other research data [193–195, 197, 198] demonstrated that, for coverings with similar particle sizes, there are considerable distinctions in thermal regime regularities. It is probable that they were caused by the technological dissimilarities [196]. The porous structure technology

TABLE 5.3

Experimental Data on Heat Transfer at Boiling on Surfaces Covered by Two Metalized Porous Coatings

Sample #	Coatings Type	Coatings and Sample Material	Layer Thickness, mm	Particle Size, μm	Porosity
1	gas thermal spraying	copper	0.15–0.2	—	30
2	gas thermal spraying	copper	0.15–0.2	—	36
3	gas thermal spraying	copper	0.15–0.2	—	23
4	gas thermal spraying	copper	0.15–0.2	—	25
5	gas thermal spraying	copper	0.4	—	25
6	gas thermal spraying	aluminum	0.6	—	20
7	gas thermal spraying	aluminum	0.8	—	19
8	gas thermal spraying	aluminum	0.4–0.5	—	20–40
9	sintered from powders	copper	0.21	—	58
10	sintered from powders	copper	0.49	—	64
11	sintered from powders	copper	0.91	40–50	64
12	sintered from powders	copper	0.72	—	74
13	sintered from powders	copper	1.34	—	70
14	sintered from powders	copper	2.20	200–250	66
15	sintered from powders	copper	1.00	—	69
16	sintered from powders	copper	2.15	—	69
17	sintered from powders	copper	1.00	250	—

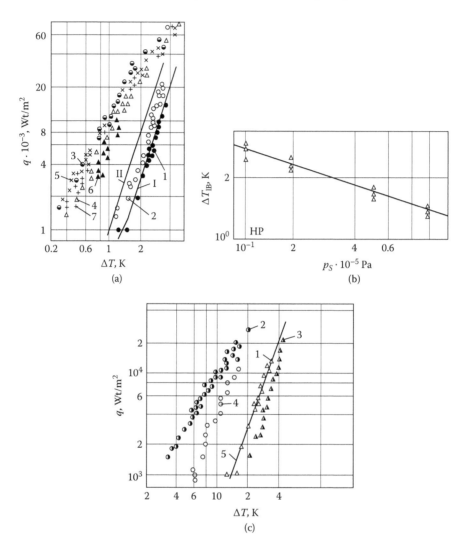

FIGURE 5.32 Heat transfer at nitrogen and R-113 boiling on surfaces coated by porous coverings brought by the gas thermal dust raising and sintering methods [193, 194]. (a) Dependence of heat flux dependence on temperature drop ΔT: I and II, calculations by the Labuntzov and Zuber correlations, respectively; 1 and 2, experimental data on smooth horizontal and vertical tubes; 3–7, experimental data on tubes coated by capillary-porous covering, δ_{CS} = 0.49, 0.52, 0.47, 0.46, and 0.40 mm, respectively. (b) Initial boiling temperature drop ΔT_{IB} (p_s); HP, the triple point. (c) Nitrogen boiling heat transfer: 1, 3, smooth surface; p = 0.098 MPa; 2–4, coated porous surface, p = 0.0196 MPa; 5, generalization of experimental data [193, 194].

has a strong influence on boiling heat transfer regularities (Figure 5.33). As shown in this figure, coverings obtained by the gas thermal raising dust method and the powder sintering method show essentially different heat transfer regularities.

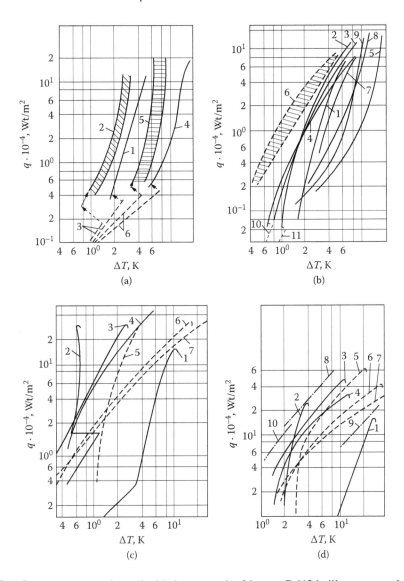

FIGURE 5.33 Heat transfer at liquid nitrogen and refrigerant R-113 boiling on coated surfaces [196]. (a) Liquid nitrogen: 1, smooth tube; 2, samples #1 and #2; 3, natural convection zone; refrigerant R-113: 4, smooth tube; 5, samples #1–4; 6, natural convection zone. (b) Liquid nitrogen: 1–4, smooth tube and samples #5–7, respectively; 5, sample #8; 6–9, the same, refrigerant R-113; 10, 11, natural convection zone. (c) Liquid nitrogen boiling on the sintered surfaces: 1, smooth surface; 2–7, samples #9–14. (d) Refrigerant R-113 boiling on the sintered surfaces: 1, smooth surface; 2–7 samples #9–14.

Heat transfer enhancement was considerably higher for sintered coverings compared to those obtained by the gas thermal raising dust. Porous coverings of similar type have considerably lower scattering; however, the range of C and m values is significantly larger.

Sintered porous coated structures with covering thickness of 0.20 mm and particle sizes of 40–50 μm are the most effective possibilities for heat transfer enhancement at the boiling of cryogenic liquids.

Results of experiments conducted on heat transfer at vaporization of liquefied gases on porous coating surfaces were reported in [199–201]. The essential influence of the technological and structure parameters on the maximum heat transfer intensity was observed. Comparative experiments were conducted for coated porous surfaces and smooth surfaces. Vertical stainless steel tubes (diameter, 25mm; length, 200 mm) were tested. The porous structure material was identical with the tube. The thickness of porous coverings was changed from 0.12 to 0.88 mm. The covering was brought by the thermal diffusion sintering method from the powders consisting from fractions with different average particle sizes.

Four types of powders having particle sizes (d) of 51.5, 81.5, 150, and 250 μm were used in the experiments. Porosity (ε) = 0.55; saturation pressure ranged from 0.125 to 0.142 MPa, when the heat flux was from 3×10^3 to 66×10^3 W/m². The influence of porous coverings thickness and the time of sintering on porous coating thermal properties were studied. It was discovered that when the sintering time was less than 8 hours, the heat transfer intensity was decreased with lowering of sintering time.

Heat transfer coefficient dependencies by butane and propane boiling on coated surfaces [201] are presented in Figure 5.34. It is seen that the optimal thickness of porous structure exists under the conditions of the experiments [199–201]. The maximum heat transfer coefficients were obtained when the porous structure thickness was in the range 0.25–0.55 mm. Unfortunately, information about the layer structure characteristics, including dependencies of layer properties on the quality of the sintering process, were not presented. Hence, supplementary analysis of nonmonotonous dependency between heat transfer intensity at boiling on coated surfaces and thickness of porous structure is required.

In addition, it was discovered that, for the working elements being considered, the dependence of the heat transfer coefficient on heat fluxes was similar to that in the case of smooth surfaces; however, absolute values of the heat transfer coefficients were significantly higher. Therefore, for generalization of experimental data on heat transfer at liquefied gases boiling on the porous structure with optimal thickness (δ_{CS} = 0.4 mm) and optimal porosity (ε = 0.55), Sirotin et al. [199–201] suggested using the known approach based on the thermodynamic similarity principle for treatment of experimental data on heat transfer at pool boiling:

$$\alpha = \frac{320 p_{CR}^{0.3}}{T_{CR}^{0.35} M^{0.15}} \left(0.62 + 3 \frac{p_S}{p_{CR}} \right) q^{0.7} \bar{d}_p^{-0.7}$$

Experiments [202] were performed on heat transfer at water, R-11, and liquid nitrogen boiling on surfaces covered by triangle pores interconnected by rectangular channels. The pitches between pores were 0.6–0.7 mm. Cross sections of

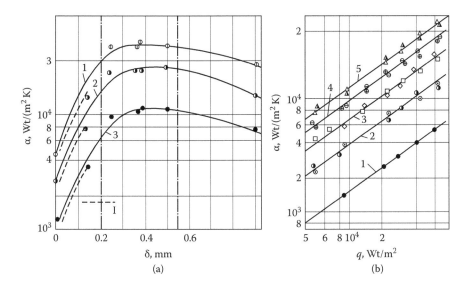

FIGURE 5.34 Heat transfer at liquefied gases boiling on surfaces covered by porous coatings [201]. (a) Influence of porous coatings thickness δ_{cs} on the heat transfer, $d_p = 0.0515$ mm; I, section of conditional dependency α on δ; 1– 3, $q = 64 \times 10^3$, 28×10^3, 9×10^3 W / m^2. (b) Heat transfer at propane boiling on the porous coated surfaces: 1, smooth tube; 2–5, sintered coated tubes.

the rectangular channels were 0.6×0.25 and 0.4×0.18 mm, and pore sizes ranged from 30 to 200 μm. The experimental results led to the following conclusions:

- The surface under consideration was extremely effective for vaporization heat transfer enhancement, especially in the range of small temperature drops. The required superheat was 10 times less than in the case of pool boiling on the smooth surface.
- Some curves for different diameter values were intersected, that is, surfaces with large values of d_P are preferable in the region of large ΔT, whereas in the region of small ΔT values higher heat transfer enhancement corresponds to surfaces with small d_P.

The study of the processes' internal characteristics is of special interest. Researchers achieved the possibility of comparing experimental heat flux values with calculations by using given values of acting pore number, vapor bubble removal frequency, and bubble departure diameter. The ratio between the latent heat flux of vaporization q_r and the entire heat flux q was determined. It was observed in experiments with refrigerants R-11 and R-113 that at $q = 10^3$ W/m², the ratio $q_r/q = 0.8$–0.9, and with the heat flux increasing to 10^4 W/m², the value q_r/q was decreased to 0.3. Hence, experiments [202] revealed the necessity of accounting for heat transferred in the latent form.

Experimental modeling of boiling on the narrow metallic perforated surface with width $b = 1$ mm was considered [202]. Two glass plates fixed the surface. Thus, single flat narrow channel connecting neighboring pores was created. This led to the following observations of phenomena accompanying vaporization inside the porous structure:

- The region occupied by the vapor exists in the channel when heat flux is absent.
- Heating of the channel causes extension of the vapor region and consequent removal of liquid, excluding only the liquid elements remaining in the corners (such liquid volumes evaporate and oscillate).
- At small superheat values (about 0.6–0.7 K), vapor phase removal was initiated and separate vapor bubbles appeared on the external surface. The frequency of bubble departure from the boiling surface was approximately 1 per 8 seconds. The appearance of oscillations in the liquid was caused by the liquid's suction from the nonactivated pores.
- Increase of superheat has no influence on the process observed in practically all of the experiments.

In fact, experimental conditions [202] corresponded to the case of vaporization inside the perforated capillaries.

Experiments on vaporization on porous coverings were considered in several studies [203–205]. Cylindrical working elements with porous coverings and internal electric heater were used in tests [203]. The sample porosity was equal to 0.55–0.60, the porous layer thickness was 0.38 mm, particle sizes ranged from 40 to 80 μm, and water and R-113 were used as heat carriers. Other experimental conditions were common. The surface described in Ref. [202] was also used in Ref. [204], and in Ref. [205] the sintered porous coverings and types of surfaces were the same as those used in Ref. [203].

Main peculiarities of experimental results [202–205] on heat transfer at vaporization were connected with low wall superheats (from 1 to 10 K), essential thermal instability, and hysteresis phenomena. Typical curves of $q = f(\Delta T)$ obtained for boiling on porous surfaces reported by Bergles [203] are shown in Figure 5.35. It is shown that at vaporization on porous surfaces, the heat flux increases with increasing wall superheat corresponding to the case of natural convection (from 1 to 8–10 K). Further increase of heat flux leads to abrupt decrease of wall superheat (approximately 10 times), that is, thermal instability of the natural convection causes appearance of a more stable thermal regime of vaporization inside the porous structure. The character of dependency $q = f(\Delta T)$ was changed in such a manner that at heat flux decrease, it does not appear to be reproducible, that is, the hysteresis of the boiling curve $q = f(\Delta T)$ occurs in such a case.

It was noticeable that these regularities appeared at small superheats of the wall opposing transient and hysteresis phenomena at high superheats (10–100 K).

Recent experimental and theoretical research on vaporization in porous structures has been directed toward the study of heat and mass transfer processes at the formation of deposits on vapor-generated surfaces.

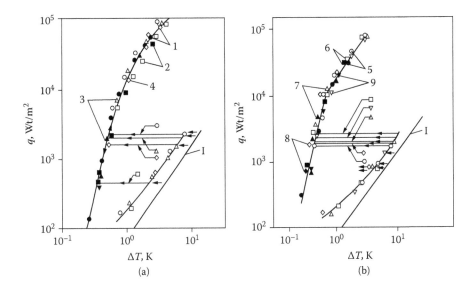

FIGURE 5.35 Hysteresis phenomena and instabilities of the vaporization process [203]. (a, b) Dependencies q on ΔT (light and shady points are heat flux increasing and decreasing, respectively); I, natural convection; 1–9, different types of perforated surfaces.

Styrikovich et al. [206] reviewed experimental research on heat and mass transfer processes in surfaces with deposits. One of the first experimental studies that considered the dashes concentration at vaporization inside the CPS was done by Picone et al. [207]. The main part of the experiments was performed under the pressure $p_s = 12.3$ MPa, mass flow rate $\rho w = 2280$ kg/m² sec, relative enthalpy $x = 0$, and maximum heat flux $q_{max} = 1250$ kW/m². Thickness of deposits of the iron oxide layer on the heated surface was changed in the range 2.5–165 μm. The concentration factor n, that is, the ratio between the dashes concentration in the heat carrier inside the CPS C and the dashes concentration in the heat carrier outside the deposits C_0 was determined in the experiments.

The drawbacks of the experimental determination of the concentration factor were analyzed by Styrikovich et al. [206]. Thus, there are doubts about the reliability of the experimental results [207]. Concentration took place only in one-half of the layer close to the heating surface and the concentration factor reached the value $n = 8000$ [207].

Other experimental results were also reported [208, 209]. The concentration factor was determined via the so-called "salt method." Regime parameters were changed as follows: pressure, from 9.8 to 16.7 MPa; specific mass flow rate, from 1000 to 3500; and heat flux, from 580 to 1160 W/m².

Reasons for immense inconsistency of experimental results for the cited reports could only be explained based on the improvement of existing physical concepts and development of newer, more reliable theoretical models. It required further research on vaporization inside CPS heat transfer.

Obvious applications of these studies are connected with heat pipes, heat transfer enhancement of processes in vapor-generating channels, cryostabilization and super-conductivity problems, etc. Therefore, not only the considered problems of boiling regimes, but also thermal regimes at drying of heating surface inside the coverings, including development of the entire crisis of boiling heat transfer, film boiling, worsened thermal regimes, transient processes, etc., should be studied extensively. Furthermore, we shall consider research results devoted to some of the mentioned problems.

5.2 EXPERIMENTS ON HEAT TRANSFER AT VAPORIZATION ON SURFACES WITH SINTERED COATINGS: THE MALYSHENKO PHENOMENOLOGICAL THEORY OF BOILING

Typical experimental results on the modeling of CPS from aluminum oxide Al_2O_3 and nichrome were published [54–57]. Plates produced from nichrome strip (size, 40×10 mm; thickness, 0.3 mm) covered by the porous structure with a volumetric porosity equal to 0.5 were used in experiments. Heating was done by electric current. Almost all of the heat was generated on the heat transfer surface for the Al_2O_3 samples, whereas a fraction of heat transferred from the wall in the nichrome samples with thickness 0.15, 0.4, 1.4, and 3.0 mm was equal to 100%, 90–98%, 88–93%, and 70%, respectively [55]. The heat transfer regularities at distilled water boiling on the mentioned surfaces under atmospheric pressure were studied in the form $q = f(\Delta T)$.

Two main boiling regimes were discovered in one study [54]. When the heat transfer regularity has the form $q \approx C(\Delta T)^m$, m is approximately equal to 2 (i.e., close to the case of boiling on smooth and polished surfaces) when $q < q^*$. If thermal load is higher than q^*, a new thermal regime appears with $m \approx 1$. This regime exists at high superheat values until heat loads close to critical values at boiling on smooth surfaces q_{CR} is reached.

It was discovered that boiling heat transfer regularities were significantly different for the thin-layer and thick-layer coverings. With the increasing in the coating's thickness, specific transient and hysteresis phenomena appear, when — at some heat flux value — significant growth of the wall superheat (about 1 order of magnitude) occurs at constant q. The essential irregularity of the temperature field is caused by this transient regime, which disappears upon reaching steady state, that is, one of the possible stable states corresponding to the correlation $q = C_0 \Delta T$, where C_0 is considerably less than that in similar thermal regimes with thin-layer coatings.

The analysis in several studies [54–57] demonstrated that transition to the mentioned thermal regimes was associated with the stable vapor film formation inside the porous structure. The vapor film remains stable in a wide range of heat flux values. Transition from the "vaporized" porous structure regime to the wetted porous structure skeleton surfaces had the obvious hysteresis form. Typical experimental results [54–57] are presented in Figure 5.36. Comparison of boiling curves for the smooth and covered surfaces justified the hypothesis that the porous structure's internal volume is filled by vapor when heat flux value q^* is reached. This implies transition to a similar film-boiling regime, when "thick" structures were used. However, when the thermal conductivity of the porous skeleton was considerably higher than that for vapor, the thermal resistance of the "vaporization" porous structure was essentially

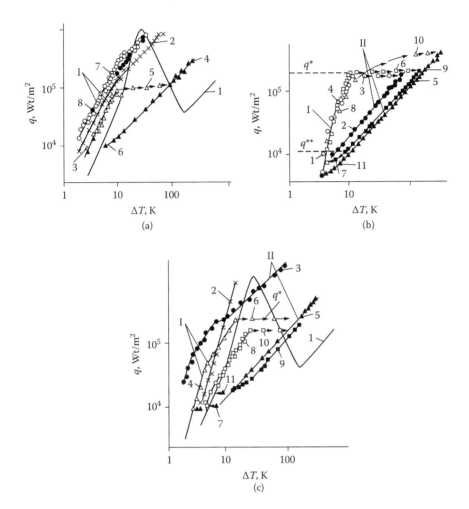

FIGURE 5.36 Heat transfer at boiling on the porous coatings from Al_2O_3 and nichrome [54–57]. (a) Dependency $q(\Delta T)$ at heat transfer boiling on surfaces: 1, polished; 2, with coverings from Al_2O_3, $\delta_{CS} = 0.4$; 3–8, with coverings from Al_2O_3, $\delta_{CS} = 1.4$ mm (3, 4, thermal regime with $q = C\Delta T$; 5, 6, $q = q^*$ and q^{**}; 7, 8, with the VRC; 7, heat flux increasing; 8, heat flux decreasing). (b) Heat transfer hysteresis at boiling on the nichrome coated surfaces, $\delta_{CS} = 1.4$ mm: transition at $q = q^* (\delta_{CS}) = 2.3 \times 10^5$ W/m^2; time $\tau_0 = 1$ hour (1, heat flux increasing; 2, heat flux decreasing); $\tau_0 > 3$ hours (4, heat flux increasing; 5, the heat flux decreasing; 6 and 7, the transient regimes). (c) Heat transfer at boiling on the following surfaces: 1, polished; 2, with nichrome coverings: $\delta_{CS} = 0.15$ mm; 3, with nichrome coverings: $\delta_{CS} = 0.4$ mm; 4–7, with nichrome coverings: $\delta_{CS} = 1.4$ mm (4 and 5, $q \approx C\Delta T$; $6 - q = q^*$; $7 - q = q^{**}$); $\delta_{CS} = 3$ mm (8 and 9, $q \approx C\Delta T$; $10 - q = q^*$; $11 - q = q^{**}$).

less than in the case of vapor film boiling. Experimental results on samples with VRC inside the porous structures revealed additional confirmation of these explanations. In such experiments, the phenomenon of vaporized structures was gone, similar to the region of hysteresis and transient regimes. One can suppose that q^*

corresponds to the moment of stable continuous vapor film formation near the wall inside the porous structure at growth of q. The value q^{**} relates to the destruction of vapor film when heat flux is decreased.

Slightly different regularities of vaporization heat transfer were obtained in other studies [36, 45] despite the fact that the experimental method and the structure parameters used were quite similar. Experiments were performed for water and R-113 boiling. Stainless steel tubes with diameters of 4 mm heated by electric current were used as samples. Coating was made from the stainless steel spherical and dendrite particles sintered to the wall. Tests were performed on seven samples with the parameters given in the table below [45].

Sample #	1	2	3	4	5	6	7
Covering's thickness, mm	1.0	1.0	0.45	0.45	0.55	0.55	1.5
Porosity	40	48	40	40	75	32	40
Particle diameter, μm	400–615	200–315	100–200	63–100	63–100	63–100	63–100

Experimental results are presented in Figure 5.37 as the boiling curves.

By comparing data between studies [45, 54–57], it is seen that heat transfer regularities at water boiling approximately concurred for the thin-layer structures with high thermal conductivities. Hence, the value of index m in the correlation $q \approx C\Delta T^m$ for sample #1 was decreased from $2 < m < 4$ to values 1.2–1.5 under small superheats. The "vaporization" thermal regime appears when $\Delta T \approx 20°C$, corresponding to comparatively high values of q and ΔT under experimental conditions [45]. The correlation view for sample #2 was approximately the same. Sample #4 revealed only one thermal regime ($m = 1.2$). Heat transfer regularities for sample #3 were typical for the transient thermal regimes.

Distinctions of the structure parameters were not essential. Experimental results for R-113, especially for samples #6 and #7, showed better agreement with the data in [54–57] for massive structures. The existence of two main thermal regimes, including transient and hysteresis regimes, is observable (see Figure 5.37). It is probable that the influence of such parameters as permeability, pore distribution, etc., on heat transfer was rather essential.

Other experiments for modeling capillary structures were devoted to regularities for concentration factor. Thus, Styrikovich et al. [206, 219] prescribed experiments on obtaining such correlation on the modeling samples of the thin-layer and thick deposits. The thin-layer deposits were modeled by filtrated screens produced from stainless steel wires with diameter $d_p = 0.2$ mm, one layer. The thick deposits were modeled by layers of balls with diameters (d) 60 and 300 μm; layer thickness = 6.5 mm. These experimental results did not provide reliable information about the concentration factor dependence on such regime parameters as heat flux, relative enthalpy (vapor quality), etc., and about the heat transfer correlation at vaporization on surfaces covered by porous deposits.

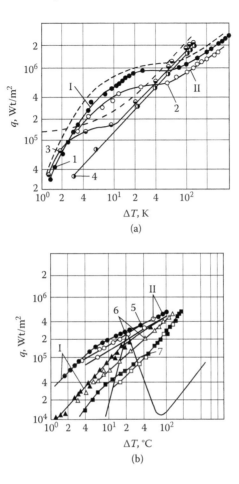

FIGURE 5.37 Heat transfer at vaporization on surfaces with stainless steel porous coatings [45]. (a, b) Water and R-113 boiling on the porous surfaces: 1–7, samples #1–7; I, boiling zone inside the porous structure; II, the drying zone; dotted lines, calculations.

Therefore, the first experiments conducted at the High Temperatures Institute (HTI) of the Russian Academy of Sciences in the wide range of heat fluxes on low-porous capillary structures (pores with characteristic size 1–10 μm) allowed the following principal conclusions:

1. There are such thermal regimes when washing of the heating surface by liquid is damaged. This causes change in heat transfer regularities associated with the filling of the porous structure by vapor. Depending on the thickness of the structure, such vapor filling could be complete or partial. In the case of partial vapor filling, stable thermal regimes are obtainable, when the formed vapor film in the porous structure has the defined thickness depending on the heat flux. The vapor film thickness remains constant with the heat flux growth inside the vaporized porous structure. It allows definite

correspondence between regime parameters and the results' reproducibility
[54]. Such a state exists until the initiation of the heat transfer crisis on the
external surface of the porous structure.

2. The vapor filling of the capillary structure was accompanied by the appearance of specific thermohydrodynamic instabilities, when the porous structure
   thickness becomes $\delta_{CS} \geq 0.8$ (Figure 5.38). It causes a strong irreversible change
   in the temperature regime at some heat flux level. It is likely associated with
   abrupt moisture content decrease starting from some critical value, when the
   relative liquid permeability still allowed liquid transport to the wall, down to
   the zero. Such a "jump" should be accompanied by drying of skeleton of the

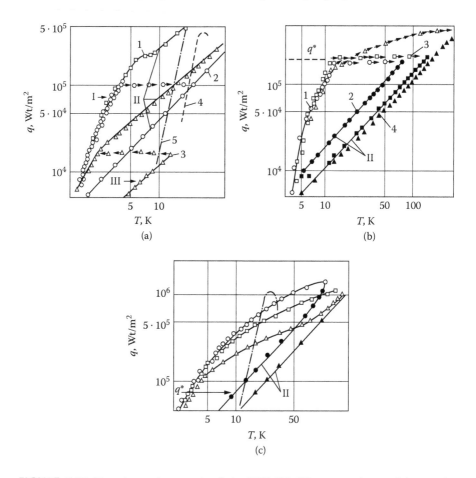

FIGURE 5.38 Experimental research of the HTI [54–57] on transient and hysteresis
processes at vaporization on surfaces with the NiCr porous coatings. (a) the boiling curves:
1, 2, δ_{CS} = 0.4, 0.8 mm, water; 3, δ_{CS} = 0.8 mm, water–ethanol mixture; 4, 5, smooth
surface, ethanol and water; I–III, different thermal regimes of vaporization. (b) Heat
transfer hysteresis at the water boiling (dark points, heat flux increasing; light points, heat
flux decreasing). (c) Influence of thermal load changing velocity on boiling curves' shape,
water, p = 0.1 MPa.

vaporized porous structure's part. The skeleton drying requires some time, hence, the transition to new steady state is extended in time [55–57]. New steady states of vaporized porous structure appear essentially more stable (curves II in Figure 5.38), and further transition to states with liquid content in accord with the capillary hysteresis regularities should accompanied by significantly less heat flux values, that is, hysteresis phenomenon in the heat transfer regularities for boiling on the porous structures occurs associated with the interlayer crisis regime.

3. Phenomena determined in [57] as the structure's memory and dependence of the hysteresis on the heat flux jump have a specific thermal nature. Explanation of these effects requires shared analysis of thermal and hydrodynamic phenomena.

Further experimental research activities at HTI were devoted to the following new problems:

- Boiling of cryogenic liquids and the cryogenic stabilization problems [224]
- Enhancement of worsened thermal regimes at crisis of the II type inside the channels via the porous coatings [227]
- Application of porous coverings operating by the "inverted meniscus" principle for cooling of surfaces under high thermal loads [228, 36]

These experiments are mostly connected with problems on critical heat fluxes. In several studies [221, 223], particular attention was also given to the hysteresis phenomena at boiling on surfaces with porous coverings. Special experiments on hysteresis phenomena at initial boiling were conducted to justify the physical concepts on the nature and regularities of thermal regimes at vaporization on surfaces with porous coatings [221].

Experimental results [221] on hysteresis at initial boiling agreed with the data of Techwer et al. [225]. In addition, new suggestions were proposed for some special hysteresis mechanisms defined by the processes, when boiling first occurred on the external surface of the porous structure and then, at heat flux growth, spread inside the pores' internal volume [225].

New shapes of the hysteresis curve and physical mechanisms were discovered for boiling on porous surfaces in [229] under conditions similar to those described in another study [225].

Many authors associated the physical phenomenon of hysteresis at boiling on porous coatings' surfaces with the following technological and design parameters:

- Degassing of liquid, heating surface, and porous structure
- Quality of mechanical and thermal contacts between the wall and the porous structure
- Heat carrier purification
- Irregularity of structure and thermophysical properties of porous structures over the thickness, area, etc.

Representative experiments devoted to the methodical study of the above-mentioned factors have not been available until now. Therefore, it was impossible to declare that systematic experimental data exist for the hysteresis phenomenon at boiling on surfaces with porous coatings. Thus, we shall only consider the analysis of hysteresis presented by Malyshenko and Andrianov [221, 223, 226]. This analysis can be considered as the phenomenological theory because Malyshenko and Andrianov combined their models and imaginations from many different problems, such as two-phase flows hydrodynamics inside the porous volumes, heat transfer, and internal heat conduction mechanisms over the wetted and drying skeleton, thermal resistances of the interfaces inside the porous structure volumes, etc. Thus, let us consider the main positions of the theory presented by Malyshenko and Andrianov [223, 226]:

1. The heat transfer regularities in the form of excess temperature $\Delta T = T_W - T_s$ dependence on the heat flux q change essentially due to the presence of the porous coverings.
2. Two boiling regimes are feasible corresponding to the mechanism of vapor bubble removal from the external surface of the covering, as opposed to one regime in the case of pool boiling on smooth surfaces. Experimental curves at the bubble boiling regime I were on the left side of the traditional boiling heat transfer curve (the heat transfer correlation in the form $q \approx C\Delta T^m$; $m >$), and the vapor generation zones were disconnected from each other inside the porous volumes (see Figures 5.36–5.38).

   Regime II was characterized by the essential growth of wall superheat and associated with the continuous vapor zone formation near the wall. The regularity was prescribed by the correlation $q \approx C\Delta T$. The transition from one of the boiling regimes to another was accompanied by the "jump" change in temperature drops. It took place at some heat flux values $q = q^*$ at transition from regime I to regime II (Figures 5.36 and 5.38), and $q = q^{**}$ correspondingly at transition from regime II to regime I, that is, the typical heat transfer hysteresis was observed in the transient region.
3. The derivative break in the dependence $q(\Delta T)$ is linked to the initial and final boiling points similar to smooth surfaces and was observed in the typical hysteresis phenomenon.

   The initial or final boiling moments on surfaces with coverings were fixed by the appearance or disappearance of the first or last vapor bubbles on the coverings' external surface.

   These moments corresponded to the following dissimilar values of temperature drops and heat fluxes: ΔT_{IB}, q_{IB} and ΔT_{FB}, q_{FB}.
4. There are two main characteristic sizes in the case of porous structures: the active gully "neck" radius on the porous structure, R_A, and the minimum radius of the pores occupied by vapor at the covering's "breaking," R_{CR}. The last one smaller from the mentioned sizes determines the initial superheat corresponding to appearance of the first vapor bubbles on the covering's external surface: $\Delta T = 2\sigma T_s/(r\rho''R_V)$.
5. The aforementioned radii were essentially larger for traditionally applied porous coverings than for smooth surfaces, because the initial boiling temperature drops of the porous coatings were essentially less than those for

smooth surfaces. Porous coverings make the processes of vapor removal from the activated pores essentially easier, because their "neck" radius curvature is determined by the first layer's particle size that is sufficiently large.

6. The minimum radius of pores occupied by the vapor can be calculated if the pore distribution and the pore net geometry are known. It can be assumed that the porous media can be considered as the nonisotropic, nonintersected cylindrical pore net with some coordinate number, Z. Using the percolation theory method yields the condition of the covering's "breaking" by the vapor as follows:

$$F(R_V) = \int_{R_V}^{R_{max}} f(R)dR = \varphi_C$$

where $\varphi_c = A/(A-1)Z$ is the percolation border for the pores' net connections and A is the net dimension scale (for three-dimensional nets $\varphi_c = 3/2Z$).

At given values of φ_c, $f(R)$, and R_{max}, we can calculate R_{CR} and ΔT_{IB}.

7. When the skeleton's wall is covered by the liquid film, the boiling final point corresponds to the changing of interface geometry inside the porous structure. This means that the cylindrical meniscus at the neck with radius R has to change to the spherical one, that is, ΔT_{CR} is in two times less than ΔT_{IB}. When the liquid film on the channel's wall is absent, then $\Delta T_{CR} = \Delta T_{IB}$.

8. After the coverings are destroyed, the system symmetry was changed by the jump. The stable open channels exist inside the porous structure and they are occupied by the vapor. The steady operating vapor-generating channels appear on the external surface of the porous structure. The average size of the heating surface part R_1, from which heat load was removed by evaporation, represents the characteristic size of the two-phase system. In Refs. [223, 226], the value of $R_1(q)$ was calculated via high-speed video capture. It was determined that $R_1 \approx \delta_{CS}$ and practically did not depend on q. The limited density of vapor-generating centers N_{max} exists for regime I:

$$N_{max} \sim \frac{1}{\pi\delta_{CS}^2}$$

The lower limit of the average distance between vaporized centers was estimated as $d_{min} \approx 2\delta_{CS}$.

9. The quantitative correlation between the vapor quality in porous media and its limit values φ_c and φ_c' (φ_c' is the vapor quality corresponding to the case when the maximum density of vaporized centers was attained) has a form:

$$N = N_{MAX}\left(\frac{\varphi - \varphi_c}{\varphi_c' - \varphi_c}\right)^{\beta} \qquad (5.1)$$

$$d = d_{min} \left(\frac{\varphi - \varphi_c}{\varphi_c' - \varphi_c} \right)^{-\nu} \tag{5.2}$$

where d is the average distance between the acting vapor channels, β and ν are the universal critical indexes of the percolation theory ($\beta = 0.35–0.47$ and $\nu = 0.82–0.92$).

By means of some additional assumptions, Equations (5.1) and (5.2) can be simplified as

$$N = N_{max} \Delta T^{\beta};$$

$$d = d_{min} \Delta T^{-\nu} \tag{5.3}$$

The heat transfer regularity for regime I can be presented as

$$\hat{q} = \Delta^{\wedge} T^{1+\beta} \tag{5.4}$$

Here,

$$\hat{q} = \frac{q - q_{CON}(\Delta T)}{q^* - q_{CON}(\Delta T)} \tag{5.5}$$

q_{CON} is the fraction of the heat flux transferred over the skeleton and removed by liquid convection.

10. Equations (5.1)–(5.5) can be used only for the infinite isotropic nets, that is, in cases of uniform coverings with large thickness [223, 226].

In the finite nets, the character of percolation transition changes and breaking of coverings by the vapor becomes possible only at some range of vapor qualities with the upper limit equal to φ_c, that is, the transition is blurred. Therefore, the value of percolation limit shifting for the coatings of finite thickness δ_{CS} is equal to

$$\varphi_c(\delta_{CS}) \approx \varphi_c - A\delta_{CS}^{(-1/\nu)}$$

$$\Delta T_{IB}(\delta_{CS}) \approx \Delta T_{IB} - B\delta_{CS}^{(-1/\nu)} \tag{5.6}$$

The critical amplitudes A and B depend on the type of covering, whereas parameter ν is universal. The incompleteness of experimental data for the empirical determination of A and B was noted by Malyshenko and Andrianov [223, 226].

11. The characteristic transition time τ_{IB} is determined by the velocity of the interface's movement inside the porous structure at the moment of its breaking under the action of capillary pressure drop $\Delta P = (2\sigma/R_{\text{CR}})\cos\gamma$.

$$\tau_{I.B.} \sim \frac{\delta_{\text{CS}}^2 \mu' R_{\Pi}}{K 2\sigma \cos\gamma} \qquad (5.7)$$

12. Both thermal regimes (I, II) and the irreversible transitions between them could not be realized for all cases of boiling on surfaces with porous structures. The nature of the interface inside the porous structure appears different for thermal regime I and thermal regime II.

    The two-phase mixture hydraulic resistance decreases due to the appearance of vapor film stabilized along the porous structure thickness. Calculations of the thermal regimes allow estimation of the average relative vapor film thickness δ'' in the porous media as $\delta''/\delta_{\text{CS}} = 0.5$–$0.8$. It was obtained in a conditional manner. The relative thickness $\delta''/\delta_{\text{CS}}$ decreased from 0.8 to 0.5 when the covering's relative thickness δ_{CS}/d_p was increased from 0 to 20.

    The entire wall superheat in thermal regime II (at the heat flux density q^*), ΔT^*, can be determined as the sum of such two terms as the superheat in regime I corresponded to the point of transition to thermal regime II, ΔT_1^*, and the temperature drop between the evaporation zone and the wall:

$$\Delta T^* = \Delta T_1^* + (d_p / 4\lambda_E) q^* \qquad (5.8)$$

13. The vapor film appearance at transition from thermal regime I to regime II was accompanied by the vapor filling of practically the entire pore volume. The stable vapor film formation in the fraction of the pore volume in the coatings small values of thermal conductivity was caused by the irregularity influence. The skeleton thermal conductivity λ_{SC} limitation is

$$\lambda_{\text{CS}} = \frac{q^* \delta_{\text{CS}}}{2(\Delta T - \Delta T_{IB})} \qquad (5.9)$$

The estimation based on Equation (5.9) yielded $\lambda_{\text{SC}} = (5$–$10)$ W/m K for the experimental conditions [223, 226]. At water boiling on surfaces with higher thermal conductivities, the transition between regimes I and II takes place at large values of q and corresponds to complete drying of the coating's surface, that is, in conditions close to the boiling crisis.

Moreover, the absence of the abrupt change in vapor quality at high heat fluxes causes smearing of front interface over the porous structure and corresponding smearing of the transition on the finite interval Δq or its transformation into the point.

14. Experimental results also revealed that close to the transient point from regime I to regime II, abrupt increase in relaxation time and transition to a new steady state occurred, and the values of such relaxation times reached several hours [221, 224].

Increasing in relaxation time near the transition caused essential differences in the boiling curves obtained in experiments at continuous changing of heat flux with different speeds (Figure 5.38).

These experiments demonstrated that it is possible to obtain various boiling curves by the heat flux speed changing. It means that influence of the value τ_W should be accounted for in the experimental data treatment.

15. The thermodynamic nature of irreversible phase transitions at boiling on surfaces with porous coverings is especially apparent due to the phenomenon of the system's memory near the transient points [56, 57]. If the transition and vapor film formation inside the porous structure will be interrupted before they come to end by thermal load, the two-phase structure remembers the initial point of the heat flux q decrease. The heat transfer regularity remains linear in the case of interrupted transition, but it corresponds to the smaller thickness of the vapor film inside the porous structure (Figures 5.36b and 5.38b). The existence of the system's memory is determined by the fact that regime II is more beneficial from the energy viewpoint in comparison to regime I, that is, it has less entropy generation at the given value of heat flux.

Several basic details were presented in Malyshenko's theoretical concept about thermal regimes and heat transfer regularities at boiling on coated surfaces flooded by liquid. Let us analyze these statements.

1. In the initial statement of the concept [223, 226], not only the temperature curve shapes correspond to the different thermal regimes at vaporization on coated surfaces, but the values of q^*, ΔT^*, ΔT_{IB}, ΔT_{CR}, and q^{**} are also accepted as specified.

2. Excluding Equation (5.8), even qualitative imaginations about the heat transfer mechanisms (local, average, or integral) are absent in the present approach.

3. The hydrodynamic phenomena accompanying the vaporization stable regimes on surfaces with porous coverings were not accounted for in the Malyshenko phenomenological theory.

4. The analysis and accounting for various temperature depressions accompanying the vapor generation processes inside the pores' internal volumes were missed by the author. Because these factors are extremely important for the correct understanding and modeling of vaporization processes in the given conditions, we shall discuss them in detail.

    (1) It is known that if the single pore skeleton's superheat is equal to $2\sigma T_s/r\rho''R = \Delta T^*$, equal to the so-called nucleation superheat ($\Delta T^*(\Delta T^* = 2\sigma T_s/(r\rho''R))$), then the evaporation inside the pore internal volume is impossible.

This means that the driven temperature drop at the vaporization inside the porous structure's internal volumes is represented not by the wall superheat ΔT, but by the difference $(\Delta T - \Delta T^*)$.

The ratio $(\Delta T - \Delta T^*)$ in regime I and in the experimental conditions [54–57, 221–226] was equal to 1 for the majority of pores. It exerts influence on the regularity of pore activation, because pores having this ratio close to 1 cannot generate the vapor. The parameter ΔT^* could be considered as some temperature depression scale caused by capillary effects.

(2) Vapor flow from the internal volumes to the porous structure's external surface was associated with the increase in essential hydraulic resistance as layer thickness grows and, in particular, with the decrease of the activated pores' diameters. The existence of excess pressure near the wall relating to the ambient pressure equal to the ΔP_0 caused the increase in saturation temperature, that is, decrease in the driven temperature drop on value ΔT_1 proportional to ΔP_0 and equal to $\Delta T_1 = \Delta P_0 (dP/dT)^{-1}$.

ΔT_1 can be considered as some temperature depression caused by hydraulic resistances. Generally, this value is rather small and its influence in comparison to even a small ΔT is also small. However, comparison with $\Delta T - \Delta T^*$ gives another result, because in many cases, especially for small pore diameters (when $\Delta T - \Delta T^*) / \Delta T^* \ll 1$, value ΔT_1 is important. Accounting for this factor shifts the zone of the activated pores into the region of larger pore sizes.

(3) The real operating conditions of porous structures generating the vapor at low pressures and having pore sizes of about 1 μm or less can be associated with the transition to free-molecular flow and corresponding appearance of temperature jumps.

(4) Special attention should be devoted to problems relating to correct accounting for temperature depressions in cases of boiling of liquid mixtures with essential differences in surface tension values or at vaporization of pure liquids with noticeable dash concentrations.

It is known that vaporization and boiling of such media are associated with the effects of considerable change in the volatile or unvolatile admixtures in the vapor and liquid accompanied by its own temperature depressions comparable with the total temperature drops.

Moreover, the dash accumulation causes their intensive concentration in the near-wall liquid layer and consequent depression of the vaporization process and hermetization of porous volumes. Such phenomena were not observed in the typical operating conditions of vapor-generating surfaces. Apparently, the correct analysis of the mixtures and solutions boiling on the surfaces with porous coverings without accounting for these factors is impossible.

(5) Malyshenko and Andrianov [223, 226], Andrianov et al. [224], and Techwer et al. [225] concentrated their attention on transient phenomena and underlined the important role of the percolation theory directed toward determining the corresponding "border" vapor quality φ_c in the porous volume. However, all models of the percolation theory models

are just approximations providing only some estimation for φ_c that are applicable only for thick uniform structures, which are different from real cases. On the other hand, values φ_c were empirically determined for many thick porous structures as the "border" values, for which the relative phase permeability was equal to zero [40]. There are doubts that models based on the percolation theory can provide results that are more accurate.

(6) The internal mechanisms and the vaporization pictures inside the porous structure based on observations of vapor removal from the external surface of the porous structure are difficult to consider as reliable because of the absence of direct observations for processes inside the structure.

(7) Calculations of different border parameters (5.3)–(5.7) were empirical, although they do not maintain limit transient reliability. For example, at $\delta_{CS} \rightarrow 0$, we could obtain from Equation (5.6) that $\Delta T_{IB} < 0$, which is not true:

$$\Delta T_{IB}(\delta_{CS}) = \Delta T_{IB} - B\delta_{CS}^{(-1/\nu)}$$

at $\delta_{CS} \rightarrow 0$, the term $B\delta_{CS}^{(-1/\nu)} \rightarrow \infty$ and $\Delta T_{IB}(\delta_{CS}) < 0$.

(8) Influence of such factors as irregularity of porous structure parameters over its thickness (ε, K, D_E, etc.) and conditions of the contact between the structure and the wall were not accounted for in the theoretical analysis [223, 226].

Therefore, the theoretical concept of irreversible phase transitions at boiling on surfaces covered by porous coatings performed Refs. [223, 226] mostly represent descriptive research. The quantitative correlations of the theory should be considered as empirical related only to the specific experimental conditions. However, the presented results and their comparison with different mass transfer models of vaporization on porous structures are very helpful for development of justified physical models.

Hence, let us consider known models of vaporization on surfaces coated by porous deposits as the first step for the analysis of existing physical imaginations about heat and mass transfer processes at boiling on porous coverings having reliable mechanical and thermal connections with the wall.

5.3 PHYSICAL IMAGINATIONS AND MODELS OF VAPORIZATION ON SURFACES WITH POROUS COATINGS

Initially, it will be correct to use Styrikovich et al.'s review [206] for main mass transfer models' prescription. The first model is based on the physical imaginations of the concentration process presented in Refs. [220, 230].

At boiling, the VRC are forming and vapor is in the process of being removed over these channels. The large pores play the VRC role; their sizes range from 1 to 10 μm [231].

In addition, a system of smaller pores with radius ranging from 0.1 to 1.0 μm also exists. The smaller pores intersect the larger vapor channels and are filled by liquid; they play the role of liquid supply channels. The model of the "wick" boiling suggested in Ref. [231] and its schematic diagram is given in Figure 5.39.

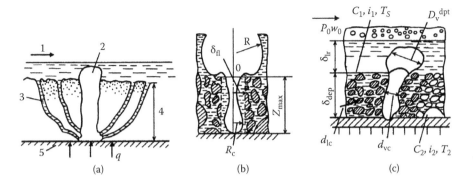

FIGURE 5.39 Vaporization inside porous structures physical models. (a) [231]; (b) [230]; (c) [208]. 1, Liquid flow; 2, vapor outlet flow from the vapor tube neck; 3, capillary channel; 4, deposit's thickness; 5, metal surface.

The mass transfer process can be prescribed with help of the diffusion equation in the form: $D\partial^2 C/\partial z^2 - \partial(w'C)/\partial z = 0$ at the boundary conditions $z = 0$, $C = C_0$, $D\frac{\partial C}{\partial z}\big|_0 = w'C_0$ [220].

Equation for w determination was delivered from the mutual consideration of mass conservation and mass transfer equations,

$$S\rho'w' = pK_s\beta\int_z^\delta \Delta T dz$$

where $K_s = (p - p_s)/(T - T_s)$, β is the mass transfer coefficient, δ is the thickness of deposit layer, and S is the cross section area.

Assuming that heat transfer was caused by the effective heat conduction from the wall to the evaporation places and the entire heat input was transformed into the latent heat, the next correlation was obtained [220] for the concentration factor:

$$n = \frac{C}{C_0}; n = \exp\frac{q_c}{D\varepsilon r\rho'}\left(z - \frac{1}{a}\frac{sh(az)}{ch(a\delta)}\right)$$

where (a^{-1}) is the thickness of the diffusive boundary layer

at $z = \delta$ $$n_\delta = \exp\frac{q_c}{D\varepsilon r\rho'}\left(\delta - \frac{1}{a}th(a\delta)\right)$$

Another principal model of mass transfer at vaporization on porous structures was suggested and developed by Styrikovich et al. [230]. It was assumed that the capillary structure represents a system of cylindrical vapor channels with capillary radius R_c. The vapor bubbles were generated on the heating wall and developed inside the vapor channel in such a manner that in the internal channel walls the thin liquid film remained. The liquid and vapor were in counter flows. The liquid supply to the heating wall was organized by the capillary pressure determined by the admixtures' concentration changing along the channel length. After the vapor bubble removal, the liquid penetrated into the main vapor channels through the lateral capillaries inside the porous structure. The liquid and vapor flows were assumed as laminar. Two basic limit cases of liquid film flow were considered: with small flow rate (long films) and with large flow rate (short films).

Calculation correlations were obtained for the admixture concentration factor [230]. According to the model, the existence of long films stabilized by evaporation is obtainable in the vapor-generating channel. These films supplied the liquid flow rate needed for the vaporization, and were determined by the admixtures concentration changing by 1 or 2 orders of magnitude over the lengths at some relative diameters. The same concentration factors were predicted even for thin structures with thickness of 10–30 μm in the model [220].

However, this contradicted experimental results [219] and the actual performance of boiling apparatus in power plants [206]. Therefore, the next step in the development of the physical imaginations of mass transfer mechanisms at vaporization inside porous structures was performed. It was the model of the admixtures concentration at boiling inside porous deposits [208]. Discussion of its content follows.

Liquid supply to the heating wall is substantially determined by the capillary hydrodynamic regularities. The vapor bubble's appearance and growth took place inside the porous structure matrix [208]. The vaporization process had largely a periodic character. The appearance of vapor phase took place mostly due to the overheated liquid heat. The following vapor bubble growth was caused by evaporation from the liquid film into the vapor channel and from the wall. The matrix skeleton was covered by the dynamic liquid film. Tangent tensions arise from the vapor side effect of the external liquid film surface. The liquid in the film should flow cocurrently with the vapor, that is, it is removed from the wall in case the dependence of the surface tension forces on temperature and admixtures concentration can be neglected. The liquid supply is caused by the action of capillary forces.

Vapor generation inside the porous structure volume was accompanied by the periodic removal of vapor bubbles [208]. For every vapor bubble formation and removal from the porous structure elementary act, there was a corresponding occurrence of liquid filling of the vapor channel and the "silence" period. Polonsky et al. formulated their mathematical model on the mentioned basis for physical imaginations. They also assumed that the vapor bubble "lifetime" was determined in comparison with the "waiting bubble time." This assumption was made with the goal of simplifying the mathematical modeling. It was an artificial assumption, but it corresponded well to the reality of the vaporization conditions observed from video observations on vaporization inside porous structures (Figure 3.29) [79].

The mathematical model can be presented by a set of equations determining the admixtures concentration factor at vaporization inside porous structures as

$$n_0 = \frac{C_2}{C_1} = \frac{G_V + G_{Fl}}{G_{Fl}}; G = \frac{q}{r\varepsilon N}$$

where C_2/C_1 denotes the admixtures concentration inside (C_2) and outside (C_1) the porous structure.

The flow rate in the liquid film was related by the material balance equation to the liquid cross section $\pi D_C^4 (1 - \varphi)/4$, liquid mass density ρ', and average liquid velocity $\bar{w}_F$. Assuming the liquid flow as laminar and annular, and taking hydraulic resistance from the vapor side as the main factor, Polonsky et al. [208] presented the following correlations:

$$G_{\mathrm{Fl}} = \frac{\pi D_C^2}{4} \rho'(1 - \varphi) \frac{\tau_V \delta_{\mathrm{Fl}}}{2\mu'}; \tau_V = C_f \frac{\rho'' w_V^{-2}}{8};$$

$$\bar{w}_V = \frac{4G_V}{\pi D_C^2 \varphi \rho''}; C_f = \frac{65 v''}{\bar{w}_V D_C \sqrt{\varphi}}$$

The assumption that operating against friction forces appears possible due to the action of capillary forces yields the following condition

$$\Delta p_V \approx \frac{\delta}{D_C \sqrt{\varphi}}$$

Thus, the concentration factor and the real vapor quality were determined as

$$n_0 = 1 + \frac{64\varphi^{3/2}}{(1-\varphi)(1-\sqrt{\varphi})} \frac{v'}{v''}; \varphi = \left(\frac{q n_0 v''}{6\pi D_k^2 r \varepsilon N} \right)^{3/2}$$

Calculations of n_0 based on the proposed model did not provide satisfactory concurrence with the corresponding experimental data. In addition, it would be right to notice that if the first model revealed the concentration degree values to be about 1000–10,000, then the experimental data were in the range 2–10; values calculated by the last model were $n_0 = 3$–30, which is considerably close to experimental data. The entire view of the model [208] corresponded well to the direct observations for vaporization inside screen structures, whereas the principal positions of the first and second models contradicted the experimental data. For example, existence of the vapor channel having simultaneously dry lateral and bottom walls was not observed in the visualization experiments. It is possible to assume the channel's bottom and lateral walls to be dry, but in such a case, the whole or the major part of the structure internal volume should be also dry.

The assumption of the model [230] that liquid and vapor flows were caused by the capillary forces associated with admixtures or soluble substance concentration differences, contradicts the numerous experimental information about the high intensity of pure liquids without vaporization of some admixtures and soluble substances on porous coverings, and in the case when surface tension gradients are absent. Therefore, the model proposed by Polonsky et al. [208] appears more realistic. However, mass transfer equations are not presented in this model.

It is necessary to note several common drawbacks of the models [208, 220, 230]:

- Did not account for the existence of different regimes at vaporization inside porous structures. Every regime needs its "own" model.
- Did not account for the interconnections of heat transfer and mass transfer processes, excluding only the most simple correlation $G = q/r$.
- Did not account for the fact that change in the regime (q and ΔT), structure, and thermophysical (λ_{CS}, λ^1, etc.) parameters causes essential change in the vapor generation mechanism inside the porous structure (vapor and liquid channels ratio, their distribution along the surface, cross section, etc.).

Therefore, in recent years, many attempts were made to improve existing models. Let us consider the most important of these models.

The theoretical model of boiling on flooded porous coverings was presented by Mankovsky et al. [88]. The heat transfer process was convective at the small wall superheat. When the wall temperature drop reached a certain value, vaporization was initiated inside the porous structure. The vapor that appeared was removed through some pores. The VRC maintained the liquid evaporation. Vaporization speed was small at low heat fluxes and vapor came out of the porous layer through the pore system as separate bubbles. Increase in thermal load led to growth in the frequency of vapor bubble formation and departure from the porous layer. Therefore, liquid movement inside the porous structure had a pulsating nature [89].

When the heat flux q reached a certain value, vaporization speed became sufficient for forming a permanent acting VRC inside the porous layer. Because the character of liquid flow inside the layer was changed, the pulsating liquid flow was terminated [89]. The interface (meniscus) was near the heating wall and the wall was flooded by the thin liquid film. According to Mankovsky et al. [88], the liquid was suctioned by capillary forces inside the layer and heated over contacts with the layer elements having considerably higher temperature than the liquid due to higher thermal conductivity. Only this thermal regime was considered [88], because the generalized correlation was performed for this case alone. The model created on these assumptions is as follows:

1. It was assumed that the local heat transfer intensity was determined in a similar manner to the liquid steady laminar flow in capillaries:

$$\alpha = C \frac{\lambda'}{d_P}$$

2. The real porous structure was replaced by the capillary system.
3. It was also assumed that every VRC corresponds to m liquid channels. It means that ($m + 1$) capillaries were operated for every single vaporization center. For ($m + 1$) channels, m were completely filled by the liquid (the heat transfer inside the VRC was neglected).

4. With the goal of determining the average heat transfer density in the single capillary q_c, the capillary wall was assumed as some fin on the heat transfer surface. The liquid temperature around the fin was supposed to be equal to the saturation temperature at the pressure with respect to the capillary pressure defined with meniscus curvature inside the capillary channel. If T_s is the saturation temperature, the nucleation superheat in capillary with the diameter D_C is

$$\Delta T^* = \frac{4\sigma T_s}{r p'' D_C}$$

5. Capillary effective diameter D_C was equal to $0.41 d_p$ (the layer natural piling) for the layer from the identical particles with size d_p, then $\Delta T^* = 9.75 \sigma T_s / (r\rho'' d_p)$. It was assumed that the temperature drop at the fin's bottom was equal to $\Delta T_0 = T_W - T_s - \Delta T^*$ [88]. This immediately drew the question: "How can we combine the first three assumptions with the final one?"

6. If heat transfer inside liquid channels plays the determining role, why was the temperature drop assumed at the nucleation superheat subtraction and the liquid temperature taken as constant and equal to the saturation one? The statement that the main heat flux is transferred in the vapor channel agrees with assumption no. 4, but opposes the first three suggestions.

7. The liquid and vapor flow rates at filtration through the porous structure are determined by conditions of complete vaporization of liquid flowing through the liquid channels and by coincidence of total hydraulic losses of liquid $\Delta P'$ and vapor $\Delta P''$ with the capillary potential $\Delta P_C = 4\sigma \cos \theta / D_C$ [88]. Then, the correlation determining the number of liquid channels m is

$$m = 1.41 \times 10^{-3} \frac{\sigma r p'' \cos \Theta}{\mu'' q_c} \left(\frac{d_p}{\delta_{cs}} \right)^2 \left(\frac{\varepsilon}{1-\varepsilon} \right)^2 - \Delta m$$

It was difficult to imagine how the liquid and channel redistribution could be independent on real pore distribution and why at the uniform pore distribution near the wall the vapor was generated in one channel but was absent in others. There were also other contradictions; for example, it was suggested to calculate the effective fin thermal conductivity as the thermal conductivity of the dry layer without accounting for the presence of liquid. It contradicted with assumptions 1 and 2.

The final correlation [88] has the following form:

$$q = \frac{64.8(1-\varepsilon)}{d_p} \sqrt{\left(\frac{1}{\sqrt{\varepsilon}} - 1 \right) \lambda_L \lambda_{cs}} (\Delta T - \Delta T^*) \frac{m}{m+1} \qquad (5.10)$$

Mankovsky et al. [88] obtained good concurrence with their own experimental data. However, results of comparison with the data of other authors (from LTIRI) were less successful. Some reasonable correlations were obtained only after the selection of thermal conductivity of skeleton equal to $\lambda_{SK} = 0.015$ W/(mK).

Although this value is even less than the vapor thermal conductivity, this could not be possible. Nevertheless, the data obtained from using the generalized Equation (5.10) were quite close, in some cases, to experimental results and coincided with calculations using the generalized equation obtained previously [71] and independently on the basis of dissimilar physical imaginations.

Furthermore, in the model analysis [71], we shall explain that in spite of the many contradictions of the model [88], the obtained equations are acceptable. Nevertheless, the simple correlation $Nu = \alpha d_p/\lambda' = 15.2$ was recommended in Ref. [89] instead of Equation (5.10). The availability of such assumption was checked in some experimental data. Liquid was oscillated inside the structure with high frequency f and amplitude ΔU proportional to the layer characteristic internal scale D: $\Delta U \approx fD$. It was assumed for the so-called the pulsation regime.

Fridgant [89], by using the differential energy equation, considered the unsteady heat conduction of the straight bar with the constant heat transfer coefficient α. Next, they left out the energy equation term accounting for the heat transfer to the liquid. Hence, the final equation has the following form

$$\bar{q}_L = \frac{2}{\sqrt{\pi}} \frac{\varepsilon_L \varepsilon_V}{\varepsilon_L + \varepsilon_V} \sqrt{f} \left(T_{W0} - T_{L0} \right)$$

where ε_L is the fraction of the pore volume occupied by the liquid; ε_V is the fraction of the pore volume occupied by the dry vapor; $\bar{q}_L$ is the average specific heat exchange between the wall and liquid at the liquid pulsating flow; and T_{W0} and T_{L0} are the wall and liquid temperatures on the skeleton surface, respectively. Parameter $\bar{q}_s$, which denotes the specific heat exchange between the wall and the structure, was also introduced. The total heat flux $\bar{q}$ was determined as $\bar{q} = A_L \bar{q}_L + A_s q_s$, where values A_L and A_s are determining experimentally.

Fridgant [89] used the scaling approach

$$f = q/rp''\varepsilon l; \quad l \sim (D_E^* l_\sigma)^m$$

where $l_\sigma = \sqrt{\sigma/(\rho' - \rho'')g}$, D_E^* is the effective pore diameter, l is the characteristic pores size, and $m = 0.5$.

Analysis of numerous experimental data revealed that m ranges from 0.25 to 0.8–0.9.

Experimental modeling was performed with R-113 as the heat carrier and using the porous structure imitation by the system of 10 parallel capillaries with diameters of 0.4 mm. These were the R-113 typical boiling conditions in the slits and capillaries, where the pulsation regimes are typical. Therefore, the regularities of the type $\alpha \approx q$ are feasible.

The unsteady heat transfer mechanism for the system liquid–wall was determined in some conditions. However, the data obtained from photo and video capture [79] did not reveal the existence of such a thermal regime inside porous structures, when two-phase mixture flows in the pores. It was difficult to imagine such a thermal regime inside typical porous structures with pore sizes from 10 to 100 μm.

It should be noted that an unsuccessful attempt was made [88] using experimental data from the Odessa Technological Institute of Refrigeration Industry (OTIRI)

[251–256] to observe vaporization at boiling inside a flooded particle bed. These vaporization processes will be studied further by special analysis. It was shown that such structures are related to the so-called movable structures, when a gap appears between the wall and structure, and boiling develops similar to slits. In addition, the determining input into the whole process of modeling is associated with the vaporization centers' density changing on the heating surface, microlayer evaporation, contact heat transfer mechanism, etc. These processes were not mentioned in the discussed studies.

The author of this book succeeded in excluding most of the drawbacks and contradictions of studies mentioned in an earlier work [71], which was devoted to the theoretical prescription of heat transfer at vaporization inside porous structures linked with the wall. It was suggested [71] that formation of interface near the wall takes place at vaporization in the CPS. The following heat transfer regimes are typical in such cases (Figure 5.40):

- At $\Delta T < \Delta T^*$ and when soluble gas is not present in the liquid, the heat is transferred through the wetted porous structure by heat conduction. When a soluble gas is present in the liquid, stable gas-vapor bubbles appear inside the structure playing the role of "thermal isolators." In the last case, the effective thermal conductivity λ_E is correlated with λ_{E0} (when soluble gas is not present) and volumetric gas content φ as

$$\lambda_E = \frac{\lambda_{E0}(1-\varphi)}{(1+\varphi/2)}$$

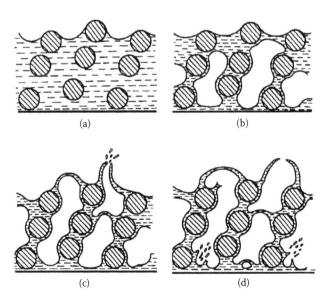

(a)

(b)

(c)

(d)

FIGURE 5.40 Model of vaporization inside CPS [71]. (a) Evaporation regime. (b) "Microheat pipe" thermal regime. (c) Thermal regime of boiling with "deepened meniscus." (d) Boiling in the microfilm thermal regime.

- At $\Delta T > \Delta T^*$, two thermal regimes are possible depending on the skeleton thermophysical properties and liquid supply conditions (flooded, capillary feeding, etc.):

- "Microheat pipes," when the vapor generated near the wall condenses completely inside the structure near the external surface (than larger values of structure thickness, temperature nucleation drop ΔT^* and λ_E, then this regime becomes more probable).
- The vapor phase filtration over the CPS.

It can be concluded from the schematic layout (see Figure 5.41) illustrating the heat transfer mechanism that heat transfer regularities for both thermal regimes "b" and "c" (Figure 5.40) are approximately similar.

Consider the model [71] for thermal regimes "b" and "c" (Figure 5.40) based on the presented imaginations.

1. The vapor phase removal realized under the action of excess vapor pressure near the wall. This excess value, $\Delta P''$, corresponded to the capillary potential $P_C = K_0(\sigma/L_0)$; $L_0 = \alpha$ is the cell size for the screen CPS (vaporization in these structures is extensively studied by the author and his team [75–79]).
2. The linear interface was formed at the structure elements' conjugations both at the places of the structure's conjugation with the wall under the excess pressure action. It provided the liquid flow to the heat transfer surface, that is, the vapor and liquid countercurrent flows.
3. It was supposed that heat was transferred mainly in the latent heat form. The main thermal resistance, as shown from the schematic diagram (Figure 5.41), is the thermal resistance to the heat transfer from the wall through the wetted structure skeleton to the interface. This thermal resistance can be determined from the solution of the heat conduction problem for the temperature field at some conditional fin covered by the liquid dynamic film with the height equal to the structure's thickness δ and specific thermal conductivity λ_{E0}. The fin's perimeter and cross section area are defined by the structure's cell geometry parameters. In the simplified case, $\delta \ll a$, it remains constant over the structure's thickness.

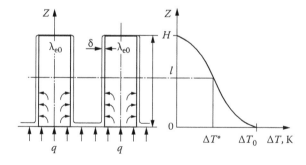

FIGURE 5.41 Model of elementary cell at boiling inside the porous structure [71].

Therefore, it leads to the problem of heat conduction over the bar with the boundary conditions:

$$\frac{d^2\theta}{dx^2} - \frac{\alpha\Pi}{\lambda_{E0}f}\theta = 0;$$

$$x = 0; \quad \theta = \Delta T_0 - \Delta T^*;$$

$$x = l; \quad \theta = 0;$$

$$x = H; \quad \theta = \Delta T^* \qquad (5.11)$$

where Π is the conditional fin's perimeter and f is the fin's cross section area.

Heat flux removal from the selected elementary cell over the heat-transferring skeleton was determined as

$$Q = \lambda_{E0}mf(\Delta T_1 - \Delta T^*)\frac{2chml}{\exp ml - 1} \qquad (5.12)$$

Here, l is the zone's size, at which evaporation from the film surface takes place. Value l was determined from the condition:

$$\Delta T^*(\exp ml - 1) = (\Delta T_1 - \Delta T^*)sh(m(H - l)) \qquad (5.13)$$

Direct calculation of parameter m was complicated by the conditional nature of value δ and its uncertainty.

Estimation of the main values determined parameter m and analysis of experimental data revealed that it could be assumed that $ml \gg 1$. Then, $2ch(ml)/(\exp(ml) - 1) \to 1$ and the heat transfer intensity would not depend on the capillary structure's thickness; consequently, $q = \lambda_{E0}m(\Delta T_1 - \Delta T^*)$. ΔT_1 in Equation (5.13) is the temperature drop at the base of heat-transferring skeleton. ΔT_2 is the temperature drop at the elementary cell bottom part covered by the liquid film and the wall average superheat $\Delta T_0 = \Delta T_1(1 - \varepsilon')$ $+ \Delta T_2\varepsilon'$. It yields the correlation for the heat flux determination with respect to the fact that $\Delta T_2 \cong \Delta T^*$

$$q = \lambda_{E0}m(1 - \varepsilon')(\Delta T_0 - \Delta T^*) \qquad (5.14)$$

Here, ε' is part of the porous structure occupied by the VRC, $\varepsilon' < \varepsilon$.

The character of liquid film thickness dependence can be defined based on the physical model with respect to the equality of the capillary potential and the hydraulic losses at the vapor and liquid film flow. Using the simplified calculation schematic of the independent laminar vapor and liquid film flow yields the following correlation for δ [71]

$$\frac{4\sigma}{a} \geq \frac{qv'}{r\delta^3}\frac{Lf}{\Pi}C_1 \qquad (5.15)$$

where L is the length of liquid transport, $L \approx H$. Then, the qualitative estimation of value l was suggested based on the assumption that more m than less l, because $ml = const$.

The joint consideration of Equations (5.11)–(5.15) leads to the following correlation for the liquid film thickness:

$$\delta \approx const \left(\frac{qv'}{r\sigma} a \right)^{0,4} \left(\frac{f}{\Pi} \right)^{0,6} \left(\frac{\lambda_{E0}}{\lambda'} \right)^{0,2} \tag{5.16}$$

The final correlation presents the dependence between heat flux and temperature drop at vaporization on coated surfaces

$$q = const^6 \sqrt{\frac{r\sigma\lambda'^3}{v'}} (1-\varepsilon')^{5/6} \sqrt[3]{\lambda_{E0}} \left(\frac{\Delta T_0 - \Delta T^*}{L_0} \right)^{5/6} \tag{5.17}$$

Here, L_0 is the characteristic size of the elementary cell cross section, $L_0 = \left[\left(\frac{f}{\Pi} \right)^4 a \right]^{0.2}$.

Generalization of numerous experiments on heat transfer at vaporization on screen structure surfaces is presented in Figure 5.42. First, thermal regimes without filling of capillary structures by vapor were considered. Generalization led to the determination of the constant in Equation (5.17) as equal to 0.094.

The considered model prescribes heat transfer regularities at vaporization on surfaces covered by porous coating, when the structure itself could be imagined as consisting of identical or approximately identical elementary cells. The elementary cells' "identification" was especially important for the near-wall zone, that is, at the first layer of the screen structure or the first layer of the layer consisting of identically sized spherical particles. These conditions were well fulfilled in several studies [75–79, 148–153].

The model comparison with corresponding experimental data showed good qualitative and fair quantitative agreement. Hence, the following were confirmed:

- Heat transfer independence on the covering thickness
- Weak dependence of heat transfer coefficients α from thermophysical properties of the skeleton and saturation pressure
- Strong dependence of α on the liquid thermal conductivity and the pore characteristic size L_0
- Depression of heat transfer enhancement with the heat flux growth

It was especially interesting and important to note that some of the most doubtful and discussed positions of the author's model [71] were confirmed in studies (see Section 3.5). It was related, first of all, to such thermal regime predictions as the existence of the "microheat pipe," when stable closed vapor–liquid loops acting on the same physical base as heat pipes appeared in the near-wall zone [69, 70].

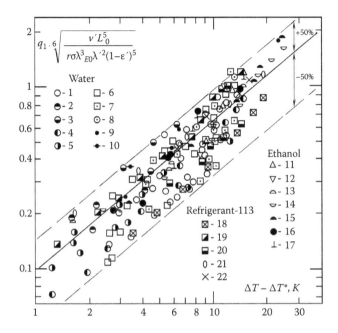

FIGURE 5.42 Generalization of experimental data at vaporization on surfaces coated by screen or particle bed coverings [71, 75–79, 146–153]. $\Delta T^* = 4\sigma T_s/(r\rho''D)$; $\Delta T_0 = T_W - T_s$; $Y = q\sqrt[6]{v'L_0^5/(r\sigma\lambda_{E0}^2(\lambda')^3(1-\varepsilon')^5)}$. water, ethanol, acetone, R-113; $0.01 \le p \le 0.1$ MPa; number of layers, 1–8; $a = 40$–125 μm; material of structures, nickel, stainless steel.

It was natural to assume that when a porous structure had some pore distribution and if it could be presented as a combination of elementary cells (it was possible, but only in the case when elementary cells are not identical), then heat transfer regularities could, and even have to, change. Therefore, comparison of experimental data on heat transfer at vaporization on coated surfaces — covered by sintered, sprayed, and other types of coverings having some distribution of pores, with calculations by Equation (5.17) — will be unacceptable. It is an essential drawback of the given model. This is because there are other drawbacks relating to the assumption of simple liquid film flow, a simple form of definition of the average temperature drop, absence of the heat transfer dependency on the structure thickness for many experimental cases, etc.

It is very important to remember that the model prescribes vaporization thermal regimes, when the heating surface is steadily washed by the liquid. It means that such thermal regimes can be considered "vaporized (or boiling) with deepened meniscus." It was interesting to analyze the limitations of this regime. Hence, it was natural to consider that the upper limit of this regime is related to heat flux value, when the heat transfer effect is depressed, that is, achievement of similarity between heat fluxes and corresponding temperatures drops in this regime and pool boiling on the smooth surface: $q = q_1$ and $\Delta T = \Delta T_1$ (see Section 3.3).

Further increase in heat flux ($q > q_1$ and $\Delta T > \Delta T_1$) can lead to one of two possible variants of "vaporization" thermal regime development:

- Filling of capillary structure by vapor accompanied by the abrupt decrease of heat transfer coefficient
- Stable boiling in the liquid film covered heating surface at the pores' bottom

The transition to one of these thermal regimes is determined by the thermal–hydrodynamic stability conditions; however, this was not considered in the author's model [71].

Main initial positions (5.11)–(5.16) of the model [71] have extremely approximated basis. It means that the model required essential improvement and development. It was correct only for conditions with reliable thermal contact between the wall and the structure. Termination of thermal contact caused a major change in heat transfer regularities. The heat transfer regularity approaches the case of boiling in the single narrow slit. Because the model presented good explanation and generalization for a number of experiments at vaporization on surfaces covered by porous layers without pore distribution in the first near-wall layer, it could be considered as the experimental determination of some general local heat transfer regularities for the above-mentioned thermal regime with single elementary cell of porous structure (the so-called "vaporization with the deepened meniscus"). This means that the approach [71] can be considered as a semiempirical model of the local heat transfer at the flooded contact zone between the wall and structure under conditions where the interface is located near the wall and separated from it only by a very thin evaporating liquid film.

Thus, good agreement between physical imaginations [71] and numerous representative experimental data allowed considering Equation (5.17) as a heat transfer local regularity for the mentioned heat transfer thermal regime (Figure 5.40b,c).

Limitations and drawbacks of the model [71] initiated the appearance of different models and physical imaginations devoted to heat and mass transfer at vaporization on surfaces covered by porous structures. For example, Antonenko [159] and Tolubinsky et al. [160–162] proved absence of boiling inside thin porous structures at high heat fluxes corresponding to conventional heat transfer regularities at pool boiling.

Using experimental modeling, Nakoyama et al. [202] proposed the following theoretical dynamic model of heat transfer at boiling on surfaces covered by porous coatings:

- It was assumed that the real porous structure could be imagined and correspondingly changed as the system of rectangular capillaries inserted into the wall and connected by unlocked pores with external liquid volume.
- The wide pores case and, consequently, regime when the horizontal channels are filled by liquid was not considered, that is, it was assumed that horizontal channels are filled by saturation vapor, because the liquid presented only in the film in the corners.

- The dynamic model treats vaporization inside the porous structure as an unsteady oscillating process consisting of the next stages:
  a) Vapor pressure growth in horizontal capillaries as result of evaporation
  b) Vapor bubbles growth out of porous covering border accompanied by pressure drop
  c) Liquid suction associated with vapor bubbles departure

At stage (a), thermal and mass balance equations yield the following correlation:

$$V'' \frac{dp''}{d\tau} + p'' \frac{dV''}{d\tau} = \frac{\alpha S(T_{SC} - T'')}{r} \tag{5.18}$$

where V is the entire vapor phase volume (it changes depending on the growing bubbles volume), α is the average heat transfer coefficient in the horizontal capillaries, and S is the heat transfer surface.

It was necessary to determine the five empirical coefficients on the general experimental and theoretical analysis base using such experimental parameters as the acting pores density, vapor bubbles departure frequencies, vapor bubbles departure diameters, etc. [202]. It was discovered that the scattering between the experimental data and calculations by using such an approach was about 300%. Nakoyama et al. [202] tried to apply their model to the determination of the optimal pore diameter. However, they noticed that positive results would be impossible without significant improvement in the corresponding determination of empirical coefficients.

It was supposed that in a simple correlation, $q = (\Delta T/C_q)^{5/3}(N_A/N)^{1/3}$, where C_q is determined from the experiment and N_A required using additional empirical coefficients. It should be noted that the model's positions could be applied only for similar porous covering geometry and may be applied for thin-layer coatings partially or completely removed from the wall. Therefore, the dynamic character of the model [202] could not be evaluated properly.

In spite of this, the calculation results and approach [202] were used in the development of models [232, 233] on steady vaporization heat transfer on surfaces covered by porous coatings. However, comparison with experimental data was not presented. The analysis in two studies [232, 233] could be based on the following correlation prescribing the boiling curve on the surfaces covered by porous coatings:

$$\Delta T = \frac{q\{1 + nth[m(L - \Delta L)]\}}{\lambda_{SL} mA\{n + th[m(L - \Delta L)]\}} + \frac{T''q\mu''(L - \Delta L)}{r^2 p'' K_V} \tag{5.19}$$

where ΔT is the temperature drop between heating surface under porous coating and liquid at pool boiling; q is the heat flux, $m = \sqrt{\alpha_L U / \lambda_{SK} A}$, $\alpha_L = \lambda'/\delta_L$ (δ_L is the microlayer thickness on skeleton surface); λ_{SK} and λ_{SL} are the dry and saturated by liquid skeleton thermal conductivities; A is the specific area of skeleton cross section; $A = 1 - \varepsilon$, $U = \sqrt{\varepsilon_v^2 / 2K_V}$ is the specific opened internal surface, ε is the porosity, $T'' \cong T_S$ is the saturation temperature, μ'' is the vapor dynamic viscosity, K_V denotes permeability;

r is the latent heat, L is the porous structure thickness; $\Delta L = L(1 - q^*/q)$ is the thickness of dry layer of porous material; ε_V is the porosity opened for vapor; q^* is the maximum heat flux at the inter-layer crisis condition:

$$q^* = \frac{2r\sigma}{a_{min}L}\left(\frac{v_L}{K_L} + v_v \frac{p_0}{p}\frac{1}{K_V}\right)^{-1}$$

where p_0 is the pressure in pool volume; p is the average pressure in the porous layer; a is pore radius $p = p_L + \Delta P/2$; $v = \alpha/\lambda_{SL}m$.

The differential pore distribution over the sizes is prescribed as

$$\frac{dV}{da} = (a - a_{min})\sqrt{\left(\frac{a_{min} - a}{a}\right)}s_V \qquad (5.20)$$

$S_V = S/V$ is the specific internal pore surface related to porous volume, m²/m³.

The opened porosity for vapor can be obtained by integrating Equation (5.20):

$$\varepsilon_V = \varepsilon \int_a^{a_{max}} f(a)\frac{da}{K}$$

where $a = \sigma/\Delta P$, K is the normalization coefficient accounting for the fact that the right side of the distribution curve vanishes at $a > a_{max}$, $K = \int_a^{a_{max}} f(a)da/K$.

The vapor and liquid permeability was calculated as K_V and K_L, respectively.

$$K_V = \varepsilon \int_a^{a_{max}} a^2 f(a)\frac{da}{8K} \quad K_L = \varepsilon \int_{a_{min}}^a a^2 f(a)\frac{da}{8K}$$

The thermal conductivity of porous material is calculated by [232]

$$\lambda_{sL} = \lambda_L \left(\frac{\lambda_s}{\lambda_L}\right)^{(1-\varepsilon)^{0.28}}$$

Liquid microlayer thickness on the pores' walls was assumed as constant and equal to $\delta = \mu$m, whereas the locked porosity was not accounted for. The correlation between the heat flux density q and permeability K_V is determined by Darcy's law: $q = \Delta P K_V^* r(\rho'')^*/\mu''L$; if $q < q^*$ [232]

$$q = \frac{\Delta p K_V^* r(p'')^*}{\mu''L}$$

where K_V^* and (ρ'') are covering permeability and vapor density at $q = q^*$.

The following values of the main parameters were used for calculations: $L = 0.4$ mm, $\varepsilon_0 = 0.3$, $\lambda_{E0} = 200$ W/(m K), $S_0 = 10^{-6}$, $a_{min0} = 1$ μm, $a_{max0} = 10$ μm; $n_0 = 1$.

The heat carrier used was refrigerant R-113. Calculations of wall superheat values at the various heat flux values (10^5, 2×10^5, 3×10^5, and 4×10^5 W/m^2) were performed depending on such characteristics of porous structures as porosity ε, coverings thickness L, and parameter n.

It was demonstrated via calculations that there were such values of the aforementioned parameters, at which the superheat ΔT reached minimum value and the heat transfer intensity attained its maximum value. The optimal porosity value was defined as equal to 0.3, the superheat for the range of $q = 10^5$–4×10^5 W/m^2 at R-113 boiling was changed from 2 to 4 K. Because the absolute values of the temperatures drops, the heat flux values did not agree with known experimental results (e.g., Ref. [196]) (see Figure 5.33). Pichlak and Techwer [232] did not present comparison with experimental data.

Let us consider the model presented in two studies [232, 233]. The main advantage of the model was the attempt to account for the interconnection between the pore distribution and the vapor and liquid permeability and its interconnection influence on the ratio between the number of liquid and vapor channels. Unfortunately, the model did not account for the pore activation phenomenon associated with the nucleation superheat:

$$\Delta T^* = 4\sigma T_s / (r\rho'' a)$$

Nevertheless, the condition of pores activation $a \geq 4\sigma T_s / (r\rho'' \Delta T)$ was extremely important for the correct prescription of the vaporization process on the covered surfaces.

The next principal drawback of the model [232, 233] is associated with the attempt to prescribe different thermal regimes of vaporization on covered surfaces by the single model. Analysis of representative experimental data of different authors demonstrated that, for example, "vaporization beginning," as a rule, was associated with wall superheat "jump" — that is, the minimal vapor layer thickness in the porous structure exists, ΔL_{min} at $q > q^*$, and the monotonous dependence of the vapor layer thickness on q predicted by the considered model was not observed in experiments.

The first term of Equation (5.19) is associated with the external thermal resistance, an essential input in the overall thermal resistance of the heat transfer at vaporization on porous covering. Such result could be justified only for thin-layer flooded structures without vapor removal from the porous surface.

Unfortunately, numerous temperature regimes at vaporization inside porous structures did not follow from the model, including those that coincided with the corresponding heat transfer regularities in pool boiling.

The hydrodynamic model of vaporization inside CPS applied to the limit heat flux determination was suggested [234, 235]. Special attention was paid to the countercurrent liquid and vapor flow inside the vapor channels. Because pressure in the liquid decreases from the external surface of capillary structures to the wall, and vapor pressure increases in the same direction, the pressure drop at the interface increases from the external surface to the capillary structure basis.

The authors considered that it was possible only in the case of vapor and liquid flow in the different channels. Therefore, in Refs. [234, 235] the steady vaporization in the porous structure was considered as the phenomenon accompanied by the liquid flow inside the channels with sizes $R_{min} < R^* < R_{BR}$ and vapor flow in channels with the radius R_V, where $R_{BR} < R_V < R_{max}$.

Such an approach of repeating the Macbeth model is correct, if the goal is to neglect the real peculiarities of the porous structure, when the local changing of the curvature of the skeleton surface could be differentiated at some point from the average curvature of the cylindrical channel with the pore's average diameter.

In many cases, the corresponding growth in the nucleation superheat $\Delta T^* = 4\sigma T_s/(r\rho^* R)$ leads to the condition $\Delta T - \Delta T^* < 0$, excluding these surface parts as the vaporization surface elements.

The condition of mechanical equilibrium in the real pulsation process and at the wave flow could not be realized. Therefore, the model of the independent existence of liquid and vapor channels requires direct experimental verification, which has not been available until now. The visual observations [75–79], probably unknown to the authors [234, 235], demonstrated at least for the screen structures, the absence of thermal regimes with independent vapor and liquid flow accompanied by the countercurrent vapor and liquid flows in the pores. However, it is possible that under some conditions, together with countercurrent vapor and liquid flow zones, zones of direct current flows also exist. The results of visual observations [75–79] at vaporization on porous structures accompanied by very strong throwing of liquid confirmed the assumption.

Nevertheless, the model [234, 235] is of some interest and could be useful for research and development of corresponding physical imaginations. The model was based on the following set of differential equations of vapor and liquid interconnected flows:

$$\rho'w'\frac{dw'}{dn} = -\frac{\partial p'}{\partial n} - \frac{\mu'w'}{K'} - \beta\rho'w'^2;$$

$$\rho''w''\frac{dw''}{dn} = -\frac{\partial p''}{\partial n} - \frac{\mu''w''}{K''} - \beta\rho''w''^2;$$

$$\frac{2\sigma\cos\Theta}{R} = p'' - p'$$

The phase permeabilities K' and K'' were calculated on the basis of known dependencies with respect to the pore distribution functions and the restriction of liquid flows in pores by the radius values, from R_{min} to R, and the vapor flows by values from R_{max} down to R, $w' = q/(r\rho''f(R))$, $w'' = q/(r\rho''(1 - f(R)))$.

The radius R and the corresponding values of the integral distribution function $f(R)$ are variables for pores separating the vapor and liquid channels. Thus, the set of equations was transformed into a single differential equation:

$$\left[\frac{2\sigma\cos\Theta}{R^2} - \frac{q^2}{\varepsilon^2 r^2}\varphi(R)\left(\frac{1}{p'f^3(R)} + \frac{1}{p''(1-f(R))^3} \right) \right]dR =$$

$$\left(\frac{q}{rK} \right)\int_{R_{min}}^{R_{max}} R^2\varphi(R)dR \left| \frac{v'}{\displaystyle\int_{R_{min}}^{R} R^2\varphi(R)dR} + \frac{v''}{\displaystyle\int_{R}^{R_{max}} R^2\varphi(R)dR} \right| +$$

$$\frac{\beta q^2}{r^2}\left(\frac{1}{p'f^2(R)} + \frac{1}{p''(1-f(R))^2} \right)dh$$

Analysis of this equation demonstrated that values R_1 and R_2 are the roots of the relation $dh/dR = 0$, where h is a local coordinate. These values restricted the physically realized region of the phase-interconnected flow inside the coating. This condition was not required at separate phase flows.

It was also obtained from the approximated thermodynamic analysis [235] that the minimum stable system state was realized at the minimum possible pore volume occupied by the vapor. This condition corresponded to the lowest location of curve $h(R)$ relating the axis R over the thickness δ_{CS} determined from searched boundary condition. Integral of the momentum equation at $h = \delta_{CS}$, $R = R_2$ yields

$$h = \delta_{CS} - \int_{R}^{R_2} \frac{\dfrac{2\sigma\cos\Theta}{R^2} - \dfrac{q^2}{\varepsilon^2 r^2}\varphi(R)\left(\dfrac{1}{p'f^3(R)} + \dfrac{1}{p''(1-f(R))^3} \right)dR}{\dfrac{q}{rK}\displaystyle\int_{R_{min}}^{R_{max}} R^2\varphi(R)dR\left[v'\left(\displaystyle\int_{R_{min}}^{R} R^2\varphi(R)dR \right)^{-1} + v''\left(\displaystyle\int_{R}^{R_{max}} R^2\varphi(R)dR \right)^{-1} \right]} + A_0$$

where

$$A_0 = \frac{\beta q^2}{r^2}\left(\frac{1}{p'f^2(R)} + \frac{1}{p''(1-f(R))^2} \right)$$

The obtained equation is a correlation between the local coordinate h and the local minimal radius R in the restricted the vapor channels zone. Awareness R and R_2 allow us to calculate the vapor quality for every coordinate h, including at $h = 0$. Moreover, it is possible to determine from this equation some q_{LT} at $R \to R_1$. This implies the beginning of the dry-out. This follows from the analysis [235] that the increasing $q > q_{LT}$ causes the monotonous flow of the liquid penetration border. Experimental results demonstrated that transition over q_{LT} has the capacity to eliminate thermal stability. This was accompanied by the temperature jump corresponding to some stable vapor layer thickness inside the porous structure. It leads to drying out of the whole structure along its height. The possibility of eliminating such a contradiction was not even discussed in the approach [235].

The physical model of heat transfer at vaporization on the porous structure coverings was suggested with the assumption that the influence of heat conduction over the skeleton can be neglected. It was supposed that vapor–liquid flow hydrodynamics has a determining influence on the heat transfer. Changing of saturation temperature and, consequently, of temperature drop inside the capillary structure, was accounted for by the corresponding change in pressure, $\Delta T = T_w - T_s - T_s \Delta P/(r\rho'')$. It revealed the dependence of the heat transfer coefficient on heat flux in the form $\alpha = f(q)/(1 + Af(q))$. Here, $f(q)$ is the heat transfer coefficient between vapor–liquid flow in the capillaries and the wall (skeleton) and A is the coefficient of proportionality in the correlation $\Delta P = Aq/(r\rho'')$. It was based on the laminar flow regime in the pores and on the existence of the proportionality between the value $q/r\rho''$ and the average flow velocity.

The next formula yields the pressure drop in the disperse-annular regime:

$$\Delta p = \Delta p_L \left[1 + x\left(\frac{p'}{p''} - 1\right)\right]^{0,8} \left[1 + x\left(3,5\frac{p'}{p''} - 1\right)\right]^{0,2}$$

where ΔP_L corresponds to the liquid filtration through the porous media:

$$\Delta p_L = 72\mu' \frac{(1-\varepsilon)^2}{\varepsilon^3} \frac{1}{a^2} w'$$

Here, w' is defined with respect to the slip conditions as $w' = w''/S = q/(r\rho'')$. Calculation of function $f(q)$ was also suggested by using one of the known dependencies for the heat transfer coefficient in the disperse-annular flow,

$$f(q) = \alpha_{bc} \sqrt{R^{0,8} + \left(\frac{\alpha_{pb}}{\alpha_{bc}}\right)^2}$$

$$\alpha_{pb} = \varphi(\pi)q^{0,75}$$

where $R = 1 + [(\rho' - \rho'')/\rho''] x$, $\varphi(\pi)$ is a certain dependence on the relative pressure, critical parameters of p_{CR}, T_{CR}, and molecular mass. Value α_{pb} was defined as the heat transfer in the particle layer. Comparison of calculations with experimental data was performed on the extremely limited experimental database. Therefore, it was difficult to evaluate the model's reliability even for the considered conditions. Nevertheless, rejecting the microlayer evaporation schematic and attempting to apply the known hydraulic and heat transfer correlations of the boiling liquid flow inside tubes for the vaporization in the capillaries could be rational.

The classification of porous coverings on the limited experimental database was presented [237]. Aleshin et al. [237] supposed that classification of the porous coverings should precede the creation of the model, and suggested to base it on the following parameters: (1) relative micro size equal to the ratio between the layer characteristic size (particle diameter, pore diameter, etc.) and some macro size (e.g., layer thickness) and (2) the effective porosity, ε_E. The latter can be determined by comparing hydraulic losses in the considered layer ΔP_E with the standard layer ΔP_{ST} with known porosity ε_{ST}. The comparison yields

$$\left(\frac{\varepsilon_{ES}}{\varepsilon_E}\right)^3 \frac{(1-\varepsilon_E)^2}{(1-\varepsilon_{ES})^2} = \frac{\Delta p_E}{\Delta p_{ES}}$$

Classification of coverings [237] is presented in Table 5.4.

Single-scale (thin-layered) coverings and two-scale (thick) coverings exist according to Aleshin et al. [237]. The first one can only be impermeable, whereas the second one can both be semipermeable and permeable. The heat transfer mechanism for pool boiling on surfaces covered by these coatings for single-scale layers appears the same as for boiling on surfaces without coverings. The special shape and size distributions of set vaporization centers could explain the existing peculiarities.

Two-scale coverings ($1/\delta_{CS} < 1$) can be nonpermeable, semipermeable, and permeable depending on the effective porosity. The layer is permeable at porosity values close to 1. Vaporization begins and circulation of vapor and liquid develops when heat flux reaches the value sufficient for the initiation of boiling.

The classification [237] and the corresponding imaginations about vaporization on porous structure mechanisms coincide with some elements of the classification system suggested in Section 5.4.

TABLE 5.4

Classification of Porous Coverings

Porous Coating	$1/\delta_{CS}$	ε_e	Permeability	Phase Border Location	Heat Removal Peculiarities
Single scale	1	—	—	on the surface	boiling on surface of the special shape with sizes distribution of set vaporization centers
		$\varepsilon_e^n \ll 1$ <1	impermeable	on the surface	boiling on surface of the special shape with sizes distribution of set vaporization centers
Two scale	<1	$\varepsilon_e^s \le \varepsilon_e \le \varepsilon_e^n$	semipermeable	in the coverings range	combination of permeable and impermeable covering's effect
		$\varepsilon_e^n \le 1$	permeable	near the covered surface	liquid film evaporation in the covering's base and the mass transfer existence in the covering

Research activities conducted in recent years have presented the most essential improvements in the physical imaginations and models of the considered processes. The physical imaginations and the heat and mass transfer models of vaporization on porous coverings submerged in liquid were performed [45, 238–240]. These theoretical and experimental studies were generalized by Kovalev and Soloviev [36].

The physical imaginations [36] are as follows. The porous structure superheat creates vaporization inside the large pores. The appearing vapor channel is filled by the vapor in such a manner that complete dry-out of walls and bottom of the vapor channel occurs. Evaporation that takes place from the meniscuses in places of small pores exits to the main vapor channel. The number of vapor channels grows with the increase in specific heat flux and velocities of vapor and liquid. Thus, liquid penetrates inside the porous structure through the small pores and vapor flows in countercurrent to liquid by the vapor channels, that is, vapor and liquid countercurrent flows occur along separate channels sustaining the dynamic phase equilibrium (Figure 5.43).

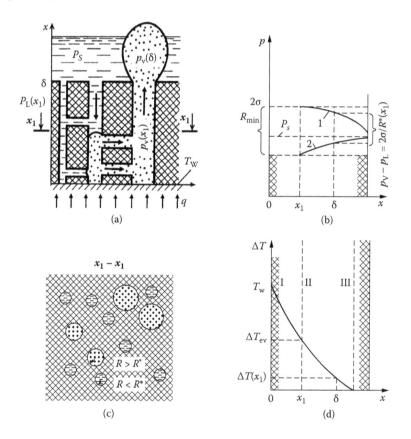

FIGURE 5.43 Model of heat transfer at boiling on CPS [36]. (a) Schematic of phase flows. (b) Pressures changing across coating's thickness: 1, vapor; 2, liquid. (c) Horizontal cross section of the porous layer $x_1 - x_1$. (d) Temperature changing of porous coating's skeleton across the thickness: I, drying layer; II, evaporation; III, saturation.

Such hydrodynamic model repeats the model [231] developing the physical imaginations of heat and mass transfer mechanism. It was assumed that heat is transferred by heat conduction over the porous structure's skeleton to the liquid's meniscus [36]. The liquid completely moistens the porous structure material.

The one-dimensional thermal conductivity problem was considered. It was suggested that the structure skeleton changed only into the wall normal direction. The pressure excess under the liquid determined the corresponding liquid meniscuses curvature and defined for every cross section X some border radius curvature $R^*(X)$ with respect to the formula: $P_M(X) - P^*(X) = 2\sigma/R^*(X)$. In the cross section X, all pores with radii $R < R^*$ are filled by liquid, but zones with $R > R^*$ are filled by vapor. The cross sections for liquid and vapor are functions of R^* and, consequently, they should change over the structure thickness.

The main correlations included the hydrodynamic and heat conduction equations [45]. The hydrodynamic equation had the following form:

$$\frac{dR^*}{dx}\frac{2\sigma}{(R^*(x))^2} - \frac{dR_{Br}}{dx}\frac{G^2(x)f(R_{Br})}{\varepsilon^2}\left(\frac{1}{p'\varphi^3(R_{Br})} + \frac{1}{p''(1-\varphi(R_{Br}))^3}\right)$$

$$= \frac{G(x)}{\varepsilon^2}\frac{d(G(x))}{dx}\left(\frac{1}{p'(1-\varphi(R_{Br}))^2} - \frac{1}{p''\varphi^2(R_{Br})}\right) + \frac{v'G(X)}{KF(R_{Br})}$$

$$+ \frac{v''G(X)}{K(1-F(R_{Br}))} + \frac{\beta''G^2(x)}{p''(1-\varphi(R_{Br}))^2}$$

where $F(R_{BR}) = \int_{R_{min}}^{R_{BR}} R^2 f(R)dR\left(\int_{R_{min}}^{R_{max}} R^2 f(R)dR^{-1}\right)$; R_{min} is the minimal radius of the pore; $G(x)$ is the vapor mass flow rate; $\varphi(R)$ is the integrated distribution function; $f(R) = d\varphi(R)/dR$; p'' is the vapor density; R_{Br} is the minimum radius of pores operating as the VRC in the given regime.

The temperature distribution over the porous structure thickness was prescribed by the heat conduction equation:

$$\lambda_{SC}\frac{d^2T_{SC}}{dx^2} + q_V - q_{EV} = 0$$

where T_{SC} is the porous structure temperature in cross section x, q_{EV} is the capacity of internal heat sources (when internal heat generation is presented), and q_V is the volumetric heat removal accounting for liquid evaporation from meniscus surfaces.

Soloviev and Kovalev [238] proposed to calculate the density of heat sources by the model of evaporation that they developed and discussed in the detail by other authors. The fundamental correlation of the model [238] is

$$q_{EV} = \left(T_{SC} - T_s\right)5.6\Pi^2\sqrt{\alpha_{EV}}\,\lambda'\left(\frac{\pi}{2} - arc\,tg\sqrt{\frac{\alpha_{EV}\delta_{**}}{\lambda'}}\right)\frac{\varphi(R)}{\sqrt{R^*}}$$

where α_{EV} is the phase transition heat transfer coefficient, R^* is surface force effective action radius, δ_{**} is the evaporated film thickness, and Π denotes perimeter.

The physical representation of this model is shown in Figure 5.44. The correlation between the hydrodynamic and heat conduction equations was created by the assumption that entire internal heat sources were transformed into the latent heat:

$$G(x) = \frac{1}{r} \int_0^x q_{EV} dx$$

The authors solved the problem on the basis of the proposed model of vaporization on porous structure coverings used in their experimentation.

Let us consider the research done on analytical heat transfer processes before its discussion. It was assumed that different zones exist corresponding to conditions of evaporation from the curved interface:

- Equilibrium film zone, $0 < \delta < \delta_0(T_W)$, where $\delta_0 = \lambda'/\alpha[1 - (T_W - T_s)r\rho''/T_s P_s]^{1/6}$ (without evaporation, the liquid film with thickness δ_0 is maintained by the adhesion forces).
- Partially evaporating film zone, $\delta_0 < \delta \leq \delta_*$ (liquid flow is realized due to the effect of disjoining pressure gradient).

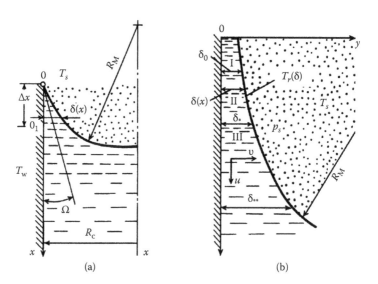

FIGURE 5.44 The evaporating meniscus model [36]. (a) Isothermal (nonevaporating) liquid meniscus in the cylindrical capillary. (b) Thin-film part of evaporating meniscus near angle point: I and II, equilibrium and evaporating films regions, respectively; III, meniscus; R_M, meniscus radius; R_C, single pore's (capillary) radius; $T_r(\delta)$, liquid evaporation surface temperature.

- Partially evaporated film zone, $\delta_* < \delta \le \delta_{**}$ (the adhesion forces influence is not essential and the liquid flow is realized due to the effect of capillary pressure gradient).
- Meniscus zone, $\delta > \delta_{**}$ (pressure drops in the liquid film are considerably less than capillary pressure, and the meniscus shape is almost spherical).

The continuity, momentum, and energy equations were used with some assumptions about the liquid films thickness determination, δ_*, δ_{**}, and δ_0. Consequently, the momentum equation for the meniscus film part appears in the form

$$\delta^4 \frac{dp}{dx} = 4v'\lambda'x\frac{T_{SC} - T_s(\delta)}{r}$$

The correlation for whole heat flux value removed due to the liquid evaporation from a single capillary was obtained [238] as

$$\Delta Q = \pi R_C (T_W - T_s)\left[3.2rp''\tilde{R}\sqrt{rp'b\frac{\delta_*^4 - \delta_0^4}{v'\lambda''(T_W - T_s)}} \right.$$
$$\left. + 3rp''\sqrt{\frac{\tilde{R}}{T_s}}\left(\frac{ro\delta_{**}^5}{v'\lambda'(T_W - T_s)}\right)^{1/4} + 2.8R_C\sqrt{\frac{\alpha_{EV}\lambda''}{R_M}}\left(\frac{\pi}{2} - arctg\sqrt{\frac{\alpha_{EV}\delta_{**}}{\lambda'}}\right) \right]$$

where R is the gas constant and R_M and R_C are the meniscus and pore radii, respectively. The capillary system average heat transfer coefficient α_S is determined as

$$\alpha_s = \frac{\Pi}{T_W - T_s}\int\frac{\Delta Qf(R_C)dR_C}{\pi R_C^2}$$

Unfortunately, the heat transfer model [238] did not obtain direct experimental confirmation.

Theoretical researches developed by several research teams [45, 238–240] and generalized by Kovalev and Soloviev [36] became extremely popular and recognized by many specialists [184–187, 241]. However, there were many serious drawbacks and contradictions in these models. It did not allow considering these models to be considered correct with respect to the physical nature of the influence of physical, geometrical, and regime parameters on heat transfer intensity at vaporization from the porous structure of the heating surface's coating. It meant that some of the aforementioned models and their physical imaginations required serious modifications. Let us consider the main critical comments relating to the models presented in Ref. [36] and devoted to vaporization on surfaces covered by porous coatings, and boiling inside the porous media.

1. Accounting for "temperature depressions" associated with capillary pressure, hydraulic losses, vapor flow continuity destruction, and other factors was absent in models [36] devoted to evaporation and boiling on porous coverings. The most serious and most important of these factors was "temperature depression" associated with the action of capillary forces. Accounting for this factor meant transition from traditional driven temperature drop ΔT (equal to $T_W - T_s$, where T_s is the saturation temperature in the liquid volume) to the value ($\Delta T - \Delta T^*$). This transition in the actual vaporization conditions in the internal volumes of the porous structure caused major changes in the theoretical models of Ref. [36] as follows:
   - Estimations of activated pores radii ($R > R_A$) are shifted to the larger pore sizes.
   - Probability of effective vaporization from meniscus surfaces appearing at intersections of liquid and vapor channels (by the model [36]) decreases abruptly due to meniscus curvature growth, significant increase in the nucleation temperature drop ΔT^*, and decrease in the difference ($\Delta T - \Delta T^*$) down to zero.
   - Vaporization from meniscus located at some distance from the wall appears quite rare due to considerable decrease in the initial temperature drop ΔT, and the driven temperature drops ($\Delta T - \Delta T^*$) become equal to zero or even result to negative values.

   Therefore, volumetric form of the vaporization assumed in [36] is physically impossible. If it becomes necessary to account for additional "temperature depression" terms, this drawback [36] will be more significant.

2. Heat supply by the elementary evaporation process is realized (by the model [36]) by heat conduction over the liquid thin film (Figure 5.44). Heat conduction over the structure's skeleton was absent.

   Such physical nature of evaporation inside porous media could be assumed only for high-conducting liquids in low-conducting skeleton, that is, $\lambda' \gg \lambda_{SK}$. However, contrary conditions when $\lambda_{SK} \gg \lambda'$ or, at least, $\lambda_{SK} \gg \lambda'$, appear typical for relevant cases. Heat transfer over the skeleton in these conditions becomes the determining process. This is because special surface thermal resistance appears inside the skeleton layer near the wall. This thermal resistance becomes more significant than smaller meniscus characteristic size. Detailed analysis of this phenomenon and the corresponding correlation for calculation have been presented previously. It is easy to understand that this disadvantage of the model in Ref. [36] is a major issue.

3. The model presented by Soloviev and Kovalev [238] contains many contradictions. It is not clear how four quite different zones could be prescribed by a single equation of the film flow. How could extensions of different film flow regimes be determined? How could border values of film thickness δ_*, δ_{**} be measured?

4. Many experiments demonstrated that in the common case, different vaporization regimes with various physical mechanisms are possible. Therefore, it is not feasible to prescribe different regimes by one model without alteration of the real physical nature.

5. Regimes of the simultaneous presence of vapor channels with internal dry-out along the height and channels filled by liquid were not observed by the visualizations. Initiation of the dry-out, if any, causes formation of some minimum stable vapor layer or complete porous structure drying along the porous structure height. These imaginations were confirmed by the HTI experiments.

6. Pressure drop $(p' - p'')$ decreases along the porous structure thickness and, consequently, R^* increases. This proves existence of the VRC having liquid in some cross section. It contradicts with the assumption [36] on the unlocked form of the vapor channels.

7. The appearance of the vapor phase and nucleation does not depend on the wall superheat and the temperature drop inside the porous structure [36].

8. The following regularities of heat transfer at vaporization on porous coatings could not be explained by the model [36]:
   - Transition from the porous structure to the pool boiling thermal regimes.
   - Filling by vapor of the whole porous structure in the moment of interlayer crisis initiation.
   - Simultaneous vaporization in every elementary cell of the porous structures.

Thus, to conclude the discussion of this model [36], it is necessary to underline that the single reason for the accounting of this model is associated with the difficulties in explaining the countercurrent vapor and liquid flow inside the single channel due to the necessity of maintaining the specific form of the interface pressure distribution. However, the authors did not account for the following conditions to remove this contradiction:

- Countercurrent flow is achievable at low velocities and low-pressure gradients, including hydrodynamic regimes with the interface elimination and throwing out of liquid.
- Liquid flow is possible not only in the liquid film, covering the skeleton, but also inside the liquid channels and liquid bridges, including countercurrent flow.
- Pulsation character of the liquid flow is feasible. Then, increase of meniscus curvature is accompanied by increase in liquid velocity and increase in liquid film thickness associated with decrease in capillary potential. The last one causes deepening of the meniscus, that is, correlation between $(p' - p'')$ and σ/R is changing.

Different types of simplified heat transfer models at vaporization on coated surfaces were performed in several studies [243–247]. These models did not account for the hydrodynamic phenomena, whereas it was assumed that local regularities of heat transfer correspond to correlation $q = A(T_W - T_s)m$, where T_W is the local temperature of the wetted skeleton. The values A and m are empirical constants depending on the structure parameters, technology, heat carrier, and skeleton's thermophysical properties.

Thermal regimes with internal boiling preceding the interlayer crisis were considered in Refs. [243–247]. This approach was developed in Refs. [243, 247].

Its principal essence is as follows. The chosen elementary cell consists of the single conditional element of the porous structure skeleton. In the boiling process, heat is removed through this element to the surrounding vapor–liquid mixture in accord with the transfer local regularity $q = A(T_W - T_s)m$. The initial energy equation for this case with respect to the known skeleton thermal conductivity λ_{sk} and porous structure specific surface S_Y

$$\frac{d^2 T_w}{dy^2} - \frac{AS_V}{\lambda_{sk}(T_w - T_s)^m} = 0$$

At the boundary conditions,

$$y \to \infty; \quad (T_W - T_s) \to 0; \quad y \to \infty; \quad dT_W/dy = 0$$

where y is the coordinate perpendicular to the heat transfer surface.

The boundary condition $y \to \infty$ corresponds to thick structures.

The third type boundary conditions were suggested in Ref. [243] for the thin-layer structures

$$\lambda_{sk} \frac{dT}{dy} = A(T_w - T_s)^m$$

For "thick" structures, Rosenfeld [243] obtained

$$q_0 = \varepsilon A(T_{w0} - T_s)^m + \sqrt{\frac{2A\lambda_{sk}S_V}{m+1}}(T_{w0} - T_s)^{(1+m)/2}$$

and for thin-layer structures

$$q_0 = \varepsilon A(T_{w0} - T_s)^m + \sqrt{\frac{2A\lambda_{sk}S_V(T_{w0} - T_s)^{(1+m)/2} - (T_{wt} - T_s)^{(m+1)}}{m+1} + A^2(T_{w0} - T_s)^2}$$

where T_{W0} is the skeleton temperature close to the bottom, $(T_{W0} - T_S)$ is the superheat of heat transfer surface covered by the porous coating, and $(T_{WT} - T_S)$ is the superheat at the external surface of porous structure.

The wall superheat could be calculated based on the common consideration of the skeleton heat conduction and eat transfer from the external surface by the correlation $q = A(T_{WT} - T_S)^m$. The value S_V for the sintered porous structure could be defined as $S_V = 6 (1 - \varepsilon) / d_p$ m^2/m^3, where ε is the structure's porosity and d_p is the diameter of sintering particles.

Selecting the value A and assuming m in the range 2–3, Rosenfeld [243] obtained good agreement between calculation results and experimental data presented in Refs. [244, 245]. This approach to heat transfer mathematical modeling combines simplicity and rationality.

However, restrictions of this approach are clear. Its main drawback is associated with its failure to account for the hydrodynamic phenomena. In this relation, the model presented by Shaubakh et al. [248] and Singh et al. [249] appears free of such weakness. It was developed to cover numerical calculations and direct comparison with the representative experimental data. Hence, it can be considered as the next significant step in the theoretical prescription of vaporization inside porous structures. Let us consider main statements of the model [248]:

1. It was assumed that liquid flows inside some artery with hydraulic resistance determined and accounted for as some function of flow rate. It was related to the wetted porous structure in the heating zone.
2. Heat flux on the heating surface was introduced as a function of coordinate and time.
3. Porous structure was divided into separate elementary volumes with respect to the finite elements method.
4. Such initial parameters as the pore size, permeability, porosity, skeleton thermal conductivity, etc., could be assumed as functions of the porous structure thickness and length.
5. It was also supposed that the vapor and liquid present inside every finite element of the porous medium and liquid flows mainly along the wall, whereas vapor moves upright to liquid flow.
6. Linear pore distribution with respect to vapor quality φ_C inside the porous medium was assumed. Correspondingly, pores defining the hydrodynamic phenomena have a radius (R_C) of $R_c = R_{max} - (R_{max} - R_{min})\,\varphi_c/\cos\Theta$, where Θ is the wetted angle (usually, $\cos\Theta = 1$).
7. The liquid and vapor flow inside the porous media correspond to Darcy's law and are independent one from each other. They are defined by the relative permeability model. The correlations for phase permeability are as follows: for vapor, $K'' = K_0\varphi^2$, and for liquid, $K' = K_0 (1 - \varphi)^3$. K_0 is the entire porous structure permeability and φ is the vapor quality inside the pore medium.
8. The mass and energy conservation equations were written for every finite element.
9. It was also assumed that the heat transfer between finite elements took place according to the linear effective thermal conductivity model.
10. The interface pressure drop $p'' - p' = \Delta p_0$ for every finite element should correspond to capillary pressure scale, $\Delta p_0 = 2\sigma\cos\Theta / R_C$.

Calculations by the model revealed satisfactory agreement with experimental data. However, computing by this model represents a time-consuming process. Such a disadvantage complicates representative checking of the model's reliability and restricts its practical application. It is also necessary to note the other principal drawbacks of this model:

- Correlations determining relative phase permeability used in the model [248] did not account that $K'' = 0$ at some small vapor quality values, and the absence of liquid permeability at some small values $(1 - \varphi)$.

- Internal heat transfer correlations are not accounted for in temperature's depression terms.
- Conditions determining phase concentration inside the porous structure are not present.
- Conditions of transition from one phase structure to another were not determined for a given porous structure, including crisis and hysteresis phenomena.
- The linear correlation between the average vapor quality and characteristic size of activated pores does not correspond to the reality of the VRC operating.

Therefore, the common drawback of all known models is associated with the lack of universal approach to the justification of extensive experimental regularities obtained for the heat transfer at vaporization on coated surfaces.

For this reason, the author, along with his team, during many years consistently developed and improved physical imaginations and models of the considered processes with the goal of achieving a consistent and free-from-contradictions explanation of experimental heat transfer regularities from the general position. Initially, the author suggested a model prescribing only thermal regimes preceding the crisis for coatings without pore distribution (with uniform pores in the first layer near the wall, such as screen and uniform size particles layers, etc.). The second-stage model improved imaginations of the first stage for porous coverings with pore size distribution, including the first layer near the heating wall.

The first-stage model was reported in Ref. [71]. Let us consider the second stage model presented in Refs. [46–48].

As had been observed in the first-stage model [71], following Equation (5.15) with respect to the problem of heat transfer through the microfin surface, the correlation between the specific heat flux q and the wall superheat $\Delta T = T_W - T_S$ at vaporization on the coated surface is

$$q\varphi(p_s) \approx const \left(\frac{\Delta T - \Delta T^*}{L_0} \right)^m , m = 0.8...1.0 \qquad (5.21)$$

Equation (5.21) does not account for the existence of different pore sizes, L_0, in the porous covering. It allowed satisfactory generalization of the experimental data only for the screen and particle layers with approximately uniform pore sizes near the wall, that is, uniform along the heating surface. For uniform structures, it was natural to suppose that the representative cell could be restricted to the single pore. Consequently, Equation (5.21) justified for screens with different pore sizes could be considered as a correlation determining some local heat transfer regularity inside the vapor-generating pores.

The other question lies in relating regularities of switching on pores for porous coatings with pore distribution. The condition for vapor-generating pores with diameter $D_i \geq 4 \ \sigma T_S / (r\rho''\Delta T^*)$ can be considered as necessary, but not sufficient. It follows from Equation (5.21) that the local heat transfer intensity depends nonmonotonously on characteristic size. Actually, if D is the pore diameter, $L_0 = kD$, where k is the proportionality coefficient depending on the representative cell's shape, and

$\Delta T^* = 4\sigma T_S / (r\rho''D)$, then $(\Delta T - \Delta T^*)/L_0$ has the maximum value for pores with diameter $D_0 = 8\sigma T_S/(r\rho''\Delta T)$. It means that in pores with diameter D_0, vaporization is possible at minimal irreversibility of superheated liquid layer in the near-wall zone.

Naturally, it could be assumed that at given value P_s, such pores will be activated, which will maintain the minimum wall superheat, that is, the maximum of thermal stability. Therefore, the necessary and sufficient condition determining the character of switching on pores is

$$\frac{\partial q}{\partial D_i} = 0 \tag{5.22}$$

In the case of the existence of some pore distributions, λ_E, f_i, and Π_i (effective thermal conductivity, skeleton cross section, and perimeter of the elementary cell, respectively) are the functions of D_i. The real discrete pore's distribution $f(D_i)$ could be prescribed by the distribution density function $f(D_i)$. Assuming that the real pore distribution is characterized by change in diameter for some of the pores ΔD_i, consider the main dependencies for values ε', λ_E, and L_0 on D_i:

$$N = \frac{4\varepsilon'}{\pi \overline{D}^2} \tag{5.23}$$

where $\overline{D}^2 = \displaystyle\int_{D_{min}}^{D_{max}} f(D_i)D_i^2 dD_i$; or for the discrete pores distribution,

$$\overline{D}^2 = \sum f(D_i)D_i^2 \Delta D_i; \tag{5.24}$$

$$\varepsilon' = \varepsilon \frac{D_i^2}{\overline{D}^2} f(D_i)\Delta D_i; \qquad 1-\varepsilon' = (1-\varepsilon)\frac{D_i^2}{\overline{D}^2} f(D_i)\Delta D_i \tag{5.25}$$

Assuming a rectangular vapor-generating cell and cylindrical vapor channel allows the following correlation for one pore with diameter D_i:

$$f_i = \frac{\pi \overline{D}^2}{4\varepsilon f(D_i)\Delta D_i}; \Pi_i = \pi D_i \tag{5.26}$$

and the characteristic size

$$L_{0i} = \left[\left(\frac{f_i}{\Pi_i}\right)^4 D_i\right]^{0,2} \tag{5.27}$$

Determination of the effective thermal conductivity of the wetted capillary structure skeleton yields the following correlation with respect to the existence of the vapor-generating cells occupying the fraction of capillary structure equal to ε':

$$\lambda'_E = \lambda_E - \lambda'\varepsilon'. \tag{5.28}$$

Substituting Equations (5.23)–(5.27) into (5.21), and accounting for condition (5.22) yields the correlation determining the character of switching on pores with diameter D_i

$$A_1 \frac{\partial}{\partial D_1}[\ln f(D_i)] + A_2 \frac{1}{D_i} = \frac{20\sigma T_s}{rp''(\Delta T - \Delta T^*)D_i^2} \tag{5.29}$$

where A_1 and A_2 are variables depending on the values λ_E, λ', and $\varepsilon(D_i)$. At small values, $\varepsilon'(\varepsilon' \to 0)$; $A_1 \to -4$; $A_2 \to -3$. At the consequent pores switching on, the value $\varepsilon' \ll \varepsilon$, that is, condition of value ε smallness is accomplished, as a rule.

Equation (5.29) can be considered as the correlation determining some pores switching on regularly.

Therefore, heat transfer regularities at vaporization in capillary structures could be prescribed in general case by Equations (5.21) and (5.29) with respect to Equations (5.23)–(5.28). Common consideration of these equations requires knowledge of pore distribution law and its border (maximum and minimum pore diameters) dispersion. Usually, this information is not available; sometimes, very limited data are given.

These characteristics for fiber–metal structures were reported in several studies [19, 171–180], allowing the assumption that real complex pore distributions could be approximated via simple functions, for example, by normal or logarithmic distributions. Therefore, corresponding solutions for Equations (5.21) and (5.29) were obtained. It is also known that, for these distributions, knowledge of the distribution's borders alone (D_{max}, D_{min}) is sufficient for prescription of the whole the distribution function $f(D_i)$.

Considering Equations (5.21) and (5.29) gives the dependence determining the heat transfer regularities at vaporization on porous structures as

$$q\varphi(p_s) = const \frac{\varepsilon^{2/3} \lambda_3^{1/3}}{D^{5/6}} \Delta T^{5/6} \psi(X, A_0) \tag{5.30}$$

where

$$\varphi(p_s) = \left(\frac{v'}{r\sigma\delta'^3}\right)^{1/6}; X = \frac{D_i}{\bar{D}}; A_0 = \frac{\bar{D}}{D_0}; D_0 = \frac{4\sigma T_s}{rp''\Delta T} \tag{5.31}$$

$$\bar{D} = \int_{D_{min}}^{D_{max}} f(D_i)D_i dD_i$$ is the most probable pore diameter.

The normal pore's distribution law is

$$\Psi(X, A_0) = \left(\frac{X - A_0}{X^2}\right)^{5/6} X^2 \exp\left[-18(0.5 - X)^2\right] \qquad (5.32)$$

The logarithmic pore's distribution law,

$$\Psi(X, A_0) = \left(\frac{X - A_0}{X^2}\right)^{5/6} \{X^{1.34} \exp[-2.3(\ln X)^2]\} \qquad (5.33)$$

Substituting the function $f(D_i)$ into Equation (5.29) yields for the normal distribution law,

$$X^3 - X^2(0.5 + A_0) = X(0.5A_0 - 0.02) - 0.014A_0 = 0 \qquad (5.34)$$

and for the logarithmic distribution law,

$$5 + (a_0 X - 1)(3 - C_1 \ln X) = 0; C_1 = \frac{36}{D_{max}/D_{min}} \qquad (5.35)$$

Therefore, general heat transfer regularities for conditions of vaporization on CPS are determined by considering Equations (5.30) and (5.34), if the real pore distribution could be approximated by the normal distribution law, and Equations (5.30) and (5.35) in case of the logarithmic law. The generalized correlation appears as follows

$$Y \equiv q\psi(p_s) = const \times \psi_0(A_0, D_{max}/D_{min}) \qquad (5.36)$$

where function ψ_0 depends on the selection of pore distribution law.

The dependencies of functions ψ_0 on the parameter A_0 at typical ratio $\gamma = D_{max}/D_{min}$, 100, and cited distribution laws are presented in Figure 5.45. At given values λ_E, ε, D_{max}, and p_s, it is equivalent to creating the correlation $q = f(\Delta T)$ in the form

$$q = C\Delta T^m \qquad (5.37)$$

Parameter A_0 determines the relative average pore diameter of the CPS. The region $A_0 \gg 1$ relates to the consequent vaporization (increase in the A_0 is accompanied by smaller pores switching on). As shown, in this region, increase in value of A_0 causes essential change in Equation (5.37). Therefore, m varies from 4 to 1.4, and when $A_0 \to 10$, $m \to 0$. It coincides well with extensive experimental data reported by research teams from Kiev National Technical University.

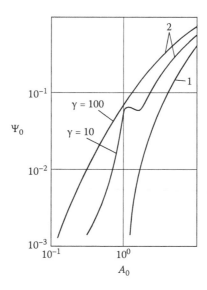

FIGURE 5.45 Dependency of parameter ψ_0 on the relative size of the most probable pore's A_0 (calculation by Equations [5.30] and [5.31]), pores distribution law, and ratio $\gamma = D_{max}/D_{min}$. 1, the Gauss pores distribution $f(D_i)$; 2, the logarithmic pores distribution.

As follows from Equations (5.36) and 5.37), such values of A_0 exist, where function ψ_0 is uncertain, that is, the hysteresis phenomenon should be observed in the related region. It is also well confirmed by experiments.

It should be noted that, in many cases, the pore distribution could not be prescribed (even approximately) by continuous distribution function — that is, there are some pores sizes at which change in value D_i is discrete. Then, varying the thermal regime of the heating surface would be accompanied by the jump transition from one region of D_i to another, and it could be considered as hysteresis effects. Analysis of Figure 5.45 also demonstrates that if pore distribution exists even in the region $A_0 < 1$, stable thermal regimes are likely restricted by $(A_0)_{min}$ depending on the distribution law and the ratio D_{max}/D_{min}. Pores with size $D_i \geq \bar{D}$ operate in the region corresponding to $A_0 \leq 1$.

With respect to the aforementioned analysis, it is interesting to note the following observations. When porous coating is positioned on the heating surface containing its own pores, some cooperative pore distribution appears, for example, screen capillary-porous coverings on the surface with roughness. Different correlations between the ranges ΔD_i of pore parameters changing in capillary-porous coatings and heating surface and common different pores distribution laws are possible.

In particular, for the woven screen structures in the near-wall zone, pore size is practically unchanging and equal to the cell size a. However, the heating surface has its own pore distribution. Therefore, in this case it will be necessary to suppose the existence at least of two different heat transfer regimes with the considerably different values of m in Equation (5.37). This conclusion coincides well with experimental data on heat transfer at vaporization on screen capillary structures presented in Refs. [75–79] (Figures 5.7 and 5.8).

Depending on the conjugation conditions, pore changes are observed for screen structures and heating surface, as both smooth and abrupt transition between these thermal regimes are possible. Such different transition forms have also been observed in experiments [75–79] (Figure 5.8). Therefore, the suggested approach and correlations obtained on its base yielded an explanation without contradictions all known peculiarities of heat transfer at vaporization inside CPS.

The most representative and systematic experimental results on heat transfer at vaporization on CPS differed essentially by manufacturing technology, structure, and thermophysical characteristics at known values of D_{max} and D_{min}, and were used for direct comparison with the suggested physical imaginations and corresponding generalized correlations. There were experimental data at boiling on the metal–fiber coverings [19, 173–180], data at boiling on the sintered coatings [43–45, 54–57], and experimental data on heat transfer on the plasma and gas-sprayed CPS [194, 196].

It was also assumed that for small values, $\gamma = D_{max} / D_{min} \leq 4$–5, pore distribution corresponds to the normal distribution law, that is, $\bar{D} = 1/2\,(D_{max} + D_{min})$. At $\gamma \geq 10$, use of the logarithmic law distribution was recommended. In cases when corresponding information on values of λ_E was not available, their estimations were obtained by suggestions presented by Dulnev and Zarichnyak [15].

Comparison of calculations by Equation (5.36) and aforementioned experimental data are presented in Figure 5.46. The present concurrence between calculations and experimental results could be assumed as satisfactory due to the following reasons: classified dependencies of heat transfer, as a rule, were described with essential scattering; CPS characteristics even manufactured by the same technology are hardly reproducible; and values of λ_E, D_{max}, and D_{min} were estimated very roughly in some cases.

Therefore, the second-stage model was created in such a manner that the first-stage model [71] was represented in its limited case, when pore distribution was absent in the near-wall zone and pores of identical size were presented on the contact surface.

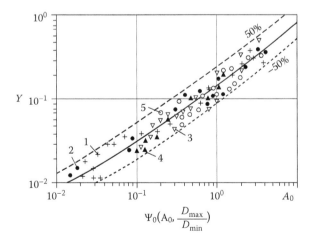

FIGURE 5.46 Generalization of experimental data by Equation (5.36). 1, Data from Refs. [54, 55]; 2, data from Ref. [19]; 3, data from Ref. [194]; 4, data from Ref. [191]; 5, by data from Ref. [195].

However, both first- and second-stage models were approved only for pre-crisis thermal regimes, when two-phase border layer inside the porous structures could be represented as a system of vapor-generating cells (channels) surrounded by the porous structure elements filled by liquid, and liquid dynamic films cover the structure skeleton down to channels' bottom, including heat transfer surface. It means that these models are not valid under conditions of intercrisis initiation and post-crisis thermal regimes. In addition, these models could not predict a major possibility of the existence of boiling regimes at different options of liquid supply: submerging, capillary feeding, and combined. The second model is not applicable for determining volumetric vapor quality in boiling regimes, possible regimes of two-phase vaporization heat transfer inside porous structure, and hydrodynamic conditions of their existence.

The third-stage model [49] eliminated the drawbacks of the second-stage model.

5.4 GENERAL MODEL OF VAPORIZATION PROCESSES ON THE COATED SURFACES (THE THIRD STAGE)

The following principles and methods, sufficiently general and covering a wide range of conditions of boiling on surfaces covered by porous coverings, were used with the goal of developing efficient calculation procedures of these processes [49]:

1. Classification of different thermal regimes and boiling conditions was suggested and justified.
2. One-dimensional models of filtration and heat and mass transfer were used.
3. Critical or maximum heat flux determination was performed for the whole range of thermal regimes at vaporization on coated surfaces covered by porous structures.
4. It was assumed that the maximum heat flux is attained when the liquid supply to some part or to the entire heating wall appears impossible.
5. The irreversibility minimum principle was used for determination of such factors as conditions of vaporizing pores, ratio of liquid, and vapor phases' concentration inside the porous media, and number of activated pores and their diameters.
6. It is necessary to classify the following types of conditions requiring their own mathematical model, that is, set of equations, boundary conditions, and constraints defining particular hydrodynamic and heat transfer regularities:
   (1) Vaporization on the heat transfer surface covered by the porous structure and submerged into the liquid. In such a case, vaporization occurs in the boiling regime (Figure 5.47a).
   (2) Vaporization on heat transfer surface covered by the porous structure under capillary feeding conditions at horizontal heat pipe's position, that is, excess liquid is absent on the external surface of porous structure (Figure 5.47b, c).

(3) Vaporization on heat transfer surface covered by the porous structure under the capillary feeding conditions at the horizontal position or at small inclinations ($\beta < 10°$) of heat pipe, when the liquid layer is presented in the foot zone of the external part of capillary structure (Figure 5.47d).

(4) Vaporization on heat transfer surface covered by the porous structure under the capillary feeding conditions at the vertical or inclined position of the heat pipe, when heat input end is located above heat output end, that is, the inclination angle $\beta < 0°$ (Figure 5.47e).

(5) Vaporization on heat transfer surface covered by the porous structure under the capillary feeding conditions at the vertical or inclined position of the heat pipe, when heat input end is located under heat output end, that is, the inclination angle, $\beta > 0°$ (Figure 5.47f).

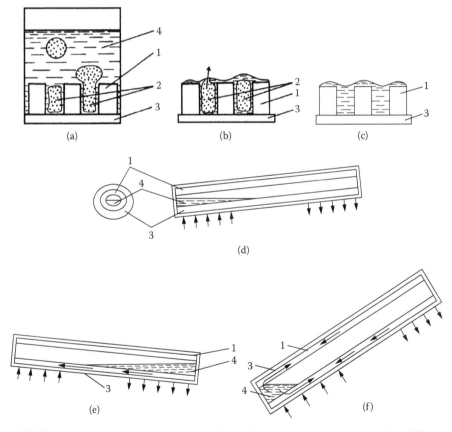

FIGURE 5.47 Classification of regimes of vaporization on surfaces covered by different porous coverings in capillary feeding and submerged conditions. 1, Skeleton; 2, vapor-generating channels; 3, heating wall; 4, liquid layer. (a) Vaporization inside submerged porous structure. (b) The same, at capillary feeding of liquid in the boiling regime. (c) The same, in the evaporation thermal regime. (d) Schematic of liquid supply to the "evaporating" or "boiling" structure. (e) The same, evaporating zone above the condensation zone. (f) The same, evaporating zone below the condensation zone.

7. In cases (6b)–(6e), as in boiling, both evaporative heat removal regimes are probable.

Vaporization inside the porous structure is possible only in the case when the wall superheat is enough for activation, at least, of some fraction of pores. The minimum superheat determining the pores stable activation was associated with the following required condition for the activation of pores with diameter D_i

$$(\Delta T)_{min} = T_w - T_s > \frac{2\sigma T_s}{rp''D_i} \tag{5.38}$$

This condition is necessary, but not sufficient, because at $T_W - T_s = (2\sigma T_s)/(rp''D_i)$, pores with diameter D_i could contain the vapor phase, although evaporation inside them is impossible. Thus, for determination of the low border of boiling thermal regime in the flooding case, it is required to identify those pores whose activation is more probable at a given wall superheat ΔT.

According to the aforementioned physical imaginations, in any stable boiling regime, those pores will be activated, an action which sustains the irreversibility minimum inside the two-phase boundary layer, minimum of the wall superheat at given heat flux, or maximum heat flux at given wall superheat ΔT.

The equation determining the correlation between the wall superheat ΔT in the boiling regime and diameter D_i of activated pores was obtained. The simplest form of this equation is

$$\frac{20\sigma T_s}{rp''D_i^2(\Delta T - \Delta T^*)} + \frac{3}{D_i} + 4\frac{\partial}{\partial D_i}(\ln f(D_i)) = 0 \tag{5.39}$$

$$\Delta T^* = \frac{4\sigma T_s}{rp''D_i} \tag{5.40}$$

In boiling regime, that is, $\Delta T \geq \Delta T^*$, substituting ΔT and thermophysical properties of the heat carrier (σ; r; ρ'', depending on the saturation temperature T_s) in Equation (5.39) yields the average diameter of activated pores, D_i. Determined values of D_i and ΔT allow the calculation of heat flux via recommendations on local heat transfer regularity as

$$q = const\left(\frac{r\sigma}{v'}\right)^{1/6}\sqrt[6]{\lambda'^3\,\lambda^2_{E0}(1-\varepsilon')^5}\left(\frac{\Delta T - \Delta T^*}{L_0}\right)^{5/6} \tag{5.41}$$

where $const = 0.094$, $L_0 \approx KD_i$.

Other correlations of this model were presented earlier by Equations (5.25)–(5.35).

The initial correlation of two-phase filtration at vaporization inside flooded porous structure is

$$\frac{4\sigma}{D_i} - \frac{v''qk_1\delta_{CS}{}^0}{K_0 f_2(\varphi)r} - \left(\frac{q}{r\varphi}\right)^2 \frac{1}{\rho''} - \frac{v'qk_2\delta_{CS}{}^0}{K_0 f_1(1-\varphi)r} \geq 0 \tag{5.42}$$

where k_1 and k_2 are the numerical coefficients accounting for thermal and hydro-dynamic irregularities along the perimeter ($k_1 = k_2 = 1$ when irregularities are not present); K_0 is the single-phase flow permeability of the porous structure; δ_{CS}^0 is the thickness of porous covering; $f_1 (1 - \varphi)$ is the relative phase permeability for liquid; and $f_2(\varphi)$ is the relative phase permeability for vapor.

The inequality (5.42) performs the necessary condition for the determination of φ. The sufficiency could be justified by minimizing hydraulic losses at two-phase filtration through the porous structure as a restricted case of the irreversibility minimization principle for stable steady states. This condition yields the following equation

$$\frac{v''k_1}{f_2{}^2(\varphi)}\frac{\partial(f_2(\varphi))}{\partial\varphi} + \frac{v'k_1 k_2}{f_1{}^2(1-\varphi)}\frac{\partial(f_1(\varphi-1))}{\partial\varphi} + \frac{K_0 q}{r\rho''\varphi^3\delta_{CS}} = 0 \tag{5.43}$$

Joint consideration of Equations (5.43) and (5.39)–(5.41) determines the values q and φ. If substituting these values to Equation (5.42) confirms the inequality, it means that the boiling regime could be realized in the present conditions (ΔT, T_s).

Stable vapor film appears in the near-wall zone of the porous structure due to the initiation of the "inter-layer crisis" at $q = q_{CR1}$. The vapor film thickness could be determined by the inequality (5.42). In this case, heat flux is given, whereas thickness of the wetted porous structure part, δ'_{CS} should be determined.

Therefore, values ΔT and q_{CR1} calculated from the equation set together with earlier determined value ΔT^* restrict the boiling regime range pre-crisis initiation. In this range, assuming superheat values, ΔT allows us to calculate heat fluxes and, consequently, to obtain the corresponding heat transfer regularities of boiling inside the porous structure.

Transition from the boiling thermal regime to the regime with stable vapor film near the wall requires changing ΔT to q and the two-zone model of wall superheat calculation.

Particular thermal regimes require corresponding changes in the mathematical model and computing algorithm:

1. In case of vaporization on a surface covered by thick porous layer, direct transition from the thermal regime with heat transfer by the effective heat conduction through the wetted porous structure and then by natural convection into the liquid layer to the thermal regime with interlayer crisis, when near-wall vapor layer is formed inside the porous structure. The initial superheat allowing pores activation in porous structure could be defined by pore diameter D_1

$$\Delta T_{min} = \frac{4\sigma T_s}{r\rho''D_1}, \quad \overline{D} < D_1 < D_{max}$$

The heat flux related to this condition is

$$q_1 = \Delta T_{min} \left(\frac{\delta_{CS}}{\lambda} + \frac{1}{\alpha_{CON}} \right)$$

Such thermal regime occurs when the corresponding superheat is less than the nucleation barrier, that is, $\Delta T < \Delta T_{min}$. It means that pore activation inside the porous structure causes the thermal regime with interlayer crisis.

2 In case of vaporization in the flooded layer at the heat flux q_{CR}, initiation of boiling crisis on the external surface of porous structure is possible. This moment could be determined according to recommendations [36] on the basis of two calculating limits.

The lower limit q_{CRmin} is determined similar to conventional surface by the Kutateladze–Zuber correlation as

$$q_{CR\,min} = 0.14 r \sqrt{p''} \sqrt[4]{g\sigma(\rho' - \rho'')}$$

The upper maximum limit q_{CRmax} is calculated by the Kovalev formula. Depending on the porous structure type, Kovalev and Soloviev [36] suggested several variants of the formula. The following correlation is preferable in the given case:

$$q_{CR\,max} = 0.52 r \varepsilon^{2.28} \sqrt{\frac{\sigma \rho' \rho''}{(\rho' - \rho'') R_{CL}}}$$

Here R_{CL} is the radius of the clamping pore equal in the first approximation to pore radius value averaged between the most probable and maximum radius values. The actual value q_{CR} is determined by the condition $q_{CRmin} < q_{CR} < q_{CRmax}$.

Theoretical calculation models for other types of conditions [(6b)–(6e)] could be created in an analogous manner with corresponding change in the correlations (5.42) and (5.43). The following thermal regimes are possible in the heat input zone: (a) boiling regime, (b) vaporization regime, (c) boiling regime accompanied by exhaustive throwing liquid out of the porous structure.

Temperature conditions of boiling thermal regime initiation ($\Delta T > \Delta T^*$) are defined by the correlation (5.38); pores activation conditions are determined by Equations (5.39) and (5.40); and corresponding heat fluxes are calculated by Equation (5.41). When the lower boundary limits of q and ΔT for boiling regimes are fixed, then the possibility of regime realization is determined by hydrodynamic equations. The basic form of these correlations is analogous to Equation (5.42), but essentially differs from it

$$\frac{4\sigma}{D_i} - \frac{v'' q k_1 \delta_{CS}{}^0}{K_0 f_2(\varphi) r} - \left(\frac{q}{r\varphi} \right)^2 \frac{1}{p''} - \frac{v' q k_1 k_2 L_{EV}{}^2}{K_0 f_1 (1 - \varphi) r \delta} - \Delta p_0 \geq 0 \qquad (5.44)$$

Here,

$$\Delta P_0 = \Delta P_T' + \Delta P'' - \rho' g L_{0E} \sin \beta \qquad (5.45)$$

ΔP_0 accounts for hydraulic losses due to liquid flow inside the porous structure along the transport and condensation zones ($\Delta P_T'$), vapor flow in vapor channel along the whole heat pipe length ($\Delta P''$), and influence of liquid hydrostatic pressure with respect to the interposition of heat input and heat output zones in the gravity field, that is, angle β. If $\beta > 0°$, the heat input end is located under heat output end and vice versa. L_{0E} is the effective length of the heat carrier transport in heat pipe; LEV is the length of the heat input zone.

Analysis of thermal regimes could be performed by accounting for the effects of other zones at given ΔP_0 and T_s. However, in any case, value ΔP_0 is a function of the total heat load Q and saturation temperature T_s, which is also dependent on the value Q. Correlation for ΔP_0 is essentially dependent on the heat pipe location in the gravity field and heat pipe filling conditions, that is, the presence or absence of excess heat carrier inside the heat pipe. In some cases, the structure of Equations (5.44) and (5.45) could be also changed.

Hence, before present analysis of the aforementioned types of conditions, it should be noted that validation of the inequalities (5.42) and (5.44) requires determination of vapor quality φ. As for the case of flooded structure, both in the capillary feeding conditions, some extremum value φ_{EXT} exists, corresponding to the minimum of hydraulic losses associated with vapor and liquid filtration along the heat input zone, and appearance of the stable steady boiling regime, when φ_{EXT} could be determined from the following equation:

$$\frac{v'' k_1}{f_2^2(\varphi)} \frac{\partial(f_2(\varphi))}{\partial \varphi} \delta_{CS} + \frac{v' k_1 k_2}{f_1^2(1-\varphi)} \frac{\partial(f_1(1-\varphi))}{\partial \varphi} \frac{L_{EV}^2}{\delta_{CS}} + \frac{K_0 q}{r \rho'' \varphi^3} = 0 \qquad (5.46)$$

In (5.46), similar to (5.42) and (5.43), values $f_1(1-\varphi)$ and $f_2(\varphi)$ are the relative phase permeabilities of porous structure for liquid and vapor, respectively. These values could be determined according to recommendations presented in Ref. [40]

$$0 \leq \varphi \leq 0.8 \qquad f_1(1-\varphi) = \left(\frac{0.8-\varphi}{0.8}\right)^{3.5};$$

$$0.8 \leq \varphi \leq 1 \qquad f_1(1-\varphi) = 0;$$

$$0.1 \leq \varphi \leq 1 \qquad f_2(\varphi) = \left(\frac{\varphi-0.1}{0.9}\right)^{3.5} [1+3(1-\varphi)];$$

$$0 \leq \varphi \leq 0.1 \qquad f_2(\varphi) = 0 \qquad\qquad (5.47)$$

Thus, according to (5.47), the boiling regime is restricted by the following range of φ values:

$$0.1 \le \varphi \le 0.8 \tag{5.48}$$

Analysis of correlation (5.44) demonstrated that such value of δ_{CS} exists, which corresponds to the local minimum of hydraulic losses at boiling on the heat input zone:

$$\delta_{CS_{ext}} = L_{EV} \left(\frac{v'}{v''} \frac{f_2(\varphi)}{f_1(1-\varphi)} k_2 \right)^{0.5} \tag{5.49}$$

If the combined solution of Equations (5.46) and (5.49) yields such value of $\delta_{CS;ext}$ that $\delta_{CS;ext} > \delta_{CS}^0$, it means that condition (5.49) is not satisfied. Then, only Equation (5.46) could be used for the calculation of φ accounting for the fact that $\delta_{CS} = \delta_{CS}^0$. When $\delta_{CS;ext} < \delta_{CS}^0$, calculation of φ is based on Equations (5.46) and (5.49) and $\delta_{CS} < \delta_{CS;ext}$. In this case, inequality (5.44) transforms to the following form:

$$\frac{4\sigma}{D_i} - \frac{v''qk_1\delta_{CS_{ext}}}{K_0 f_2(\varphi)r} - \frac{v''qk_1 \left(\delta_{CS}^0 - \delta_{CS_{ext}} \right)}{K_0 r}$$

$$- \left(\frac{q}{r\varphi} \right)^2 \frac{1}{\rho''} - \frac{v'qk_1 k_2 L_{EV}}{K_0 f_1(1-\varphi)r\delta_{CS_{ext}}^0} - \Delta p_0 \ge 0. \tag{5.50}$$

From the physical viewpoint, transition from δ_{CS}^0 to $\delta_{CS;ext}$ is a sequence of escalation of throwing liquid out until the liquid will fill the porous structure with the layer thickness equal to $\delta_{CS;ext}$. Effects of the third and fourth terms of the correlation (5.50) are usually insignificant in comparison to other terms.

Validation of the inequality (5.50) or (5.44) appears possible after determination of the value ΔP_0. When $\beta = 0$ and excess liquid on the external surface of the porous structure is absent, the value ΔP_0 could be calculated as

$$\Delta p_0 = L_{EV}q \frac{1}{rK_0\delta_{CS}^0}(L_{TR}v'(T_s) + 0,5L_C v'(T_1))$$

$$+ C_f''\{0,5(L_C + L_{EV}) + L_{TR}\} \frac{\Pi''\Pi_1^2 L_{EV}^2 q^2}{8r^2 S'''^3 \rho''} \tag{5.51}$$

where C_f'' is the friction factor in the vapor channel; L_C and L_{TR} are the length of the heat pipe condensation and transport zones, respectively; Π'' and Π_1 are the perimeters of the vapor channel and the porous structure inside the heat pipe; S'' is the cross section of the heat pipe vapor channel.

Neglecting the injection and suction effects within the boiling and condensation zones, both external surface roughness of the wetted structure yields the following formula to determining the friction factor:

$$C''_f = const/(Re'')^M \tag{5.52}$$

where $Re'' = qL_{EV}\Pi_1/r\mu''\Pi''$; $const$ and M are dependent on Re''.

In the case of heat pipe operating, the value of T_s could be obtained only from the experimental data, whereas it could be treated as a function of heat flux in other cases (e.g., in design calculations). Hence, it is natural to assume the heat removal temperature, T_R, and conditions of internal and external heat transfer by corresponding heat transfer coefficients α_1 and α_2 of the heat removal zone of the heat pipe or to calculate these values.

The correlation determining the value of T_s at given α_1 and α_2 is

$$T_s = T_R + \frac{q\Pi_1 L_{EV}}{\Pi_2 L_C}\left(\frac{1}{\alpha_1} + \frac{1}{\alpha_2} + \frac{\delta_w}{\lambda_w}\right) \tag{5.53}$$

where δ_W and λ_w are the wall thickness and thermal conductivity of heat pipe wall, respectively; Π_1 and Π_2 are the perimeters of the heat input and heat output zones of heat pipe, respectively; T_R is the temperature of the ambient cooling of the heat pipe. The value $\overline{T}_1$ is

$$\overline{T}_1 = 0.5\left(T_s + T_R + \frac{q\Pi_1 L_{EV}}{\Pi_2 L_C \alpha_1}\right) \tag{5.54}$$

Naturally, correlation of T_s and q requires the use of the iterative procedure for their calculation.

If the combined solution of correlations (5.44)–(5.54) yields that at $\Delta T = \Delta T_1$ inequality (5.44) or (5.50) is not accomplished, this means that the boiling thermal regime is impossible when $\Delta T \geq \Delta T_1$, that is, at $\Delta T = \Delta T_1$ and consequent value of q, the "boiling limit" is reached and capillary structure is completely filled by the vapor.

In addition, it means that at $\Delta T < \Delta T_1$, the evaporative thermal regimes could be realized. The correlation between q and ΔT in the evaporative thermal regime is as follows

$$q = \lambda_E \Delta T / \delta_{CS}^0 \tag{5.55}$$

The capillary heat transport limit of heat pipe could be examined as

$$\frac{4\sigma}{D_E} - \frac{v'qL_{EV}^2}{K_0 r \delta_{CS}^0} - \Delta p_0 \geq 0 \tag{5.56}$$

If at given $\Delta T = \Delta T_1$ and q determined by (5.55), inequality (5.56) is satisfied, it means that the "boiling limit" is critical in the present case. If inequality (5.56) is not satisfied, then thermal regime, rewarding the inequality (5.56), could be determined by the consequent decrease of value ΔT. Afterward, the values q and ΔT obtained through such a procedure determine the capillary heat transport limit of heat pipe in the evaporative regime.

The correlations (5.44)–(5.56) are valid in other cases (Figure 5.47d–f), but with some peculiarities.

Let us consider the model for conditions shown in Figure 5.47d–f, when excess liquid is presented inside the heat pipe.

The excess liquid plays a role in the liquid artery. When there is a lack of information about overfilling, while the essential excess liquid layer is presented (about several millimeters), then it is possible to neglect hydraulic losses throughout the heat input and the transport zone liquid path. It is equivalent to the abrupt reduction in L_C and L_{TR} and related terms in the correlation for ΔP_0.

In such cases, substitutions $L_{EV} = \Pi_1/2$ and $L_{EV}^2 = \Pi_1^2/4$ are required for correlations (5.44) and (5.50). In contrast, value of $2L_{EV}$ should be used instead Π_1 in Equation (5.51). If heat pipe operating conditions are typical and $\Delta P'' \ll \Delta P'$, other correlations remain unchanged.

Some peculiarities are associated with the calculation of ΔP_0. The initial correlation (5.51) accounting for the hydrostatic term is

$$\Delta p_0 = \frac{L_{EV} q}{r K_0 \delta_{CS}^0} [L_{TR} v'(T_s) + 0.5 L_C v'(T_1)]$$
$$+ C_f'' \left[0.5(L_C + L_{EV}) + L_{TR} \right] \frac{\Pi'' \Pi_1^2 L_{EV}^2 q^2}{8 r^2 S'''^3 \rho''}$$
$$- k_L (L_C + L_{TR} + L_{EV}) \rho' g \sin \beta$$

If liquid overfilling is absent, and liquid removal from the structure could be neglected, then $k_L = 1$. In other cases, $k_L < 1$, and could be calculated if the overfilling volume is known. If such information is lacking, but it is known that overfilling is considerable, then it could be assumed that $k_L = 0.5$ and the hydraulic losses of the liquid transport through the porous structure from the condensation zone to the evaporation zone could be neglected. Then, only hydraulic losses of the liquid transport along the perimeter of the heat input zone should be accounted.

Subsequently, calculation of ΔP_0 could be realized by the following correlation:

$$\Delta p_0 = \frac{0.25 \Pi_1^2 v'(T_s) q}{r K_0 \delta_{CS}^0} + C_f'' [0.5(L_C + L_{EV}) + L_{TR}] \frac{\Pi'' \Pi_1^2 L_{EV}^2 q^2}{8 r^2 S'''^3 \rho''} + D_0 \rho' g \cos \beta$$

where D_0 is the characteristic size of the heat pipe cross section.

Heat transfer regularities in all cases are based on Equation (5.41), which defines the local heat transfer law in some elementary cell that is impermeable for liquid and vapor skeleton. The skeleton is covered by the liquid film moving under the action of capillary forces in countercurrent with vapor flow. This simplified model created by the author in 1976 has satisfactory concurrence with numerous experimental results obtained mainly for the screen and particle bed structures. As mentioned earlier, any pore distribution is absent in the near-wall zone of these structures. Later, it was shown that this model could also be used in numerous other cases of boiling on coated surfaces via the combined treatment of Equations (5.39)–(5.41).

In the case of pool boiling on smooth surfaces, it is also usually associated with the existence of some pore distribution on the solid surface, and the concept of the representative cell relating to the single vaporization center on the heating surface could likewise be applied for this case. It could be assumed that in this case, the local heat transfer law is presented as $q \approx \text{const}(\Delta T - \Delta T^*)^M / (L_0^N)$, where N and M are constants and L_0 is the characteristic size of the representative cell.

It is not difficult to see that in this restricted case, the suggested approach leads to major concurrence with experimental observations.

Experimental data presented by experts from the Thermacore Inc. for heat transfer on coated surfaces covered by sintered porous aluminum at pool boiling conditions and results of experiments performed in the Kiev National Technical University (KNTU) [19] for FMPS inside heat pipe were used in direct comparison with calculations performed by the third-stage model.

The difficulty of comparison between experimental data and calculations by the third-stage model was associated with incomplete information about the main parameters of porous structures $(\lambda_E, D_E, D_{\max}, D_{\min}, K_0)$, and lack of important details concerning experimental conditions.

This problem was resolved for the above-mentioned conditions. Results of the comparison presented in Figures 5.48 and 5.49 supported the suggested theoretical model, in spite of some essential deviations for several experiments.

Such conclusion was based on the following analysis:

1. Thermophysical parameters of CPS $(\lambda_E, D_E, K_0,$ etc.) have a statistical nature with essential scattering.
2. Experimental investigations on thermophysical parameters were performed by using samples having essentially larger sizes (in several times) than the cross section size of typical porous coatings. Although it is known [35] that thermophysical parameters of porous structures with small thickness are characterized by the essential irregularity within the thickness, usually this factor is not determined quantitatively.
3. Heat transfer characteristics of the porous structure in the near-wall layer $(\lambda_E, D_E, K_0,$ etc.) are essentially dependent on the quality of contact between the wall and the structure. As a rule, the information about the quality of contact is not available and, in many cases, not obtainable.
4. Information about the reproducibility of different parameters of CPS is also lacking.

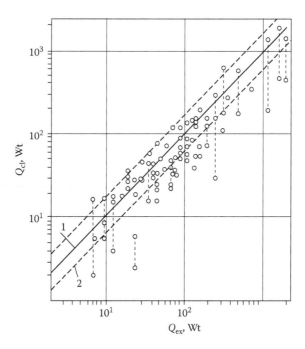

FIGURE 5.48 Comparison between calculations Q_{cl} based on the third stage model and experimental maximum heat fluxes Q_{ex} transferred inside the heat pipes with fiber–metal wicks [19]. 1, $Q_{ex} = Q_{cl}$; 2, 50% Q_{ex}.

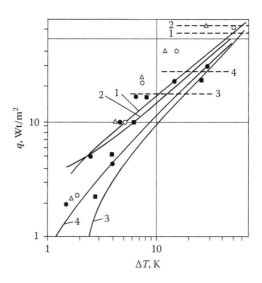

FIGURE 5.49 Comparison between calculations and experimental data at ammonia and acetone boiling on surfaces coated by the porous coverings manufactured by Thermacore Inc., $\delta_{CS} = 0.75mm$. 1, ○ ammonia, $T = 20$ K; 2, Δ ammonia, $T = 40$ K; 3 — acetone, $T = 20$ K; 4 •— acetone, $T = 50$ K; ------, calculation $(q_{CR})_{max}$ by [36]; _____, calculation by the third-stage model, $q = f(T_W - T_s)$ at $0 < q < q_{CR,max}$.

Hence, the following important conclusions could be formed:

1. The present approach and theoretical models of the first, second, and third stages developed on its basis allow us to make a general explanation — without internal contradictions — about the entire amount of experimental facts and heat transfer regularities at vaporization on surfaces covered by CPS.
2. This approach ensures the sequence at transition from the simpler model to the more complex one.
3. Effective models of the theoretical prescription of the heat transfer at vaporization on the more complex CPS types with VRC or with arteries, etc., could be created based on this approach.
4. As follows from the third-stage model analysis, the set of physical imaginations suggested and developed by the author allows understanding of the potential capabilities of the subsequent problems:
   • Establishment of the general theory prescribing heat transfer regularities, crisis phenomena, and formation of two-phase structures at vaporization on capillary-porous coated surfaces.
   • Determination of such cases when heat transfer regularities at vaporization in CPS, capillaries, and slits are the same.
   • Development of justified classification of vaporization conditions and forms on surfaces coated by porous coverings.

The problems mentioned in the last two items are significantly important for the analysis of hydrodynamic and heat transfer regularities at vaporization on surfaces with movable porous structures.

5.5 HEAT TRANSFER AT VAPORIZATION ON SURFACES COVERED BY MOVABLE CAPILLARY STRUCTURES

A relatively small number of studies have been devoted to experimental and theoretical research on vaporization on surfaces covered by movable porous structures. All these structures consist of particles and represent particle layers; therefore, only vaporization inside these structures was considered.

The first heat transfer experimental results at vaporization on movable bed particles covered surfaces were presented in Refs. [84–86]. Experiments devoted to boiling heat transfer on the wire submerged into the sand layer were prescribed by Gorbis et al. [84]. These experiments were conducted for water at atmosphere pressure under saturation and subcooling conditions in flooded and capillary feeding cases.

It was discovered that in the heat fluxes region typical for pool boiling on conventional surfaces, heat transfer regularities are the same as for usual conditions. Significant decrease of critical heat fluxes with growth of porous layer thickness was observed. The visual observations and photo capture of the vaporization process allowed us to discover the existence of the following thermal regimes:

• Natural convection inside the porous structure without phase transition
• Removing the layer from the heater
• Formation of channels

Siromyatnikov et al. [261] obtained temperature dependencies at vaporization on structures submerged into the sand layer wire. A quantitative analysis of these dependencies was not performed, but they have qualitative concurrence with data presented in Ref. [84]. The experiments on heat transfer at vaporization on real elements of heat exchanger surfaces were presented in Refs. [85–87]. These researches were developed and generalized in Refs. [251–256].

Nichrome strip with thickness 0.2 mm and width 15 mm glued to the textolite plate was used as the heating element in experiments [86]. The working chamber had the rectangular cross section shape and size $0.13 \times 0.20 \times 0.18$ m. The lateral chamber walls restricted the particle layer in the boiling container. The walls had glass illuminators, and the distance between them was almost equal to the working element width. The particle layer was pressed at the top by a special screen and had a height sufficient for gravity stability.

The fractions of particles with an average size of 0.62, 1.33, and 2.28 mm were used in the experiments. During the experiment, the layer height was changed from 5 to 150 mm, and vaporization took place in all cases. Two thermal regimes with different dependencies of heat transfer coefficient α on heat flux q were observed. The experimental data [86] at heat fluxes 6×10^3 to 7×10^4 Wt/m^2 for the first regime are presented in Figure 5.50, and were defined as satisfactory by the correlation $\alpha = (1.4 + 470d_p)q^{0.55}$. Significant temperature irregularities were discovered at $q > 7 \times 10^4$ Wt/m^2. Under these conditions, in some experiments the wall superheats increased at 100–150°C. It was shown that separate dry places appeared. This regime is characterized by an essential decrease of the heat transfer coefficient with heat flux growth. This regime was almost not analyzed.

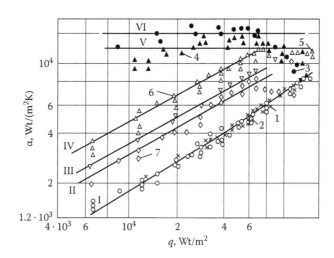

FIGURE 5.50 Heat transfer at vaporization inside the dispersed layer submerged in the distillate [86]. 1, Calibration tests [148]; 2, calibration tests [86]; 3 and 4, boiling inside the monel particles layer, $d_p = 3.58 \times 10^{-4}$ m and 4×10^{-4} m [148]; 5–7, boiling inside the glass particles layer, $d_p = 0.62 \times 10^{-3}$, 1.33×10^{-3}, and 2.28×10^{-3} m; I, the Rosenow's correlation for pool boiling heat transfer at the same conditions; II–IV, calculation by Equation (5.57); V, VI, calculation [148].

In a study conducted by Berman and Gorbis [86], the heat transfer regularities were considered. At vaporization inside the particle layer, the heat transfer, as a rule, was higher than at the pool boiling under the same conditions. It would be correct to agree with assumption presented in Ref. [148] that at vaporization inside the layer, the thin liquid microlayer was formed near the wall. Thermal resistance of this microlayer determined the heat transfer intensity. The authors attempted to explain essentially different heat transfer regularities at the same conditions and heat transfer mechanisms presented in several studies [86, 148, 149]. The presence of particles submerged into the film in the near-wall layer should affect the film's average thermal resistance. Such an effect is stronger when particles and liquid thermal conductivities are more different one from each other.

Berman and Gorbis [86] choose as the representative cell the plain element of the heating surface with width equal to one particle size, with it joining the single cylinder having a diameter equal to the particle diameter d_p and covered by the liquid film. The cylinder had a linear contact with the wall and its thickness was equal to δ. The cylinder surface modeling the corresponding particle row had contact with the surrounding vapor–liquid mixture at extremely high heat transfer intensity [85].

Such imagination might be justified. Unfortunately, the electric model accepted for the determination of the cell's average thermal resistance dependency on the main factors δ, d_p, λ_L, and λ_{SK} is dissimilar with this position. According to Berman and Gorbis [86], the particle surface outside the liquid film is not covered by the liquid film, and at the same time, its temperature is equal to the saturation temperature. Such schematic representation appears very contradictive, because particles sustain the intensive heat removal, and the more than higher their thermal conductivity, the less their diameter. According to Berman and Gorbis [86], the heat transfer coefficient α can be determined as

$$\alpha = \frac{\lambda'}{\delta} + \pi \frac{\lambda'}{d_p}\left(1 - \frac{\delta}{d_P}\right) \tag{5.57}$$

With the goal of achieving better agreement with the experimental data, it was assumed in Ref. [86] that δ is dependent on the heat flux as $\delta = 6.3 \times 10^{-2} q^{-0.45}$. It is not difficult to see that this approach does not explain the essential increase of heat transfer coefficient with the increase in particle diameter.

Significant dissimilarity in the regularities of the heat transfer at boiling inside the porous layer between experimental data [86, 148] is associated with the substantial differences in the particles' thermal conductivity values. Particles with very high thermal conductivity were used some in experiments [148]. In this case, particles play an important role in heat removal, thereby reducing the influence of q on α. It is known that in some experiments [147–149], spherical particles (from glass) with low thermal conductivity were applied. Hence, the heat transfer coefficients were almost not dependent on heat flux q. The absolute values of α were slightly different from each other, even when particle sizes were the same.

In the experiments [69–70] conducted on vaporization inside the glass particle layer, the heat transfer law $q \approx C\Delta T^{0.93}$ was attained, approximately corresponding to the insignificant dependency between α on q. A reliable contact between the heating surface and the porous layer was obtained via the special pressing plate.

In the later publications [85, 87, 252–256], the ranges of main parameters were widened and the physical imaginations were strengthened. It was pointed out by Gorbis and Berman [85] that vaporization inside the porous layer was initiated at lower temperature drops than at pool boiling. Moreover, it was noticed that, depending on the heat flux and the layer characteristics, the following types of layers are possible:

- Sense layer with the vapor filtration
- Porous layer with the stable and periodic channel formation regimes
- Uniform fluidization layer

The value $\beta^* = G_P / (g\rho_P F_H)$ was suggested in Ref. [85] as a parameter effecting regimes of vaporization inside the layer, where β^* is the length scale and determines the effective layer height and G_P/g denotes the mass of a layer.

At $3 \times 10^{-3} \le \beta^* \le 10^{-2}$ m, with q growth independent of d_p, Gorbis and Berman [85] observed the transition from the vapor filtration regime to the pseudo-fluidization regime. At $\beta^* > 3 \times 10^{-2}$ m, transition from the filtration regime to the channel formation regime was detected. The boiling inside the particle layer with $d_p < 10^{-4}$ was accompanied by strong foaming.

It was suggested to substitute parameter β^* by the layer's height H_{Lim}, at which point further increase in H does not influence the enhancing effect of the layer on the boiling heat transfer [87, 252–254].

As shown in Figure 5.51, at $H < H_{\text{Lim}}$, increase in H causes enhancement of the heat transfer. Beginning from some β^* value, that is, H_{Lim}, heat transfer regularities appear autonomous on the layer's height. Various research groups [87, 252, 254] performed the correlation $H_{\text{Lim}} \cong 6\sqrt{\sigma / (g\rho_L)}$. Classification of regimes of dispersed system parts and phases' movement at boiling is presented by Berman and Gorbis [254] (Table 5.5).

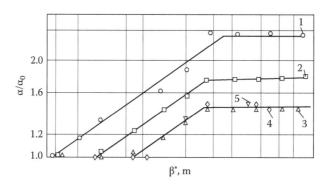

FIGURE 5.51 Dependency of heat transfer at water boiling on surface covered by dispersed particles multilayers on the effective height of layer β^* [85]. 1–3, Glass spheres, $d_p = 2.3, 1.3, 0.62 \times 10^{-3}$ m; 4, 5, electro-corundum particles $d_p = 10^{-4}, 5\times10^{-5}$ m; $\beta^* = G_P / (g\rho_P F_H)$.

TABLE 5.5

No.	Experimental Conditions	Particle Diameter, mm	Layer's State	Vapor Flow Type	
				Through the Layer	On the Surface
1		>2	motionless		
2	single heating element	0.2–1.5	unstable thermal fluidization	isolated bubbles	
3	$H < H_{lt}$	<0.2	suspension		
4		1–2	motionless	(a) $q < q^*$	
5	single heating element	0.2–1	partial removal from the heating surface by the vapor	periodically appearing vapor channels	two-phase layer from joining vapor bubbles
6	$H < H_{lt}$	<0.2	the whole layer removal	(b) $q > q^*$	
7			$w < w_f = 0.3$ m/c — motionless state, periodically appearing vapor channels		
8	tubes bundle	$d_{P.Opt}$	$w_{bf} - w_{ns} = 1.4$ m/c — beginning of the bed particles layer's thermal fluidization		
9	$H < H_{lt}$	0.8–1.3	$w > w_{ns}$ — stable circulating thermal fluidization, lifting movement in the center, flow down near the walls. The vapor is lifted through the center		

Typical local dependencies of heat transfer at vaporization inside the flooded particle dispersed layer at $H > H_{Lim}$ are given in Figures 5.52 and 5.53 [85]. These results and data from another study [86] (Figure 5.50) restrict the available direct experimental results. As seen from Figure 5.54, the dependencies of Figure 5.53 appear rather typical for conditions of boiling on a single working element. The same relates to boiling on the tubes bundle (Figure 5.52).

According to Berman and Gorbis [254], general heat transfer regularities at vaporization in the flooded dispersed layer are as follows. The influence of pressure, liquid properties, thickness of the heating surface, material properties, boundary conditions, etc., were qualitatively analogous to the pool boiling under similar conditions, but without the particle layer, that is, at $H = 0$. As follows from Table 5.5, thermal regimes 4, 5a, and 7 correspond to boiling curves $\alpha \approx Cq^{2/3}$, thermal regimes

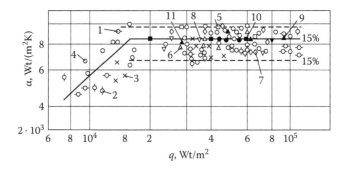

FIGURE 5.52 Heat transfer dependency on heat flux at boiling inside vacuum evaporator with dispersed layer at the tube bundle [85]. 1–3, Distilled water, $p = 0.117, 0.298, 0.460$ bar; 4–11, $p = 0.182$ bar.

4 and 6b reveal the law $\alpha \approx Cq^{1/2}$, and in the circulating thermal fluidization conditions, $\alpha \approx Cq^{0}$. The influence of particle size on heat transfer regularities has a nonmonotonous character at $q < q^{*}$. It has optimum level of the heat transfer enhancement at $\bar{d}_{p} = 0.8$–1.3 mm. When $q < q^{*}$, heat transfer intensity is independent of $\bar{d}_{p}$.

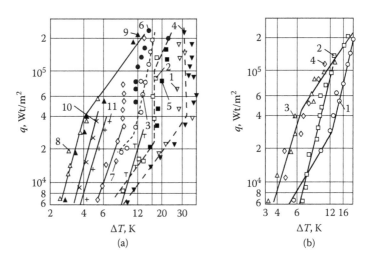

FIGURE 5.53 Heat transfer at vaporization inside flooded dispersed multilayered bed [85]. Pool boiling: (a) horizontal position, glass particles, $d_{p} = 2.3 \times 10^{-3}$ m: 1–3, $p = 0.035, 0.2,$ 1.0 bar; 4–6, $p = 0.036, 0.2, 1.0$ bar; 7–9, $p = 1$ bar; $\beta^{*} = 5.6 \times 10^{-3}, 3 \times 10^{-2}, 3.1 \times 10^{-1}$ m; 10–12, $p = 0.5, 0.2, 0.036$ bar; $\beta^{*} = 1.1 \times 10^{-1}$ m. (b) Horizontal position, electro-corundum, $d_{p} = 10^{-4}$ m; $p = 1$ bar: 1–4, $1.8 \times 10^{-3} \leq \beta^{*} \leq 7.2 \times 10^{-3}$ m. (c) Vertical position: 1, pool boiling, smooth surface; 2, glass balls, $d_{p} = 2.3 \times 10^{-3}$ m.

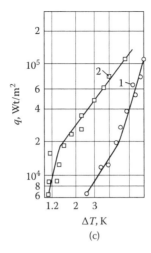

(c)

FIGURE 5.53 *Continued*

Visual observations [252, 254] allow the conclusion that at vaporization inside the porous dispersed layer, the two-phase layer occurs between the wall and the layer. Hence, the relative part of the heating surface occupied by the two-phase layer increases with q growth and approaches 1. It was observed during the experiment that high intensity heat transfer, low-amplitude temperature pulsations and stable (without superheat) operation of thin wall heating surfaces at $q = const$ are typical for the channel formation regimes. All these facts provide a basis to consider that under the two-phase layer between the dispersed structure and wall, a thin nonsteady liquid film exists. This two-phase layer may or may not have contained some particles (at the layer's removal regimes) [254].

In several studies [87, 252, 254], empirical correlations were formulated that generalized the obtained experimental data for the optimal viewpoint of heat transfer

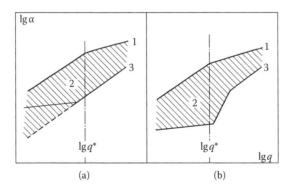

(a)   (b)

FIGURE 5.54 Boiling on heating surfaces submerged in the dispersed layer [254]. (a) $p = 1$ bar; (b) $p = 0.035$ bar; 1, at $H > H_{im}$ in the channels formation regime; 2, thermal pseudo-fluidization regime; 3, pool boiling.

enhancement, at conditions $H \geq H_{\text{Lim}}$. The whole layer removal took place at small particles sizes, d_P, when the average diameter, relating to the Laplace constant $\sqrt{\sigma/(\rho'g)}$, was less than 0.2. Then this phenomenon developed even for very small heat fluxes, and was moving into the pseudo-fluidization of the whole vapor–liquid mixture. When particle sizes exceeded $\bar{d} = d_P/\sqrt{\sigma/(\rho'g)} > 2$, then the dispersed layer boiling heat transfer would be similar to that observed under pool boiling conditions. The determined influence of size on heat transfer intensity was suggested to account for the following empirical correlation:

$$\varphi\left(\frac{d_P}{\sqrt{\sigma/(\rho'g)}}\right) = C_1 + C_2 \sin\left(\frac{C_3 d_P}{\sqrt{\sigma/(\rho'g)}} - C_4\right)$$

where C_1, C_2, C_3, and C_4 are empirical constants, equal correspondingly to 0.785, 0.235, 2.82, and 1.55 (the function view was selected in such a manner so as to reach an agreement between experimental and calculation results at the $\bar{d}_P$ range 0.2–2).

To account for the influence of the wall's material properties, known imaginations about this factor's influence on pool boiling were used, and the correction function was inserted in the following form:

$$\psi = 1 + C_5\left(\varepsilon_c^{-\kappa} - 1\right)(1 + th(C_6 \bar{P} + C_7)). \tag{5.58}$$

Here, $\bar{\varepsilon}_W = \varepsilon_W/\varepsilon_0$; $\varepsilon_W = \sqrt{\lambda_W \rho_W c_W}$ is the wall material thermal–assimilability coefficient; $\bar{\varepsilon}_W$ is the relative thermal–assimilability coefficient; ε_0 is the standard value of the thermal–assimilability coefficient; $\bar{P} = p/p_{\text{CR}}$ is the nondimensional pressure; $\aleph$, C_5, C_6, C_7 are empirical constants ($C_5 = 0.5$, $C_6 = 3.5 \times 10^3$, $C_7 = 1.58$); if $\aleph > 0$, it had to be for the thick wall, because the heat transfer coefficient α is independent of the layer's thickness, δ_W and $\kappa = 0$ by $\delta_W \ll \delta_{W.\text{Lim}} \cdot \delta_{W.\text{Lm}} = 10^{-3}$ and by $\delta_W > \delta_{W.\text{Lm}}$, $\kappa = 0.3$ (where δ_W is the heating wall thickness).

Correction functions f_1 and f_2 were used to account for the influence of saturation pressure:

$$f_1 = C_8 - \frac{C_9}{C_{10}\tilde{p} + C_{11}}; \quad f_2 = C_{12} - \frac{C_{13}}{C_{10}\tilde{p} + C_{14}}$$

where C_8–C_{14} are empirical constants ($C_8 = 0.34$, $C_9 = 1.54$, $C_{10} = 10^4$, $C_{11} = 5$, $C_{12} = 0.015$, $C_{13} = 0.098$, $C_{14} = 8$). The extreme quantity of the empirical constants was associated with the authors' desire to achieve good comparison between the generalization and experimental data (see Table 5.5). In Refs. [85–87, 252, 254], the following classification of thermal regimes was proposed:

1. The small-size particles were motionless inside the whole structure's volume. It was recommended to obtain data for this thermal regime from Ref. [148]; aside from this, however, Ferell and Alexander passed up on the chance to use of the some above-mentioned experiments with small particles.

    Unfortunately, the authors [85–87, 252–254] did not provide their own analogous data, but only gave the generalized equation,

$$\alpha = A\bar{\lambda}_E \left(\frac{3(1-\varepsilon)}{2md_P} \right)^{a'}$$

2. When capillary, viscosity, and inertial forces would be approximately 1 order and $q < q^*$, then it would be necessary to use the next formula:

$$\alpha = A_1 \Psi \varphi f_1 \left(\frac{(\lambda')^2}{v_L \sigma T_s} \right)^{1/3} q^{2/3}$$

3. When capillary forces action can be neglected and $q < q^*$, then

$$\alpha = A_2 \Psi f_2 \left(\frac{(\lambda')^2 T_{CR}}{v' \sigma T_s} \right)^{1/2} q^{1/2}$$

The boundary values q^*, which separated the thermal regimes (2) and (3) could be defined from the corresponding value α equality:

$$q^* = \frac{(\lambda')^2 T^3_{CR}}{v' \sigma T_s} \left(\frac{A_2 f_2}{A_1 f_1 \varphi} \right)^6$$

At given correlations ($A_1 = 0.62$; $A_2 = 0.45$) for walls with thickness 0.1–0.2 mm and $q = const$, $A_1 = A_2 = 1$ for wall with the thickness exceeding 1 mm and by convective heating.

The proposed concept of experimental data generalization was based on 15 empirical constants and five empirical function applications. Naturally, such an approach to the theoretical prescription of heat transfer when vaporization inside the dispersed porous structure did not allow verification of the physical imaginations, and their correct use for generalization. Therefore, some contradictions and drawbacks of this attempt would be required as soon as they were sequenced in relation to the necessity of improving physical imaginations. The main points of these issues were developed [85–87, 254] as follows:

1. As assumed by the authors, the first thermal regime of the influence of capillary forces was realized by the small heat fluxes and small particle sizes. However, in one experiment [148], the independence of the heat transfer intensity from the heat flux was observed by using same conditions described in another study [86] (including heat flux and particle size scale, the heat carrier type, and pressure level). In this set of experiments [86], however, a strong dependence, $\alpha \approx Cq^{2/3}$, was found.

The attempt to explain the boiling heat transfer in the dispersed layer pecullarities by using the influence of the effective thermal conductivity λ_E of the near—wall liquid film with submerged particles is associated with the considerable contradictions and discontented. It is necessary to account that the authors' results [86] were obtained by means of the electroanalogy method with quite doubtful boundary conditions.

2. The statement about the increase in inertial forces with the decrease in heat flux did not agree with the known position that inertial forces increased with the increase in heat fluxes.

3. The statement about the heating surface being a factor determining the size was not found in generalized correlations [87, 252–254]. In these experiments, two types of working elements were used: 100×15 mm horizontal plates and horizontal tubes with a diameter of 10 mm. These sizes did not change during experiments, therefore the experimental data [85–87, 252–254] could not point out any influence on size. However, experimental data from Ref. [87] noted that differences for the plate and the tube did take place.

   Experimental results on the wire, submerged at the dispersed sand layer, showed essential differences compared with previous data [84]. The position [254] about the independence of heat transfer regularities on the corresponding determined size was presented to be doubtful. It is likely that the independency of the dispersed layer heat transfer on heater size is possible in some are as. However, under what conditions could these results be possible? The experimental results did not provide an answer to this question.

4. There was no satisfactory explanation [254] to the following question: In the experiments [85–87, 252–254], why was the essential heat transfer enhancement field typically a dispersed layer "removal" from the wall, but not the pseudo-fluidization?

   Similar vaporization conditions at the dispersed layer, for example, in Ref. [148], did not only point to the removal of a layer, but it was also observed as quite another character of α dependency on q.

5. Boiling on the submerged inside dispersed layer tube bundle heat transfer [85, 253] was associated with the independence of heat transfer intensity from heat flux that was not sequenced from the suggested correlations [87, 252–254] for separate elements.

The cited contradictions and absence of correspondence between experimental and generalization results of vaporization on the flooded dispersed layer prompted this book's author to suggest and develop his own approach to the discussed problem of mathematical modeling. The main positions of the model were as follows:

1. The principal difference between the experiments [148–149] and conditions [85, 86] was connected with the fact that in Ref. [148, 149], H/b $\ll$ 1 by $\rho_s/\rho' \gg 1$, whereas in Ref. [85, 86] H/b $\gg$ 1 by $\rho_s/\rho' = 1.5$–2.0.

2. It is possible to assume that in the study by Ferell et al. [148, 149], conditions took place in the gravity dispersed layer pressing to the heating wall. In contrast, in another study [85–86], by H/b $\gg$ 1, the near-wall layers were considerably unloaded on the layers located above, which led to their

significant removal from the wall. This phenomenon was confirmed by the shooting results reported by Gorbis et al. [84]. The layer's removal for most cases was also confirmed in other studies.

3. Assuming that at boiling inside the dispersed layer removed from the wall, the main physical mechanism is similar to the slit boiling. Thus, analogy between the heat transfer regularities in the layer and in the slits exists with respect to accounting of some equivalent slit gap approximately equal to the particle's size.

4. At the same time, by the boiling inside the dispersed layer, in comparison with slit boiling, differences became more pronounced, associated with simultaneously two-phase mixture filtration processes developing at the mentioned conditions. It was similar to the fluidization process. Due to the smaller particle sizes and layer height, and larger heat flux, the fluidization mechanism was developed more strongly. Fluidization under the boiling process can be defined as "thermal-pseudo-fluidization." By the whole "thermal fluidization," the particles are retained as suspension particles. The main heat transfer mechanism will be the same as in pool boiling. The particles' influence could be produced over the suspension thermophysical properties changing of new dispersed media liquid + particles (effective viscosity, μ_E, ν_E; thermal conductivity, λ_E ; heat capacity, C_{PE}, etc.). Figure 5.55 illustrates this physical imagination as a schematic diagram.

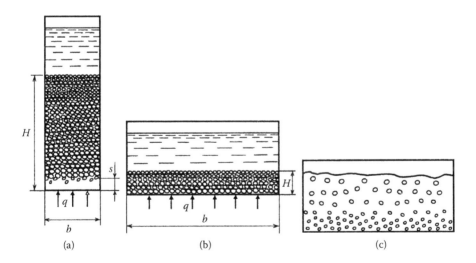

FIGURE 5.55 Pool boiling on the surface, submerged at the flooded dispersed layer, typical hydrodynamical regimes. (a) $H \gg b$; $s - d_p$; $U \ll U_{CR}$; the partial or whole dispersed layer removal from the wall. (b) $H < b$; $\rho_p \gg \rho'$; $U \ll U_{CR}$ the gravitational dispersed layer pressing; (c) b, $U \gg U_{CR}$, the whole thermal-pseudo-fluidization; U, the pool boiling determined velocity; U_{CR}, the thermal-pseudo-fluidization velocity.

5. Therefore, if they are to be used, the known criteria of the boiling heat transfer equations and the semiempirical correlations, which were generalized in the slit boiling experimental data obtained by author with his coworkers [105–108], then it would be possible to obtain the model for heat transfer by removing dispersed layer boiling regularities' prescription. However, it would be important to find some criteria that could account for every mechanism input in the general process. This criteria could be obtained if the traditional boiling velocity $q/r\rho''$ could also be compared with some known critical fluidization velocity $U_{CR} = C_W g d^2 (\rho_P - \rho') / \mu'$, where C_W is the numerical constant accounting for the particle movement regime. Therefore, for the whole heat load, by boiling at the dispersed layer, the relation can take the form

$$q = q_1(1 - \Psi) + q_0 \Psi$$

Here q_1 is the heat flux transferred by the conventional boiling heat transfer at the horizontal slit; q_0 is the heat flux transferred by the conventional pool boiling with some effective thermophysical properties; ψ is the interpolated multiplier that defines the velocities ratio U/U_{CR}, and with respect to the limit transition, we obtain:

$$\Psi = \frac{U}{U + U_{CR}}$$

The heat flux q_0 dependency on thermophysical, geometrical, and regime parameters was known. The semiempirical correlation for q_1 could be obtained from experimental data generalization on the horizontal slit boiling process. It contains the following basic positions:

1. The combined heat transfer mechanism took place via the horizontal slit boiling. When the vapor phase ensued, heat was transferred over the thin liquid microlayer to the phase border by thermal conductivity. When the liquid phase existed near the wall, the contact or convection heat transfer mechanism had to be developed.
2. The thermal conductivity over thin microlayer liquid film by the thickness δ mechanism was the main factor when small gaps in the horizontal slit took place. Then, the entire specific thermal resistance could be defined by $R_M = \delta / \lambda'$.
3. The initial local formed microlayer thickness and the local hydrodynamic parameters (the phase border movement velocity $\overset{*}{R}$, the pressure gradient dp/dR, and the vapor bubble local radius R) had the following approximate connection:

$$\delta \approx C_M \sqrt{\frac{\mu \dot{R}}{dp/dR}} \tag{5.59}$$

4. Hydrodynamics of the liquid micro-layer at boiling inside the horizontal slit is associated with the phase border movement in the system of simultaneously growing vapor bubbles. The density of growing vapor bubbles is defined by the same regularities as at pool boiling:

$$n_F \approx const \left(\frac{r\rho'' \Delta T}{\sigma T_s} \right)^m \tag{5.60}$$

where $1 \leq m \leq 2$.

5. The average pressure excess inside the vapor bubble volume was mainly defined by the liquid removal from the slit conditions

$$p'' - p_s \approx c_w \rho' w^2 \tag{5.61}$$

6. The liquid average outflow from the slit, by width b, velocity can be determined from the mass conservation equation:

$$w = bn_F \pi R \dot{R} \tag{5.62}$$

7. The vapor bubble velocity $\dot{R}$ at the horizontal slit with the gap "s" definition simplified correlation can be stated on the energy model schematic base by the condition q-$const$ in the following form:

$$\dot{R} \approx \frac{\dot{C}_R R q}{r\rho'' s} \tag{5.63}$$

The common consideration (5.59)–(5.63) with respect to the condition $s = Cd_p$ leads to the following equation

$$\frac{\delta}{\lambda'} = R_M = \frac{const}{b\lambda'} \sqrt{\frac{r\rho'' d_p v'}{q n_F}} \tag{5.64}$$

If, in the case of horizontal slit boiling, it could be possible to assume that — as with pool boiling — it would be right to take $m = 2$, then the vaporized centers' density dependency on the main factors for slit boiling at the removed layer case would be likely weaker than for pool boiling, which was $m < 2$. Therefore, the transition to the undimensional form and excluding the dimensional coefficient at (5.60) and (5.64), we could use the transformation of Equation (5.58) to exclude the dimensional coefficients at (5.60) and (5.64) at the next form

$$n_F L^2 \approx const \left(\frac{r\rho'' \Delta T}{\sigma T_s} L \right)^m \tag{5.65}$$

It would be right to account for the fact that by the pool boiling, L is the defined size depending on pressure and heat carrier properties, and it is possible to take

$$L^2 = const \frac{\sigma}{(\rho' - \rho'')g} \qquad (5.66)$$

With respect to (5.60)–(5.66), it is possible to obtain for, q_1,

$$q_1 = const(b\lambda')^2 \left(\frac{r\rho''}{\sqrt{\sigma(\rho' - \rho'')gT_s}} \right)^m \frac{g(\rho' - \rho'')\Delta T^{2+m}}{\sigma r\rho'' d_p v'} \qquad (5.67)$$

When the heat transfer intensity for boiling at the dispersed layer had to be essentially different from this one for the pool boiling conditions, which was $q \gg q_0$ and $q_1(1 - \psi) \gg q_0 \psi$, then it would be possible to obtain

$$q = q_1(1 - \psi)$$

With respect to (5.66) and (5.67), and values U and U_{CR}, the following could be obtained:

$$q = const(b\lambda')^3 \left(\frac{r\rho''}{\sqrt{\sigma(\rho' - \rho'')gT_s}} \right)^m \frac{g(\rho' - \rho'')\Delta T^{2+m}}{\sigma r\rho'' d_p v'}$$

$$\times \left[1 + C_q \frac{qv'\rho'}{r\rho'' g d_p (\rho_P - \rho')} \right] \qquad (5.68)$$

Solving (5.68) and relating q, we obtain the experimental data generalization equation.

At $m = 1$:

$$Z = C_1 \left[\sqrt{1 + C_2 Y^3} - 1 \right] \qquad (5.69)$$

where

$$Z = \frac{qv'}{r\rho'' g d_p^2} \frac{\rho'}{\rho_P - \rho'}; \quad Y = \left(\frac{1}{\sigma T_s} \frac{\rho'}{\rho_P - \rho'} \frac{(b\lambda')^2}{g r\rho'' d_p^3} \sqrt{\frac{(\rho' - \rho'')g}{\sigma}} \right)^{1/3} \Delta T \qquad (5.70)$$

At $m = 2$:

$$Z = C_1'\left(\sqrt{1 + C_2'(Y')^4} - 1\right); \quad Y' = \left(\frac{1}{\sigma T_s}\frac{\rho'}{\rho_P - \rho'}\frac{(b\lambda')^2}{gr\rho''d_P^3}\sqrt{\frac{(\rho' - \rho'')g}{\sigma}}\right)^{1/3}\Delta T \quad (5.71)$$

Experimental data [85, 86] on heat transfer boiling at the dispersed layer generalization resulting to the form (5.70) are presented at Figure 5.56; they were compared with formula (5.69) by coefficients $C_1 = 7.7\times10^{-5}$; $C_2 = 0.44$. The suggested model not only allowed experimental data generalization, but also the influence of many important factors on heat transfer (when boiling at the dispersed layer), to be explained without contradictions.

Therefore, it follows from formula (5.69) that, for a given pressure, liquid properties, and temperature drops, an optimal value of particle diameter, when the heat flux is at maximum, exists. It was in a good agreement with experimental facts [85–87]. It is also seen from (5.69) that index n at the heat transfer intensity dependency decreased with growth in temperature drop. So, at the dependency $q = C_T\Delta T^n$,

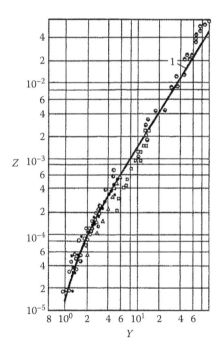

FIGURE 5.56 The experimental data of pool boiling on submerged surface by the flooded dispersed layer surfaces by the suggested previously presented physical model base (Equations 5.69–5.70) and its comparison with the experimental data published in Refs. [85, 86]. Heat carrier, water; $\rho_P = 2700, 4000$ kg/m^3; $2.3\times10^{-3} \le p_s \le 1$ bar; $0.1\times10^{-3} \le d_P \le 12\times10^{-3}$ m; Z and Y from formula (5.70); 1, the line calculation on formula (5.69) base by $C_1 = 7.7 \times 10^{-5}$; $C_2 = 0.44$.

by small superheat values, $n = 3-4$; by larger superheat values, $n = 1.5$. This was also in good agreement with the experimental results [85–87]. The model prescribed the following important experimental facts:

- The heat transfer enhancement diminishes with heat flux growth and particle size decrease.
- Independence of the heat transfer intensity from dispersed layer height by $H > (2-3)b$;
- The heat transfer enhancement scale decreasing with saturation pressure decreasing, $p_s < 0.1$ bar.

From the suggested model viewpoint, by boiling inside the dispersed layer porous structure, heat transfer regularities had to be the same on the horizontal tube external surface as on the horizontal flat plate having a width equal to the tube's diameter. This position was well confirmed by experiments [85–87]. The heat transfer regularities would be the same for tube's bundle inserted at the dispersed porous layer by small heat fluxes as on the single tube's external surface. Forced circulation associated with joint operation of tubes situated in the same vertical row is increased with the heat flux growth. The circulation associated with the tubes bundle action (N tubes located at the common vertical row) would enforce the fluidization process; the influence of circulation could be accounted for by the multiplier of type $(1 - \psi)'$ with the single difference that, instead of the determined velocity, it had to use the mentioned circulation velocity w_0. At first approximation, $w_0 \approx Nq/r\rho''$, which was

$$(1 - \Psi') \approx \left(1 + N \frac{qv'}{r\rho'' g d_p^2} \frac{\rho'}{\rho_p - \rho'} C_N \right)^{-1}$$

Then, for the tubes bundle with respect to (5.67), we would obtain

$$q = const(b\lambda')^2 \left(\frac{r\rho''}{\sqrt{\sigma(\rho' - \rho'')g T_s}} \right)^m \frac{(\rho' - \rho'')g\Delta T^{2+m}}{\sigma r\rho'' d_p v'}$$

$$\times \frac{1}{1 + C_q \dfrac{qv'\rho'}{r\rho'' g d_p^2(\rho_p - \rho')}} \times \frac{1}{1 + C_N N \dfrac{qv'\rho'}{r\rho'' g d_p^2(\rho_p - \rho')}} \quad (5.72)$$

An analysis of formula (5.72) shows that for the tubes bundle by q growth, the dependency $q = f(\Delta T)$ was becoming the next correlation form $q \approx C\Delta T$. It also had good concurrence with experimental results.

The optimal particle size — from the viewpoint of heat transfer enhancement — can be defined as the simplified theoretical correlation, which prescribes the vaporization heat transfer at the flooded dispersed porous layer regularities in the following form:

$$q = \frac{C_1}{A_1} d_p^2 \left(\sqrt{1 + (C_2/d_p^3)A_2\Delta T^3} - 1 \right)$$

where

$$A_1 = \frac{r\rho''g}{v'} \frac{\rho_p - \rho'}{\rho'}; \quad A_2 = \frac{1}{\sigma T_s} \frac{\rho'}{\rho_p - \rho'} \frac{(b\lambda')^2}{r\rho''g} \sqrt{\frac{(\rho' - \rho'')g}{\sigma}}$$

Then, using the extreme condition at the form $\partial q/\partial d_p = 0$, we obtain the formula for the optimal particle diameter as

$$d_{P;OPT} = \sqrt[3]{C_2 A_2 / 8\, \Delta T}$$

It was obtained for the atmospheric pressure water boiling inside glass particle structure case and for typical heat transfer enhancement regimes wall superheat values (by order 3–5 K) $d_{P;Opt} = 0.7$–1.3 mm. It showed good agreement with experimental data [85, 87].

Heat transfer by vaporization at the dispersed, flooded layer worked out in Ref. [257], in which the author's own model could be used and also developed for the thermal and hydrodynamic combined fluidization conditions, which were studied experimentally and presented in several papers [251, 255, 256, 258]. In correspondence with the author's model [257] in the case of q value definition, it would be necessary to use at the correction function $(1 - \psi)'$ or ψ', not two defining velocities (the determined boiling velocity $q/r\rho''$ and pseudo-fluidization critical velocity U_{CR}) but three, adding to the one mentioned above, the forced circulation velocity w_0. Then, function $(1 - \psi)'$ has to account for the influence of hydrodynamic pseudo-fluidization on the heat flux [257] and could be presented as

$$(1 - \Psi_w)' \cong \left(1 + C_w \frac{w_0 v'}{gd_p^2} \frac{\rho'}{\rho_p - \rho'} \right)^{-1}$$

It would be possible to imagine other methods of "thermal and hydrodynamic pseudo-fluidization" common action on the heat transfer by vaporization inside the flooded dispersed layer. Let us input the determined velocity for the case of common thermal and hydrodynamic pseudo-fluidization action, using the energy approach and defined the velocity w_{0q} by the following method

$$w_{0q} = \sqrt{K_1 \left(\frac{q}{r\rho''}\right)^2 + K_2 w_0^2}$$

where K_1 and K_2 are experimental constants.

In the next analysis, the value w_{0q} could be used in the comparison with the pseudo-fluidization critical velocity. The real choice of the cited approach or maybe some other prescription of the combined action on the considered process boiling and the external forced circulation would require other, more effective model working out and further experimental developments. However, together with the suggestion outlined in Ref. [257], a theoretical model could be considered as the first positive step, allowing researchers to explain, without serious contradictions, all main experimental facts defining the heat transfer inside the dispersed layer vaporize regularities.

This model has undergone a strong critical analysis by Berman and Gorbis [254]. The main comment was associated with the technical errors at the definition of empirical constants (it was originally written: $C_1 = 6.6$, $C_2 = 0.36$; but should have been corrected to read as $C_1 = 7.7 \times 10^{-5}$, $C_2 = 0.44$). Correspondingly, the values of Z order had to be changed (at Figure 5.56). The next critical comment was devoted to the author's [254] position about the possible independency of the heat transfer intensity on the heater's size. The comment was discussed above and it was shown that this position did not agree with experimental facts.

However, it would be necessary to note that the heat transfer independency on heater width could exist at the wide heater case. At the "channel formation" this thermal regime was considered the most typical for boiling inside the dispersed layer structure; the vapor phase was removed mainly at the channels' outlet, independently of the heater's size (presented in Ref. [254]). It meant that with the distance between the channels, L_{CH}, their sizes had to have an influence, by some conditions on two-phase flow near the wall and consequently on the liquid microfilm formation at places, and as the result of heat transfer regularities. If heater sizes were essentially greater than the distance between the channels, then the independence on heater size has to exist. In another case, the heater width or diameter had to have an influence on the heat transfer intensity.

The research on channels' formation regularities were likewise not produced as the determination of channel sizes and distances between them. The preliminary estimations, suggested by Gorbis et al. [84], gave the values L_{CH} equal by order 1 sm and more. It meant that the heaters used in several experiment [85–87, 252–254] could be assumed as "narrow," which corresponded with the model [257]. In the other case, it was necessary in formulas (5.69), (5.70), and (5.71) to insert b instead as the determined size of the average distance between channels L_{CH}. It would be possible to make as the case when $L_{CH} - const$; in both cases, the value L_{CH} will be dependent on the geometry, regime, and the heat carrier thermophysical properties.

Berman and Gorbis [254] also noted that in Ref. [257], only a limited part of the experimental results was used. It was discovered that only in Refs. [85–87] could it be possible to obtain the entire necessary information for the corresponding generalized treatment, following the model Ref. [257]. In relation to Berman and et al.'s [254] other publications [251–256], it was also basically impossible to obtain the necessary information from them. However, the simple consideration and comparison of experimental results (Refs. [85–87] with Refs. [251–256]) did not yield essential differences. Therefore, the generalization on experimental data [85–87] base could be considered sufficiently representative for the theoretical prescription of the heat transfer when boiling inside the flooded dispersed layer.

The whole heat transfer regularities were principally confirmed in later publications [251, 255, 256], but the main focus was on scale deposits formation regularities by boiling inside the dispersed particles layer. Some very important peculiarities of these processes by boiling inside dispersed layer were discovered: it marked the first possibility of the solution boiling without the scale deposits and even the self-cleaning phenomenon for some conditions. Unfortunately, this particular branch of cited research had no prolongation.

The vaporization heat transfer experiments at the flooded by subcooled liquid dispersed layer at hydrodynamic and "thermal" pseudo-fluidization conditions were conducted and presented in several papers [258, 259, 261]. Special visual observations (video and photo capture) were conducted on separate experimental setups with the goal of obtaining reliable information about the dispersed layer structure at the pseudo-fluidization process under the subcooled boiling conditions. The authors' attention was concentrated on the pseudo-fluidization layers near the wall zone. The authors noted the following hydrodynamic phenomenon peculiarities, which have taken place in this region:

- The pseudo-fluidized layer structure's unevenness.
- The relatively stable liquid film existence near body surface that was periodically destroyed by the local hydrotransport flows and solid particles' vertical movement.
- The subcooling boiling under the liquid pseudo-fluidization could be pressed. The turbulent mixing-up level inside the liquid flow at the dispersed layer had to be very high; it was leading to the momentum and energy transfer enhancement between the center and the near-wall liquid layer and, consequently, to more intensive vapor bubble condensation and subcooling boiling suppression.

Research on heat transfer was conducted under the conditions described in several papers [258, 259, 261], by atmospheric pressure on the presented design as the closed-loop experimental setup. The water flow fluidized the granular layer at the rectangular cross section with the sizes 32×200 mm and vertical channel at the height of 800 mm. The heating elements were stainless steel horizontal tubes with diameters of 3.0, 5.44, and 6.26 mm, and length of 200 mm. Materials used consisted of granular materials from particles of the same size and with approximately spherical form: from aluminum, $d_P = 0.95$, 1.34, 1.55, and 2.43 mm; from glass, $d_P = 0.91$, 1.34, and 1.72 mm; alundum, $d_P = 3.54$ mm. The experiment was conducted at the following parameters ranges: $20 \leq q \leq 280$ kW/m^2; $19 \leq \Delta T \leq 81$ K; $0.633 \leq \varepsilon \leq 0.914$; $0.048 \leq w \leq 0.235$ m/s.

The authors were trying to organize the vapor bubbles activation suppression with pseudo-fluidized layer help, but these efforts were unsuccessful. The experimental data by boiling absence was generalized by the next criteria equation:

$$Nu_{E.L} = 0,11Re_{E.L}^{0,72}Pr_L^{0,58}\left(\frac{Pr_L}{Pr_W}\right)^{0,25}$$

Vapor channel effective diameter was selected as the determined size, the average fluidized media temperature as the determined temperature, and velocity was defined by relating it to the narrow part of the channel's cross section. The equation was right at the following determined criteria ranges:

$$80 \le Re_{E.L} \le 5000; 3 \le Pr_L \le 10; 1.13 \le Pr_L / Pr_W \le 1.7$$

The heat transfer coefficient by the subcooling boiling was defined over the temperature drop calculation as the difference between the wall temperature and the saturation temperature. The next empirical formula recommended for developed boiling conditions is as follows:

$$\alpha_{D.B} = 0.285q^{0.95}\Delta T^{-0.63}$$

At the transition regime from convection to the developed boiling, the heat transfer coefficient was suggested by Tuchnin [259] to be calculated using the next interpolation ratio:

$$\alpha = \sqrt[n]{\alpha_{S.B}{}^n + \alpha_{D.B.}{}^n}; n = 4$$

The heat transfer intensity by boiling at the pseudo-fluidized dispersed layer comparison with pool boiling heat transfer one by forced convection, was made by one and the same ΔT and w, and it shows that heat transfer at the pseudo-fluidized layer by small heat fluxes q was 2–2.5 times more intensive than by pool boiling. When q increased, this ratio decreased, and with larger q, it contained 10–15% of corresponding value at conventional pool boiling. Therefore, by subcooling boiling, the vaporization heat transfer organization at the dispersed layer would be essentially less effective.

It is be possible to point some other studies devoted to heat transfer by vaporization at the flooded dispersed layer [262, 263]. In Ref. [262] theoretical and experimental results are presented. It was a study on seawater evaporator–distiller characteristics. For this apparatus, the solid particles' "saturation" at seawater was used, with the goal of increasing the heat transfer coefficient between the wall and the solution. The working part was the vertical tube by the diameter 38/35 mm and the length $l = 2$ m; heating was carried out via electric current. The solid particles were the glass spheres with 3 mm diameter. Comparison of experimental results with those reported in the literature show that the best agreement was reached by using the Richardson and Smith formula.

$$\alpha = \frac{\lambda'}{d}\left[0,023Re_d^{0,8}\ Pr^{0,4}+1,49\left(1+35,4c_{p0}^{2,12}\right)(1-\varepsilon)^m\left(\frac{w}{\varepsilon}\right)^{1,15}\right];$$

$$m = 0,079\left(\frac{\rho'wd_p}{\eta'}\right)^{0,36}$$

where C_{p0} is the solid phase heat capacity, w is the liquid velocity, d is the tube's internal diameter, and ε is the layer porosity. The last value changed at the experiment from 0.65 up to 0.94.

In Ref. [263] heat transfer intensity was studied by experiment from the horizontal cylinder up to the fluidized layer filled by the glass spheres with an average diameter of 0.69 mm. The experimental element — the tube (quadratic cross section, 280×280 mm; height, 500 mm); the measuring cylinder, which was produced from stainless steel tube, had a diameter 15×15 mm and was heated by electric current. Experiments were conducted with the water inlet temperature equal to 15.5°C and 66°C at the velocity range $5.7 \times 10^{-3} \leq w \leq 64 \times 10^{-3}$ m/c (by the layer porosity changing from 0.43 up to 0.82), and by heat fluxes from 1.5×10^4 to 140×10^4 W/m².

The comparative analysis of experimental data obtained for boiling at the pseudo-fluidized dispersed layer and without it, shows the significant heat transfer enhancement by boiling at the layer. The heat transfer coefficient dependencies for boiling at the layer at the convection heat transfer field and by the subcooling boiling beginning have two maximum values: by $w = 0.01$–0.02 m/s and $w = 0.04$–0.05 m/s.

The authors tried to generalize experimental data, using only α_{max} values in the following form:

$$Nu = 0.202 \cdot (d_p/d)^{0.175}\ Ar^{0.3}\ Pr^{0.28}$$

The corresponding maximum heat transfer coefficients and velocities could be prescribed by the next correlations: $Re = 0.16 \cdot Ar^{0.474}$ (first maximum) and $Re = 0.24 \cdot Ar^{0.568}$ (second maximum). These correlations also prescribe the data for vertical heating surfaces. At the conclusion, it was necessary to agree that the approach suggested in Ref. [257] for building of reliable theoretical models of vaporization on surfaces covered by movable porous structures rather than a perspective for the upcoming research and development. Further development of the present approach is extremely significant for the improvement of corresponding physical imaginations.

5.6 MODELS OF HEAT TRANSFER INSIDE EVAPORATORS OF LHP AND CAPILLARY PUMPED LOOPS

Design peculiarities and most of the typical experimental data for LHP and capillary pumped loops (CPL) evaporators were considered in Chapter 3. This section is devoted to the corresponding physical concepts and mathematical models suggested and developed by this book's author. It is known that the principle of vapor and liquid transport lines separation and using CPS mainly as locking wall providing pumping of working fluid was realized in LHP at a high technological level. Famous experts in the field such as Goncharov, Gerasimov, Maidanik, and Kiseev [183, 187, 356–360] were among those who have notably contributed to the development of the LHP technology. Designs of LHP and CPL evaporators allow transferring of thermal power OF ~1 kW or even greater, heat flux density of up to 100 W/cm², heat transfer coefficients in the range of 5×10^4–10^5 W/m² K, vapor and liquid lines length of more than 10 m, etc.

Development of reliable models of LHP and CPL evaporators requires special accounting for their peculiarities:

- Low heat and mass transfer intensity at small heat flux density
- Noticeably nonmonotonous character of thermal resistance dependency for some evaporators' designs and existence of a wide zone of constant thermal resistance for others
- Experimental data on heat transfer coefficients inside the LHP and CPL evaporators
- Dependency between the heat transfer intensity inside evaporators and saturation temperature
- Influence of the evaporator position, its position in relation to the condenser, transport lines geometry, and length, on the heat and mass transfer inside evaporator. etc.

The physical principles and experimental data performed by LHP and CPL researchers demonstrated that they are capable to applying such technological designs, in which a system of VRC is located just near the evaporator wall providing effective vapor generation process. It is the so-called "reverse meniscus" thermal regime.

Typical schematics of LHP and CPL evaporators are presented in Figure 5.57. Typical modes of heat transfer occurring inside the evaporators are shown in Figure 5.58.

Hence, the following thermohydraulic modes of circumferential channels operation exist:

1. Channels and porous structure are filled with a working fluid and heat is transferred from the LHP evaporator wall to the surface of compensation chamber and to the edge the vapor-generating surface of the evaporator mainly due to the effective heat conduction (Figure 5.58, mode 1). This thermal regime occurs when initial boiling is not possible even inside the near-wall circumferential channels, that is, temperature drop is less than $4\sigma T_s/ (r\rho''R_{ch}) = \Delta T_{min}$, where R_{ch} is the inscribing radius of the corresponding near-wall channel.

2. Increase in wall temperature drop causes small vapor bubbles to appear inside the channel and initial boiling to develop. The boiling regularities are similar to the case of boiling inside narrow slits of identical size. Thus, such a thermal mode is similar to boiling in narrow slits and it will be developed until wall superheat will attain a temperature drop sufficient for vapor appearance inside the porous structure, that is, $4\sigma T_s/(r\rho''R_{ef}) = \Delta T_{min2}$.

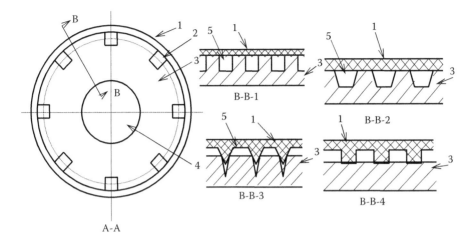

FIGURE 5.57 Design schematics of the LHP evaporators. A–A is the cross section of the cylindrical LHP evaporator; B–B-1 to B–B-4 are the typical circumferential vapor generation channels: 1, evaporator wall; 2, axial VRC; 3, main porous structure; 4, compensation chamber; 5, circumferential vapor generation channels.

3. Porous structure, including the contact surface between the porous insert and evaporator wall, is saturated with a liquid. Vaporization occurs only on those porous structure parts that are liquid-free due to heat transfer through the contact surface to the vapor–liquid interface (Figure 5.58, mode 2).

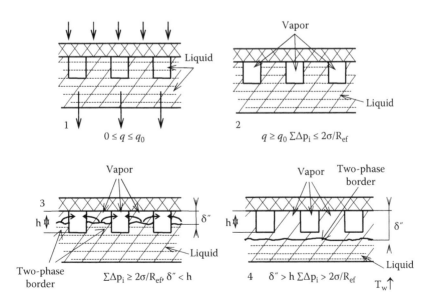

FIGURE 5.58 Typical thermal modes of heat transfer inside the LHP evaporator.

Curvature of the vapor–liquid interface is dependent on the vapor–liquid pressure drop determined for each steady-state mode by the sum of hydraulic resistances of LHP or CPL. The value of curvature remains constant within a single elementary cell of circumferential channel (Figure 5.58, mode 2). At the same time it is possible to develop here not only the clear vapor zone but also the zone with exhaustive two-phase as is shown in Figure 5.59.

4. Further rise in heat load maintains the increase in the sum of the above-mentioned hydraulic resistances. When this sum exceeds $2\sigma/R_{max}$, local deformation of evaporation surface occurs at some decrease of thermal resistance in the evaporation process. When heat load continues to increase and the sum of hydraulic resistances exceeds $2\sigma/R_{ef}$, where $R_{min} < R_{ef} < R_{max}$, then contact between the wall and the liquid in the porous structure will be eliminated and the two-phase boundary layer will appear between evaporator wall and the vapor–liquid interface (Figure 5.58, mode 3).

This mode exists until the entire porous structure around the circumferential channels is occupied by the two-phase layer. Then, vapor phase filtration through the porous structure occurs in the radial direction and the balance between the capillary pressure and the sum of hydraulic resistances appears to be irreversibly ruined.

5. The following thermal mode occurs when the main porous structure remains in superheat state, vapor occasionally enters the compensation chamber, and capillary pressure becomes steady state. The present mathematical model is not valid for such operating conditions (Figure 5.58, mode 4).

The basic thermal modes of LHP operation are presented in Figure 5.58.

When the thermal regime corresponds to temperature drop in range $\Delta T_{min} \leq \Delta T \leq \Delta T_{max}$, where $\Delta T_{min} \cong 2\sigma T_s/r\rho''R_0$; $\Delta T_{max} \cong 4\sigma T_s/r\rho''D_{ef}$; R_0 is a pore radius, D_{ef} is the channel effective diameter, and heat transfer intensity is determined experimentally for boiling heat transfer inside narrow slits as $\alpha \approx Cq^m s^{-n} p^k$, where $0 < m < 2$, $0 < n < 1$, and $0 < k < 1$. The accurate values of C, m, n, and k should be determined experimentally.

In case of intensive vaporization, recommendations presented by Labuntzov and Krukov [352, 353] and approximation performed by Krukov [351] allow us to determine the dependency of specific mass flow rate j_z on coordinate z. Hence, if

$$j_z = 0.6\sqrt{2\bar{R}_g T_s}\,(\rho_s - \rho_0)\sqrt{\frac{\rho_0}{\rho_s}} \qquad (5.73)$$

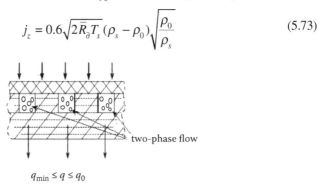

$$q_{min} \leq q \leq q_0$$

FIGURE 5.59 Schematic of thermal regime mode at narrow slit boiling.

then

$$j_z = 0.84\varepsilon \frac{1}{\sqrt{\overline{R}_\partial T_s}} \frac{dP_s}{dT} \Delta T_0(z) \tag{5.74}$$

where $\Delta T_0(z) = T_s - T_0 = (\Delta T(z) - \Delta T^*)/(1 + 2\sigma/r\rho''R)$, $\Delta T^* = 2\sigma T_s/r\rho''R$. Specific super-heat of the evaporator wall ν is determined by solving the heat conduction equation:

$$\vartheta = \vartheta_0 + \frac{2\sigma T_s}{r\rho''R} = \frac{q_0(a+b)}{\sqrt{\lambda_{\text{eff}} bB_0}} \exp\left[-z\sqrt{\frac{B_0}{\lambda_{\text{eff}} b}}\right] + \frac{2\sigma T_s}{r\rho''R} \tag{5.75}$$

where

$$B_0 = 0.84r\varepsilon \frac{dP_s}{dT_s} \frac{1}{\sqrt{R_\Gamma T_s}}; \quad \text{at } z = 0, \quad \vartheta = \frac{q_0(a+b)}{\sqrt{\lambda_{\text{eff}} bB_0}} + \frac{2\sigma T_s}{r\rho''R} \tag{5.76}$$

Superheat of evaporator wall is required for providing certain heat flux through the porous structure saturated with liquid. The total superheat of the evaporator wall can be determined by analyzing the heat transfer to contact surface between the evaporator wall and porous structure through a sequence of thermal resistances.

Assigning the total specific thermal resistance of the contact as R_{k0} yields the following correlation for total average superheat of the evaporator wall in mode 2:

$$\Delta T_0 = \frac{q_0(a+b)}{\sqrt{\lambda_{\text{eff}} bB_0}} + \frac{2\sigma T_s}{r\rho''R} + \frac{q_0(a+b)}{b} R_{k0} + \frac{3}{32} \frac{q_0}{\lambda_M \delta_M} \frac{a^3 + b^3 + ab^2}{a+b} \tag{5.77}$$

Comparison between Equation (5.74) and Equation (5.76) gives

$$j_z = \frac{B_0}{r} \frac{q_0(a+b)}{\sqrt{\lambda_{\text{eff}} bB_0}} \exp\left[-z\sqrt{\frac{B_0}{\lambda_{\text{eff}} b}}\right] \tag{5.78}$$

The one-dimensional filtration equation is $-dp/dz = (\nu'/K_f)j_z$; consequently,

$$-\frac{dp}{dz} = \frac{\nu'}{rK_f} \sqrt{\frac{B_0}{\lambda_{\text{eff}} b}} q_0(a+b) \exp\left[-z\sqrt{\frac{B_0}{\lambda_{\text{eff}} b}}\right] \tag{5.79}$$

Then, the pressure drop in capillary structure is

$$\Delta p_m = \frac{\nu' q_0(a+b)}{rK_f} \exp\left[-z\sqrt{\frac{B_0}{\lambda_{\text{eff}} b}}\right]$$

$$+ c_1 \quad \text{and} \quad \text{if } z = 0 \rightarrow \Delta p_m = \nu' q_0(a+b)/rK_f \tag{5.80}$$

In thermal mode 2, value Δp_m can be considered as supplementary hydraulic resistance due to concentration of heat and mass fluxes in the zone of vapor generation according to the reverse meniscus principle.

When specific design of the LHP allows assumption of all hydraulic resistances (except Δp_m), much less than capillary potential developed by porous structure $2\sigma/R_{ef}$; it is possible to obtain a correlation of the maximum heat flux $(q_0)_{max}$ that can be developed depending on certain type of the working fluid, saturation pressure, and basic design parameters of the near-wall capillary structure:

$$(q_0)_{max} < 2\sigma r K_f / R_{ef} v'(a+b) \tag{5.81}$$

As seen from Equation (5.81), accomplishment of $(q_0)_{max}$ requires both increasing the near-wall zone permeability K_f and reducing the radius of pores involved in the vaporization process. It is known that bidisperse pore structures provide maximum heat flux densities at the highest efficiency [354–356]. Reduction of circumferential channel dimensions (a and b) also causes $(q_0)_{max}$ to increase; however, it also leads to an increase in the sum of the hydraulic resistances inside the LHP. Under certain conditions, it can play the main role in heat flow Q and heat flux q_0 limitations before the value $(q_0)_{max}$ is achieved.

Thus, besides Equations (5.76), (5.77), and (5.80), the known steady-state thermohydrodynamic equations of the LHP should be also involved in the model. The necessity of such mutual consideration appears obvious in the case of determining transition conditions from thermal mode 2 to thermal mode 3, and for analysis of heat transfer regularities of thermal mode 3. Assuming that vapor channel length L_{vch} is independent of heat supply, the following equation yields a dependency between the curvature radius and surface of vapor–liquid interface:

$$\frac{2\sigma}{R} \ge \frac{Q}{rL_{vch}} \left\{ 4v'' \frac{\Pi_n (a+h)^2 (a+b)}{n_0^2 (ah)^3} + 0.16v' \sum \frac{1}{K_{fi}} \ln \frac{R_{i+1}}{R_i} \right\}$$

$$+ \left(\frac{Q}{r}\right)^2 \frac{1}{\rho''} \left[\frac{8.1\cdot 10^{-3} L_{vch}}{n_0^2 d_0^5} + \frac{3.2\cdot 10^{-2} L_{vch}}{D_k^5} \right] + \frac{v'Q(a+b)}{\Pi_n L_{vch} r K_f} \tag{5.82}$$

The substitution of following terms of Equation (5.82)

$$\frac{4v''\Pi_{v.ch.}(a+h)^2(a+b)}{n_0^2 r(ah)^3 L_{v.ch.}} = A_1; \quad 0.16 \frac{v'}{rL_{v.ch.}} \sum_{i=1}^{i=n} \frac{1}{K_{fi}} \ln \frac{R_{i+1}}{R_i} = A_2;$$

$$\frac{1}{r^2 \rho''} \left[\frac{8.1\cdot 10^{-3} L_{v.ch.}}{n_0^2 d_0^5} + \frac{3.2\cdot 10^{-2} L_k}{D_k^5} \right] = A_3; \quad \frac{v'(a+b)}{\Pi_{v.ch.} L_{v.ch.} r K_f} = A_4 \tag{5.83}$$

allows the next correlation

$$2\sigma/R \geq (\{A_1 + A_2 + A_3 Q\} \times Q + A_4 Q \tag{5.84}$$

Simultaneous consideration of Equations (5.84) and (5.77) and introducing specific thermal resistance at evaporation section of the LHP as $\Delta T_0 / q_0 = R_{vch}$ in the thermal mode 2, yields

$$R_{vch} = \frac{\Delta T_0}{Q} \Pi_{v.ch.} L_{v.ch.} = \frac{a+b}{\sqrt{\lambda_{eff} b B_0}} + \frac{\Pi_{v.ch.} L_{v.ch.} T_s}{r\rho''} \{[A_1 + A_2 + A_4] + A_3 Q\}$$

$$+ \frac{a+b}{b} R_{k0} + \frac{3}{32} \frac{1}{\lambda_m \delta_m} \frac{a^3 + b^3 + ab^2}{a+b} \tag{5.85}$$

As seen from Equation (5.85), in thermal mode 2 the specific thermal resistance (heat transfer coefficient) is practically independent of heat supply value. It is also justified by existing experimental data. Equation (5.85) provides a proper description of the following parameter interrelations:

- Decreasing dimensions a and b causes reduction of R_{vch}, that is, increasing heat transfer coefficients; increasing the heat transfer parameters of evaporator wall, δ_m and λ_m, maintains the decrease of R_{vch}.
- When the LHP is operating in the gravity field, the extra term $\rho'gL_0 \sin\varphi$ should be added to the right-hand side of the correlation (5.84). The angle φ accounts for the LHP evaporator position in relation to the condenser. If $\rho'gL_0 \sin\varphi > 0$, the value R_{vch} is increased, whereas the negative values of this term uphold the reliable operation of the LHP at low heat supply until the correlation (5.84) becomes valid.

Moment of transition from thermal mode 2 to thermal mode 3 is defined by the correlation (5.84). Thus, if (5.84) is not valid at increasing heat supply and $R \rightarrow R_{min}$ in the contact zone between the evaporator wall and porous structure saturated with liquid, it creates the conditions encouraging vapor–liquid interface displacement from the evaporator wall and generation of vapor layer between the evaporator wall and vapor–liquid interface inside the porous structure. As shown by Yatscenko [52] and Demidov and Yatsenko [361], the actual geometric shape of the interface is quite complex.

Generation of vapor layer causes vaporization over the whole surface of the vapor–liquid interface separating the porous structure saturated with the liquid from the evaporator wall. It means both that the first term of the right-hand side of Equation (5.85) could be neglected and increasing the value of R_{k0} in the third term. In other words, vaporization of liquid in the contact zone requires introducing an extra thermal resistance term δ/λ_{ef} corresponding to the heat transfer from the evaporator wall to vapor–liquid interface through the porous structure occupied by a vapor layer with thickness δ.

Depending on hydraulic resistance ratio, the rate of vapor layer thickness increasing changes with respect to the increase in the heat supply, but, as a rule, $d\delta/dQ > 0$. Available experimental data justifies the increase in specific thermal resistance when heat supply rises in thermal mode 3. Thus, the following equation is valid for thermal mode 3:

$$R_n = \frac{\Pi_{v.ch.}L_{v.ch.}T_s}{r\rho''}\left\{A_1 + A_2' + A_3Q + A_5\right\} + \frac{a+b}{b}R'_{k0}$$

$$+ \frac{3}{32}\frac{1}{\lambda_m\delta_m}\frac{a^3 + b^3 + ab^2}{a+b} + \frac{\delta}{\lambda_{ef}} \tag{5.86}$$

Vapor layer thickness δ is determined by solving the corresponding hydrodynamic equation at boundary condition following from the correlation (5.84). When $R = R_{min}$, the border between the thermal modes 2 and 3 ($Q = Q_0$) could be determined. Irreversible failure of the LHP operation occurs during further displacement of the vapor–liquid interface when $Q = Q_{max}$ and thickness δ becomes equal to h, that is, increase of vapor filtration hydraulic resistance is not equalized by the decrease of hydraulic resistance of liquid filtration inside the porous structure anymore. The value Q_{max} corresponding to $\delta = h$ determines the maximum value of heat supply allowing reliable operation of the LHP.

The present approach cannot be considered a completed theory or model prescribing accurate values of corresponding thermal and hydraulic resistances. A number of improvements can be made to theoretical descriptions at different stages of the heat and mass transfer process including more accurate two-dimensional or three-dimensional models of heat and mass transfer. However, the present concept should be considered as the basis for further investigations devoted to the following issues:

• Determination of optimal geometrical and technological parameters (a, b, h, d_0, etc.) by using the minimum value of thermal resistance $\Delta T_0/Q$ at q given value of heat supply.
• Calculation of value Q_{max} in constraint of $\Delta T_0/Q$ (a_{opt}, b_{opt}, L_{vch}, etc.).
• Determination of LHP optimal parameters at a given value of Q_0 with respect to minimum mass value, maximum reliability, etc.
• Investigation of the physical nature of various thermal and hydrodynamic instabilities.

Using the current concept, especially in modeling the LHP dynamics, requires paying special attention to conditions of transition between thermal modes, uncertainty of hydraulic parameters due to flow mode changing, position of the vapor–liquid interface and its stability, as well as inconsistency of factors determined in Sections 4.1, 4.2, and 4.3.

6 Heat Transfer Crises at Vaporization inside Slits, Capillaries, and on Surfaces Covered by Capillary-Porous Structures

6.1 PHYSICAL EXPLANATIONS AND SEMIEMPIRICAL MODELS OF BOILING HEAT TRANSFER CRISIS

Various regimes and geometric factors play an important role in the development of physical explanations when a boiling heat transfer crisis occurs in restricted spaces.

The application of conventional models of heat transfer crisis represents the simplest way of resolving such a problem. Prof. Kutateladze [266, 268] developed a known correlation of the hydrodynamic theory of boiling crisis with the accuracy of one empirical constant basing on explanations of the hydrodynamic nature of the crisis and using both the similarity between the boiling and barbotage processes and the dimensional approach.

Soon after, various models of boiling heat transfer hydrodynamic crisis were suggested, including the phenomenon of hydrodynamic stability failure due to jet-shape vapor phase removal.

For example, the initiation of a crisis was associated with the loss of vapor jet (columns) stability [267]. Accounting for the loss of stability using the Helmholtz model was used for the analysis of the limit stability of the vapor (liquid) jet. Column density was estimated using Taylor's interface stability correlation. Borishansky and Gotovsky [269] and Linchard and Dhir [270] developed similar explanations. In particular, thickness of the stable near-wall vapor layer was anticipated via the vapor bubble removal diameter [270]. Kirichenko and Chernyakov [271] attempted to determine a correlation of another type of boiling crisis via the vapor bubble junction condition at the increase in density of vaporization centers.

Such type of crisis occurs at boiling in restricted spaces, when $q \ll q_{CR}$. Kirichenko and Chernyakov [271] and Labuntsov [272] observed that calculations of pool boiling critical heat fluxes (CHFs) do not always correspond to the correlations of the hydro-dynamic crisis theory, according to which an increase in heat fluxes is constrained by the conditions of vapor phase removal. Although thermal or thermodynamic boiling crisis occurs when vapor phase removal conditions are well below the limit, the wall temperature exceeds the value corresponding to the restricted metastable liquid state.

Labuntsov [272] suggests that restricted heat fluxes at thermal crisis can be estimated as follows:

$$q_{CR} \geq \alpha(T_W - T_s) \tag{6.1}$$

Calculations of pool boiling heat transfer coefficient α account for its dependence on heat flux. As noted by Labuntsov [272], thermal crisis can be achieved at the given wall temperature (boundary conditions of the first type), that is, by using a condensing vapor wall heating or by means of high-temperature gas heating at the external wall side. The hydrodynamic crisis occurs when the boundary conditions of the second type are realized along the heated wall [269–271]. However, as noted by Yagov and Pusin [273], a deviation from the hydrodynamic crisis theory was observed even at heating by the constant heat flux.

In general, a CHF is defined by the joint action of hydrodynamic factors and drying conditions of liquid microfilm in the vapor bubbles' feet [272]. Accounting for the liquid microfilm drying helps explain the influence of saturation pressure on the regularities of the CHFs [272, 273].

Calculations presented by Tolubinsky [277] are based on the pool boiling internal characteristics determined experimentally. Koshkin et al. [13] and Grigoriev et al. [131] showed the significant effect of the thermophysical properties of the heater's wall and its thickness on the bubble crisis at pool boiling. In agreement with [272, 273], initial physical statements explained that the analogous effects are associated with microfilm drying processes and heat flux spreading from the drying spots determined by boiling process relaxation times and thermophysical properties of the wall.

Analysis of the pool-boiling crisis regularities based on the position of thermal instability accomplishment was presented in several studies [274, 275]. Such an approach is based on the determination of q and ΔT matched to the case when a transition from one boiling form to another occurs along the wall of the given geometry. Conditions of the stable coexistence of different boiling forms were established.

Moreover, joint dimensionality analysis and criteria treatment of the experimental data based on the scaling method [136] and generalization of experimental data via characteristic scales obtained from statistical treatment of internal boiling parameters [277] were used for simplification of experimental data on boiling heat transfer crisis.

Various models of heat transfer boiling crisis are based on the statement that different forms of near-wall liquid microfilm evaporation play a major role in initiating the crisis.

The general outlook of different approaches to the problem relating to the generalization of experimental data on critical boiling heat fluxes can be obtained from the reports of several leading experts in the field; aside from presenting models, these studies also include an analysis of other concepts. Thus, Yagov and Zudin [278] presented a model of crisis phenomena and discussed models developed in Refs. [279, 280].

Determination of CHF was associated with evaporation of the liquid microfilm in the vapor slug base [279, 280]. It was supposed that vapor slug diameter is equal to the critical wavelength of the Taylor instability. Liquid microfilm thickness was determined as a quarter of the Helmholtz instability wavelength; in addition, CHF occurs when liquid microfilm drying time becomes smaller than a vapor slug existence time in the given near-wall location.

This approach yields a correlation — which is similar to the Kutateladze formula — that also accounts for the influence of geometric factors. Yagov and Zudin [278] cited the following drawbacks of the model:

- Iterative character of the CHF calculations.
- Application of empirical correlations for such hardly determined parameters as flow velocities of vapor layers, local near-wall velocities of two-phase flow, etc.

Professor Yagov and his team developed a model of the CHF determination at saturation and subcooled liquid forced flow in the channels [273, 278, 282], which was considerably improved from the viewpoint of the above-mentioned disadvantages of the methods [279, 280]. According to the Yagov model, vaporization at forced boiling in channels and in case of pool boiling takes place inside thin liquid layers formed in vapor slug feet on the heating surface. Evaporation of liquid microlayers along the drying spot borders is the main contributor to the vaporization intensity.

At given size values of the drying spot, L_D, and liquid film average thickness, $\bar{\delta}$, the limited steady heat flux, which could be removed from the heat transfer surface, is equal to $q_D \approx (r\sigma/v') \times (\bar{\delta}/L_D)$. The crisis occurs when the heat flux $(q) \geq q_D$.

Assuming that the drying spot size L_D is proportional to the vapor slug equivalent diameter before its departure from the wall, and using a correlation between the liquid microlayer average thickness, $\bar{\delta}$, and the heat flux, Yagov obtained two equations of boiling heat transfer CHFs. One is valid for the low-pressure region ($p_s/p_{CR} < 10^{-3}$) and another one for the high-pressure zone ($p_s/p_{CR} > 0.05$).

The combined use of these correlations yields an interpolation formula that coincides well with numerous experimental data of different authors, and avoids the principal drawback of the crisis hydrodynamic theory connected with significant quantitative and qualitative deviation of experimental data and calculations at pool boiling in low-pressure region, especially for liquid metals [278].

A similar approach was developed [273, 278, 282] for boiling of saturated and subcooled liquid in channels. The influence of heater sizes on CHFs at pool boiling is not accounted for by the present generalized correlations. According to Yagov and Zudin [278], at forced flow conditions, an effect of the characteristic size d_0 can be estimated as

$$q_{CR} \approx B_1 d_0^{-0.13}/\left(1 + B_2 d_0^{-0.25}\right) \tag{6.2}$$

Hence, in the correlation $q_{CR} \approx C d_0^m$, m is in the range $-0.13 < m < 0.12$, and B_1 and B_2 are empirical constants. Such appearance of dependency deviates from numerous experimental data on boiling in restricted spaces. Therefore, the models by Yagov et al. [273, 278, 282] are inadequate for the given case.

Celata [283, 284] presented a comprehensive analysis on CHF at forced liquid flows in channels. According to Celata [284], the majority of theoretical models can be classified into five groups by different physical mechanisms of crisis initiation:

1. *Liquid layer limit overheating model.* The near-wall bubble layer hampers the liquid transport to the wall. Hence, overheating of the near-wall liquid layer increases and initiates a heat transfer crisis.
2. *Boundary layer removal model.* Kutateladze [268] suggested the model based on the analogy between the dynamic action of the vapor flow blown into the liquid and similar action of gas flow at barbotage. Such a phenomenon decreases a velocity gradient in the near-wall zone until the complete removal of the liquid layer. Drawbacks of the current approach were presented in the references to the paper [284].
3. *Liquid flow-blocking model.* It was supposed that the crisis is initiated when the normal-to-the-wall liquid flow is broken by the counter vapor flow. It corresponds to critical counter phases' flow (vapor–liquid) along the interface [286]. Smogalev [287] suggested the energy schematic of the CHF achievement, that is, at the specific value of the vapor flow kinetic energy, liquid flow stable removal from the wall can be achieved.
4. *Model of vapor removal restriction or bubble cloud formation near the wall.* According to the model, the CHF appears when the near-wall vapor bubble cloud blocks sufficient liquid transportation to the wall [288, 289]. It was assumed that the CHF occurs after the actual vapor quality near the wall reaches some critical value. The authors assumed this value as equal to 0.82. Experimental investigations [290] showed that the CHF initiates upon reaching critical vapor quality, from 0.3 to 0.95. Celata [284] examined the third and fourth models and demonstrated that calculations have deviated significantly from the experimental data. Taking into account the drawbacks of the first and second models, Celata suggested the fifth model.
5. *Model of liquid sublayer drying.* The CHF emerges at complete drying of thin liquid microlayers located under vapor slugs (vapor blanket) moving in the near-wall zone. It should be noted that such an approach has some similarities with the Yagov model [273, 278, 282].

The basic principles of the current model were developed in several studies [283, 284, 291]. The CHF value can be estimated as

$$q_{\mathrm{CR0}} = (\bar{\delta}\rho'r/L_B)U_B \qquad (6.3)$$

where $\bar{\delta}$ is the average thickness of evaporating liquid sublayer, and L_B and U_B are the characteristic size and characteristic velocity of the vapor slugs located near the wall, respectively. Distinctions in correlations suggested by different authors [283, 291, 292] are mostly related to the designation of values $\bar{\delta}$, L_B, and U_B.

In all known cases, the estimations of $\bar{\delta}$, L_B, and U_B are not associated with geometric parameters [284]. However, at boiling in restricted spaces, both experimental data and physical explanations confirmed a considerable effect of geometry on CHF, which is due to the significant increase in hydraulic losses during the two-phase flow filtration.

Hence, it seems helpful to use the fundamentals of hydrodynamic theory of boiling crisis. On the other hand, a disagreement between this theory and experimental data should be accounted for.

Moreover, existing critical comments to the crisis hydrodynamic theory, including statements suggested by Zuber [267], should be considered carefully. The model [267] is associated with the specific two-phase boundary layer structure and eliminates an opportunity to account for specific surface effects such as roughness, thermal accumulation properties of the wall, etc.

However, the known energy form of the crisis hydrodynamic theory [268] is not associated with the specific structure of the two-phase boundary layer. Thus, it provides definite possibilities for development and improvement of the theory. These circumstances were not considered in the analysis of the crisis hydrodynamic theory [273, 277, 278].

Therefore, application of the crisis hydrodynamic theory to boiling in restricted spaces is promising and reasonable.

It is also important to note that authors of recent models criticized components (parameters) of the crisis hydrodynamic theory used in their models such as hydrodynamic loss factor (C_f), Reynolds number, characteristic velocities, etc.

In addition, insufficiency of CHF experimental data for many cases of boiling in restricted spaces and the absence of data on boiling internal characteristics and CHF mechanisms should be accounted for in the development of effective models for generalization of the experimental data on CHF in restricted spaces. The first step toward developing such an approach was taken by Smirnov [265]; the fundamentals were presented by Bologa et al. [264].

6.2 MODIFIED HYDRODYNAMIC THEORY OF BOILING CRISES IN RESTRICTED SPACES

Accounting for boiling heat transfer crisis as a result of one type of two-phase boundary layer stability failure and transition to another means that analysis of the crisis regularities should be connected with the problem of two-phase boundary layer stability loss.

The simplest solution to the problem is based on the approach suggested by Kutateladze [268]. This model assumes that two-phase boundary layer stability loss occurs when the specific kinetic energy of blown vapor flow E is sufficient for formation of the stable vapor layer with specific potential energy Π, that is,

$$E \geq k_1^2 \, \Pi_g \qquad (6.4)$$

where k_1 denotes the two-phase boundary layer stability criteria [268].

Applying such an approach for boiling in restricted spaces, it was assumed that the specific potential energy of the stable vapor layer will be reduced by the term, which is proportional to a work required for overcoming the friction forces acting at two-phase flow filtration through restricted spaces (slits, capillaries, capillary-porous structures, etc.), that is,

$$E \geq k_1^2 \prod g - k_2 \prod_{HL} \tag{6.5}$$

where $\prod_{HL}$ assigns a specific work required for overcoming the action of friction forces, which is proportional to two-phase mixture filtration hydraulic losses, ΔP_{HL}.

Similarly, the initial correlation for treatment of boiling crisis experimental data accounting for the possible influence of the wall material's thermophysical properties, its thickness, liquid subcooling, etc., can be obtained. During a boiling heat transfer crisis, heat supply can be partially accumulated by the wall when the thickness of the heating surface significantly exceeds the thickness of the relaxation layer (temperature oscillations occur at boiling within such a layer). Assigning E_a to represent the decrease in kinetic energy of the blown vapor flow corresponding to the accumulated heat yields the following equation

$$E \geq E_a + k^2 \prod g \tag{6.6}$$

At boiling under the action of non-Archimedean forces with density f, the generalization of initial experimental data on the crisis appears in the form (6.4). The mutual action of other volumetric forces should be accounted for by the specific potential energy value $\prod_g$ as

$$\prod g \quad ((\rho' - \rho'')g + f)\delta \tag{6.7}$$

When the effect of forces f on stable vapor film thickness is similar to the action of gravitational forces, then

$$\prod g \quad \sqrt{\sigma(\rho' - \rho'')g} + f\delta \tag{6.8}$$

Specific features of calculations of boiling CHFs appear for long slits, capillaries, and under restricted circulation conditions (narrow channels, inlets, nonorganized circulation loops, etc.) [129, 130, 132]. From the hydrodynamic viewpoint in such cases, CHFs are not determined by microcirculation stability conditions in the direction normal to the wall. On the other hand, stability of vapor removal along the heating wall surface influences the values of CHFs. Hence, crisis phenomenon in such cases has a hydrodynamic nature, and it is similar to barbotage or flooding processes in chemical technology systems [293].

Under such conditions, CHFs calculated under the assumption of stability loss along the channel could be at the same order of magnitude as the fluxes determined from the two-phase boundary stability loss in the direction normal to the wall. Thus, it is necessary to account for the two following correlations:

$$E_y > k_1^2 \prod_{gy}; E_z > k_1^2 \prod_{gz} \tag{6.9}$$

Under such conditions, E_z is proportional to E_y with the proportionality coefficient determined by the channel geometry, that is,

$$E_z (F_z/F_y)^2 \quad E_y \tag{6.10}$$

Assuming that the z axis coincides with the heating surface plane, F_z is the heat input surface and F_y is the cross section area of channel S. Hence, as a replacement for (6.9) and (6.10), we obtain

$$\sqrt{E_y} \frac{1}{\sqrt{k_1^2 \, \Pi_{gy}}} > 1; \quad \sqrt{E_y} \frac{F_z}{F_y} \frac{1}{\sqrt{k_1^2 \, \Pi_{gz}}} > 1 \tag{6.11}$$

Substituting of E_y with E allows us obtain the joint condition of crisis as an interpolating correlation

$$\sqrt{E} \left(\frac{1}{k_1 \sqrt{\Pi_{gy}}} + \frac{F_z}{F_y} \frac{1}{k_1 \sqrt{\Pi_{gz}}} \right) > 1 \tag{6.12}$$

Equation (6.12) concurs well with the generalization of numerous experimental data on critical fluxes in two-phase thermosiphon [343]. Hence, a consistent explanation for different regularities of boiling heat transfer crisis in slits observed in experiments [129, 130] can be suggested. Therefore, assuming the hydrodynamic nature of the boiling crisis in restricted spaces, the initial correlation of the hydrodynamic theory can be easily modified as

$$\sqrt{E} > \frac{1}{\left(k_1^2 \, \Pi_{gy} - k_2 \, \Pi_{TPy} \right)^{-0.5} + (F_z/F_y) \left(k_1^2 \, \Pi_{gz} - k_2 \, \Pi_{TPz} \right)^{-0.5}} \tag{6.13}$$

The action of volumetric forces should be accounted for by Π_{gy} and Π_{gz}, similar to Equation (6.7).

The present analysis of boiling crisis regularities can be developed for such complicated cases as flow in channels, thermal-hydraulic crisis, etc. Complete prescription of the two-phase layer should comprise equations determining the dynamic or steady characteristics of the liquid microlayer and the microlayer drying conditions, and account for heat conduction inside internal surface layers of the heating wall.

However, lack of experimental data on the process internal characteristics restricts estimation of the model's correctness. Hence, generalized correlations of the modified hydrodynamic crisis theory can be applied as the first approximation for the experimental data treatment. Such correlations require the identification of two-phase flow hydraulic losses, ΔP_{HL}. Its structure depends on the nature of restricted spaces. Obtained correlations should be checked by comparison with experimental data. Specific applications of the suggested approach to experimental data generalization at boiling in restricted spaces are further considered.

6.3 EXPERIMENTS ON BOILING HEAT TRANSFER CRISIS AT FORCED LIQUID FLOW IN SLITS AND CAPILLARY TUBES

As noted earlier, experimental data on boiling heat transfer crisis in restricted spaces are very limited. It is probable that the study conducted by Kozhelupenko [140] is one the most comprehensive experimental investigations devoted to this problem. A thorough review [4] confirms this statement.

Detailed experimental results presented by Kozhelupenko [140] were cited in several reports [294–297]. A review of the problem, experimental methods, and results, including generalization of experimental data presented by Kozhelupenko [140], remain up-to-date.

CHF experimental data at forced convection inside capillary tubes [140] are presented in Table 6.1, and those for annular channels are shown in Table 6.2.

6.3.1 Boiling Crisis at Forced Flow in Capillaries

In general, CHF depends on channel geometric parameters (diameter, thickness, length, roughness, shape, and position in gravity field), regime parameters (pressure, subcooling, heat carrier velocity, and heat input conditions), and thermophysical properties of heating surface and heat carrier. Boiling heat transfer crisis at forced flow, as a rule, appears complicated because of hydrodynamic and thermal flow irreversibility, which is especially noteworthy at large subcooling and high heat fluxes.

In spite of numerous experimental investigations of CHFs at forced flow, only some of them were conducted at low pressures. The diagram presented in Figure 6.1 illustrates the allocation of experimental works by two independent factors: channel diameter and outlet pressure. It is shown that experiments were completed mostly in

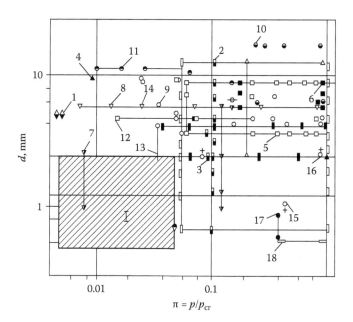

FIGURE 6.1 Experimental research on the CHF inside cylindrical capillary channels. Heat carrier—water: 1, [301]; 2, [303, 304]; 3, [307]; 4, [326]; 5, [327]; 6, [328]; 7, [298]; 8, [346]; 9, [347]; 10, [330]; 11, [335]; 12, [309] (ethanol, acetone, ethylenglycol, mixtures); 13, [306] (ethanol); 14, [329] (ethanol); 15, [345] (helium). Heat carrier — helium and nitrogen: 16, [312]; 17, [300]; 18, [311] (helium). I, Region of regime parameters studied by Kozhelupenko [140].

TABLE 6.1

Experimental Data on Boiling Heat Transfer in Capillary Channels

Reference	The Channel's Parameter/ Heat Carrier	Regime Parameters				Method of Crisis Fixation	Formula
		p, MPa	ρw, kg/ (m² sec); w, m/sec	ΔT, K	q_{CR}, MW/ m²		
[307]	Horizontal tube: $d = 2$–3 mm; copper, stainless steel/water	2–20	$(2$–$40) \times 10^3$	20	5.2–81.2	burnout, wall temperature	$q_{CR} = q_0(1 + B\Delta T)$ $(1 + w/w_0)^{0.8}$ $q_0 = f_0(p/p_{CR})$ $B = f_1(p/p_{CR})$
[303, 304]	Vertical tube: $d = 2$–3 mm; $l = 40$; alloyed copper	1–7.5	$(5$–$30) \times 10^3$	100–200	0.58–0.81	burnout	$q_{CR} = 3 \cdot 10^4$ $(\rho w)^{0.44} \Delta T$ $1 \le p \le 5$ MPa
[313]	Vertical tube: $d = 1.29$–4.78 mm; $l/d = 25$–250	0.1–7		0–78	2.83–41.4	burnout	At $G/(l/d) < 150$ $q_{CR} = 270d^{-0.2}$ $(l/d)G^{0.85}$ At $G/(l/d) > 150$ $q_{CR} = 1400d^{-0.2}$ $(l/d)^{-0.15}G^{0.5}$

(Continued)

TABLE 6.1   (Continued)

Experimental Data on Boiling Heat Transfer in Capillary Channels

Reference	The Channel's Parameter/ Heat Carrier	Regime Parameters				Method of Crisis Fixation	Formula
		p, MPa	ρw, kg/ (m² sec); w, m/sec	ΔT, K	q_{CR}, MW/ m²		
[305]	Water (further information not available)	0.1–20	$(0.8{-}8)\times10^3$ 0.7–45	0–80 0–240	1.6–5.5 1.16–52.2	burnout	$q_{CR} = 1.45\varphi(p)$ $\sqrt{1+\left(\dfrac{2.5}{\varphi(p)}\right)}\,w^{2}\Theta(\Delta T)$ $\varphi(p) = p^{1/3}$ $(1-p/p_{CR})^{4/3}$ $\Theta(\Delta T)=1+(15/\sqrt{p})(c_p/r)\Delta T$ channel size influence starts at $d<2$ mm
[306]	Horizontal tube: $d = 2{-}3.5$ mm; $l = 40{-}210$, stainless steel; bronze; 96% solution of ethanol; water	0.2–6	$50{-}500\times10^3$	0–220	18.6	by the wall temperature growth	For water: $q_{CR}=q_0(1+(9.5\Delta T/T_s))$ $(1+Bw^{0.8}))$ For C_2H_5OH: $q\kappa_p = q_0(1+$ $4.6\Delta T/T_s)$ q_0, B — coefficients depending on p

[312]	Vertical tube: $d = 2$ mm; $l = 30$ mm; 96% solution ethanol	0.5–55	$(5–34) \times 10^3$	0–160	9.5	by changing tube color; bridge schematic	$$\frac{q_{CR}\sqrt{\rho'/\rho''}}{r\rho'_w 0\left(1-\dfrac{\Delta i}{r}\sqrt{\dfrac{\rho'}{\rho''}}\right)}$$ $= 0.0013 + 0.14 Fr^{*-1}$ $$Fr^* = \frac{\rho'_w}{\sqrt{\rho''(\rho'-\rho'')}\,\sqrt[4]{\dfrac{\rho'-\rho''}{g\sigma}}}$$ corresponds to the data [306, 307]
[301]	Vertical tube: $d = 0.5–12$ mm; $l = 2–300$ mm; water	1–21	$(0.75–90) \times 10^3$	0–220			influence of d, ρw, ΔT on the critical heat flux
[302]	Vertical tube: $d = 0.5–2$ mm; $l = 140$ mm; brass; copper; water	1.1–7.1	$(10–90) \times 10^3$	60–160	175–230	Burnout	At $d = 0.5$ mm $q_{CR} = [1.4 \times 10^6 p^{-0.3} + 10^{-9}(\rho w)^3 \Delta T]$

(Continued)

TABLE 6.1 (Continued)
Experimental Data on Boiling Heat Transfer in Capillary Channels

Reference	The Channel's Parameter/ Heat Carrier	Regime Parameters				Method of Crisis Fixation	Formula
		p, MPa	ρw, kg/ (m² sec); w, m/sec	ΔT, K	q_{CR}, MW/ m²		
[299]	Tubes: $d = 2$–12 mm; $l/d = 10$–120; water	0.49–19.7	$(0.5$–40) $\times 10^3$	25–220			At strong subcooling $$\frac{q_{CR}}{rp''w_0} = 18.25 \left(\frac{c_p \Delta Tp'}{rp''}\right)^{0.35} \sqrt{\frac{\chi'}{w_0 d\rho' c_p}} \left(\frac{r}{c_p' T_s}\right)^{1.2}$$
[310]	Tube, rectangular channel, annular: $d = 3.63 \times 16.8$ mm; aluminum; water	0.16	$(1.2$–16.5) $\times 10^3$				$q_{CR} = \alpha_{CR}(T_{wCR} - T')$
[314]	Vertical tube: $d = 3.25$–9.9 mm; $l = 150$–300 0 mm; water	0.1–14	$(40$–850) $\times 10^3$	15–220			At small ρw $q_{CR} = 16.5(r + \Delta i)\rho w$ $[d^{0.1}.\rho w + (4l/d)]$ small and big velocities range defined by charts

Ref	Description					Method	Equation
[300]	Vertical tube: $d = 0.67$ mm; $l = 120$ mm; stainless steel; helium; $d = 0.315$–1.29 mm; nitrogen	0.1	0.025–0.22	0	up to 0.088		$q_{CR} = 1/4$ $[rp''w/(l/d)]$
[311]	Vertical tube: $d = 2.13$ mm; $l = 100$ mm; stainless steel; down flow; helium	0.1–0.2	0.04–0.65	$x = 0.3$–0.3	up to 0.01	not specified	$\dfrac{q_{CR}}{r\sqrt{g\sigma(\rho'-\rho'')}} = 0.031 + 0.078$ $(1-x^{3.92})$
[308]	Vertical tube: $d = 2.05$ –10.2 mm; $l = 40$ mm; water	5.14	$(1.6$–$7) \times 10^3$	0		by using tube color changing	$q_{CR} = \left\{6.0 - 3.05 \left[1+(1-x)/x \times (\rho''/\rho')\right]^{-1}\right\} \times d^{0.4} \times 10^5$
[298]	Vertical tube: $d = 0.97$–4.0 mm; $l = 120$–1220 mm; stainless steel; water	0.1–0.18	$(0.2$–$18) \times 10^3$		0.12–1.9	by using tube color changing and wall temperature	$q_{CR} = 2.4 \times 10^7$ $(w^{3/2} d/\sqrt{l})^{0.8}$

(Continued)

TABLE 6.1 (Continued)

Experimental Data on Boiling Heat Transfer in Capillary Channels

Reference	The Channel's Parameter/ Heat Carrier	Regime Parameters				Method of Crisis Fixation	Formula
		p, MPa	ρw, kg/ (m² sec); w, m/sec	ΔT, K	q_{CR}, MW/ m²		
[309]	$d = 4$ mm, $l = 60$ mm; solutions: water + ethanol; acetone + water; ethanol, water + ethylenglycol: 0–100%	0.33–1.32	2500–10,000	10–110		burnout	q_{CR}^{lb} $= \left[\begin{array}{c} q_{CR}^{hb}\,(1-N) \\ +q_{CR}^{lb}\,N \end{array} \right]$ $\rightarrow \left[\begin{array}{c} 1+1{,}5\Delta N^{1.8} \\ +6{,}8\Delta N \\ \times \dfrac{\left(T_s^{Sl} - T_s^{Sl}\right)}{T_s^{HK}} \end{array} \right]$

N is concentration; lb and hb are the low and high boiling components, correspondingly; Sl is solution

TABLE 6.2

Crisis Boiling Heat Transfer at Forced Flow in Annular Channels [140]

Reference	Geometric Parameter/Heat Carrier	Regime Parameters	Recommended Calculation Correlations
[315]	$\delta/h = 12$–14 (h is slit height; water)	$p = 0.1$ MPa $w = 2.6$ m/sec $2 \leq \Delta T \leq 35$ K $1.6 \leq q_{CR} \leq 6$ MW/m²	$q_{CR} = 0.1r\sqrt{\rho''}\sqrt[4]{g\sigma(\rho' - \rho'')}$ $+ 0.013\rho' c_p w (T_s - T')$
[321]	$d_h = 0.008$–0.03 m $\delta_{sl} = 0.0005$–0.0025 m $l = 0.1$–0.8 m eccentricity $0 \leq \varepsilon \leq 1$ vertical flow, degassed, and dissolved water	$5 \leq p \leq 20$ Mpa $250 \leq \rho w \leq 2000$ kg/(m² sec) $20 \leq t_{fn} \leq (T_s - 20)°C$ $x < x_{BR}$	$q_{CR} = 0.82 q_{cr1} \left[\sqrt[4]{\rho w} \right] \left(1 - \dfrac{p}{p_{cr}} \right)^{0.1}$ $\left(\dfrac{d_{eq} l_h}{d_h} \right)^{0.2} \left[1 - 0.06 \sqrt{\rho w} \left(\dfrac{d_{eq}}{d_h} \right)^{0.2} x \right]$, MWt/m^2; $q_{cr1} = 7r\sqrt{a'jp'\rho''} \times 10^{-6}$
[317]	$D = 0.014$ m; $d = 0.01$ m; $\delta_{sl} = 0.002$ m; $l = 0.9$ m; $l_h = 0.03$–0.9 m, water	$9.8 \leq p \leq 13.7$ MPa $500 \leq \rho w \leq 2000$ kg/(m² sec) $20 \leq t_{fn} \leq 280°C$	According to the analysis, increasing the power has no effect on the outlet conditions: $\Delta N_{CR} = N_{CR}$ $- \left[\Pi_h \left(\int_0^{z_1} q(z)dz + \int_{z_1}^{z_2} q_{tb}(z)dz \right) \right]$ where Π_h is the heated perimeter and q_{tb} is the initial boiling heat flux

(Continued)

TABLE 6.2 (Continued)

Crisis Boiling Heat Transfer at Forced Flow in Annular Channels [140]

Reference	Geometric Parameter/Heat Carrier	Regime Parameters	Recommended Calculation Correlations
[316]	$d = 0.01$ m; $deq = 0.003$–0.004 m; $l = 0.1$–0.9 m; $1h = 0.03$–0.9 m; water	$7.35 \leq p \leq 21.1$ MPa; $500 \leq \rho w \leq 3500$ kg/(m² sec); $30 \leq tin \leq (250 - 300)°C$; $1 \leq (qmax/qmin) \leq 4.4$	$\dfrac{\Delta i_{CR}^{n}}{i' + r} Fr^{*0.3}\left[1 + \dfrac{190}{(F/s)}\right]$ $\sqrt[4]{\dfrac{q_{max}}{q_{min}}}\,\dfrac{2l_{max} - l_h}{l_h} = 1.8 - 0.8Fr^{*0.15}\dfrac{i_{in}}{i'};$ $Fr^* = \dfrac{\rho w}{\rho'}\sqrt[4]{\dfrac{\rho' - \rho''}{g\sigma}}$
[324]	$d = 1.5$–3.7 mm; $\delta_{sl} = 0.5$–3.6 mm; $l = 0.5$–2.87 m; vertical channel; water	$p = 0.098$ MPa; $3 \leq w \leq 12$ m/sec; $60 \leq tin \leq 136°C$; $6.8 \leq q_{CR} \leq 12$ MW/m²	$\delta_{sl} < 2.5$; $p \leq 0.1$ MPa; $\Delta T < 11\,K$; $q_{CR} = 2.61 \times 10^6 \delta_{sl} w^n (1.23 + 0.01 t_{out})$; $n = (1.06 + 0.42\delta_{sl})/(1 + 1.7\delta_{sl})$
[322]	$D = 0.02$ m; $\delta_{sl} = (0.1$–$1.3) \times 10$ m; $l = 0.02$–0.15 m; $1h = 0.02$ m; vertical channel; distilled and degassed water	$5 \leq p \leq 21.5$ MPa; $50 \leq \rho w \leq 800$ kg/(m² sec); $0 \leq \Delta i \leq 600$ kJ/kg; $0/8 \leq q \leq 4.5$ MW/m²	$\rho w > 200 kg/(m^2\ sec)$; $q_{CR}^{sub} = q_{CR}\left[1 + 0.012\Delta Tpw\left(\dfrac{th\delta_{sl}}{l}\right)^2\right];$
[320]	$d = 6.3 \times 10$ m; $\delta_{sl} = 2.3$–6.6 mm	$0.196 \leq p \leq 0.589$ MPa; $0.3 \leq w \leq 3.94$ m/sec; $11 \leq \Delta T \leq 83$ K; $6.99 \times 10^5 \leq q \leq 58 \times 10^5$ W/m²	$q = C\Delta T_{in}$; $C = 2.3$–6.0

Ref.	Conditions	Parameters	Correlation
[323]	Vertical channel, $\delta_{Sl} = 0.0001-0.0002$ m; single-side and two-side heating; distilled and degassed water	$9.8 \le p \le 21.6$ MPa $50 \le \rho w \le 800$ kg/(m² sec) $\Delta i = 420$ kJ/kg	Within the researched range of parameters $q \ne f(\Delta T)$, $q \sim \delta$
[318]	$d = 10$ mm; $D = 14$ mm, $\delta_{Sl} = 2$ mm, $l = 0.35$ m; two parallel cylindrical and annular channels; up-flow water in vertical channel	$9.8 \le p \le 17.6$ MPa $500 \le \rho w \le 1000$ kg/(m² sec) two-side cooling of heat generating element by water	The total critical power, N_{CR} $$N_{CR} = \Delta i_{CR1}\rho w_1 S_1 + \Delta i_{CR2}\rho w_2 S_2;$$ $$\frac{\Delta i_{CR}}{i'} Fr^{*0.3}\left[1+\frac{190}{(F/s)}\right] = 1.8-0.8\frac{i_{in}}{i'} Fr^{*0.5},$$ here $Fr^* = \frac{\rho w}{\rho''}\sqrt[4]{\frac{\rho'-\rho''}{g\sigma}}$
[325]	$\delta_{Sl} = 0.50-11.1$ mm; $d = 1.5-96.5$ mm, vertical channel	$0.017 \le plt \le 0.19$ MPa $100 \le \rho w \le 1000$ kg/(m² sec) $-3.1 \le x_{in} \le 0.0$ $-2.8 \le x_{CR} \le 0.74$	$q_{CR}/G = f(x, l/d, Y);$ $$Y = \frac{G\bar{D}}{\lambda}\left[\left(\frac{G}{\rho}\right)^2 gD\right]^{0.6}\left[\left(\frac{\mu'}{\mu''}\right)\right]$$
[319]	$d = 10$ mm; $\delta_{Sl} = 1.25, 3.0, 4.5$ mm; $l = 70, 120, 170$ mm; water, antifreeze-65, ethylenglycol	$0.10 \le p \le 0.49$ MPa $\rho w = 1000, 3000, 5000$ kg/(m² sec) $30 \le tin \le 100°C$	$q_{CR} = (\rho w)^{0.3}\left[A+B\times 10^{-3}\Delta T_{av}\right]$, MW/m² where A and B are empirical constants

the high pressure range (typical for power engineering conditions, where pressure exceeds 2 MPa). This allowed developing recommendations on heat flux calculations [331], which can be used for tubes with diameter exceeding 4 mm.

Despite some variations in the CHF quantitative estimations presented by different authors, the qualitative nature of different parameters' influence on CHF is clear. It is known that the influence of channel cross section parameters on CHF becomes evident when their sizes appear comparable with the vapor bubble departure diameter. Figure 6.1 shows that the majority of experiments were performed at large channel diameters.

For many years, researchers from the National Technical University (Ukraine) conducted representative and systematic experiments on CHF regularities inside capillary channels. The influence of subcooling, velocities, and channel diameter was investigated [303]. Experimental data were treated by the empirical correlation:

$$q_{CR} \quad C_1 w^n \Delta T^m$$

Analysis of the influence of geometry factors on CHF revealed that q_{CR} is almost independent of the length l when relative length $l/d = 8$–10. Increasing the channel diameter from 1 to 4 mm caused q_{CR} to decrease twofold. Experiments were performed at pressures exceeding 1.0 MPa.

A brief review of experimental and analytical investigations as well as CHF calculation correlations for channels with diameters less than 4 mm is presented in Table 6.1. Analysis of the contents of Table 6.1 yields the following conclusions:

- Calculation correlations of q_{CR} show that most of the experimental data [298, 300, 302, 303, 306–308, 311, 313, 318] were generalized empirically.
- Several studies [299, 300, 310, 313, 314] presented generalized correlations accounting for the influence of geometric parameters on CHF, whereas estimates of the effect of channel size were different. Hence, in Refs. [299, 301, 331] the CHF appears proportional to d (channel diameter), whereas in one study [298] it was observed that there are such thermal regimes when immediate liquid boiling in small-diameter tubes causes the CHF, that is,

$$q = 0.121 p (\sqrt{w^3 d^2 / l})^{0.8}$$

which does not concur with the experimental data of other authors.

- Extensive experiments [300, 311] on CHF at helium-forced flow inside capillary channels under low pressures (~0.1 MPa) were initiated by the development of cooling devices for special magnetic systems. At such pressure levels, the ratio of liquid and vapor densities (ρ'/ρ'') was equal to 7 and the ratio of helium pressures (p/p_{CR}) was 0.4. It is similar to water properties at saturation pressure, ~12 MPa. Consequently, when water is used as heat carrier, such helium values made it impossible to apply the recommendations obtained from experiments on helium flow inside capillaries.

- The CHF at high velocities of ethanol flow were experimentally investigated by Povarnin [306] and Shtokolov [312]. Povarnin [306] discovered the typical CHF thermal regimes caused by the liquid film drying on the wall (the so-called crisis of the second type). It was supposed that q_{CR} depends on the joint effect of average flow velocity and mixing action of bubbles, that is,

$$q_{CR} \quad \left(1 + \frac{w}{w_0}\right)^{0.8}$$

The majority of authors noticed that increasing the mass flow velocity caused the CHF values to increase; however, the magnitude of such dependency was estimated in varying methods. For example, in Refs. [299, 301, 303, 313] index n in correlation $q_{CR} \approx w^n$ varied from 0.44 to 1.00. In one study [302], such an index was equal to 3, but the arrangement of correlation eliminates its influence at $\rho w < 30{,}000$ kg/(m² sec). It is probable that it was caused by the flow "blockage" inside a channel with a diameter of 0.5 mm.

In the case of noteworthy inlet subcooling and high flow velocities, experimental data [312] on CHF at 96% pure ethanol flow were generalized via the Kutateladze hydrodynamic theory of boiling crisis. The correlation concurred with experimental data [303, 307]. However, the attempt to apply such a correlation in the case of low velocities failed.

The region of high subcooling was studied by Gluschenko [299]. Accounting for the influence of vapor generation, condensation of vapor slugs, and conditions of heat removal from the wall on the CHF value, the generalized correlation based on experimental data [303, 307, 324] was obtained by using a dimensionless set of parameters. It was recommended for tube diameters between 2 and 12 mm at high subcooling and at pressures exceeding 0.5 MPa [299]. The effect of subcooling was analyzed at high pressures (more than 13.8 MPa).

Macbeth [314] suggested a generalized correlation for tube diameters between 1.1 and 19 mm at low mass velocities. However, its application was restricted when a special chart was used to determine the regime regions.

Labuntsov [305] offered a correlation for the calculation of water flow CHF at $0.1 \le p \le 20.0$ MPa, $0.7 \le w \le 45$ m/s, $0 \le \Delta T \le 240$ K, $1.8 \le q_{CR} \le 5.4$ MW/m², which represent generalized experimental data for channel diameters exceeding 2 mm.

Calculations of CHF based on recommendations [299, 303, 305, 310, 332] at regime parameters considered by Kozhelupenko [140] are presented in Figure 6.2.

Calculation correlations for the CHF performed in several studies [299, 305, 307, 308, 310–312, 314, 332] account for flow subcooling in the crisis cross section. Calculation of subcooling based on the experimental data treatment appears uncomplicated. However, the application of CHF correlations accounting for subcooling caused significant disagreement with the experimental data. It is known that maximum scattering in values of CHF at boiling in channels reached 15–40% or even greater. This is attributable to the statistical nature of the process. Calculation of subcooling in the crisis cross section is of the same order of error. Because subcooling in crisis cross section can be determined from experimental data as $\Delta T = T_S - T_{IN} - q_{CR}(PL_{CR}/G)$, the inaccuracy of q_{CR} calculation by 1.15–1.4 times causes substantial error (up to several times) in the calculation of ΔT_{CR}. Thus, a major

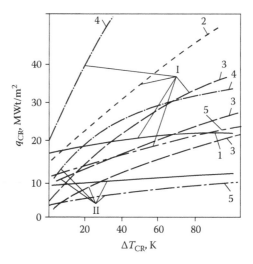

FIGURE 6.2 Comparison of correlations for CHF calculations at $l = 100$ mm, $d = 2$ mm, $p = 0.4$ MPa. I, $\rho w = 16 \times 10^3$ kg/(m^2 sec); II, $\rho w = 4 \times 10^3$ kg/(m^2 sec). 1, [305]; 2, [332]; 3, [303]; 4, [299]; 5, [310].

opportunity to obtain a reliable empirical correlation between q_{CR} and ΔT_{CR} is eliminated. Therefore, it was required to modify the initial correlation for the CHF experimental data generalization by switching from parameters in the crisis cross section to really independent variables, that is, inlet flow parameters.

Liquid mixture boiling depends on both the content and physical nature of components. It is connected with local (close to the heating wall) decreasing of light component concentration in the boiling mixture. An experimental study on the internal characteristics of different liquid mixtures [277] demonstrated a significant influence of concentration on bubble average velocity and boiling heat transfer intensity.

Gluschenko [299] reviewed a number of experimental and theoretical studies devoted to pool boiling heat transfer, the CHF, and their principal mechanisms. In contrast, research on the boiling heat transfer crisis in small-diameter channels is practically absent, except for one study [309], which considered the CHF problem at forced mixture flow inside tubes with diameters less than 4 mm. Boiling of water mixtures of ethanol, benzyl, acetone, and ethylenglycol was studied. In these experiments, a concentration was changed from 0% to 100% excluding the water–ethylenglycol mixture. Because experimental data treatment was based on empirical correlation, its application in case of boiling heat transfer at forced flow in capillaries is not possible.

6.3.2 BOILING HEAT TRANSFER CRISIS AT LOW-VELOCITY FLOW INSIDE ANNULAR CAPILLARY CHANNELS

Available experimental data on boiling heat transfer crisis in annular channels are different from the quantitative viewpoint because of the application of various experimental methods. However, the qualitative influence of regime and geometric parameters on CHF values remains the same.

Extensive experimental data [316–318, 321–323, 325] were obtained at pressures corresponding to operational conditions of the annular heat generating elements of power plants. Correlation between the CHF value and pressure in annular channels is essentially different from the corresponding dependency for tubes. As a rule, it contains a maximum value of the CHF.

Briefly, the distinctiveness of experiments in annular channels with characteristic size $d_{eq} < 4$ mm are summarized in Table 6.2. Experimental research [317, 321] on the dependency between channel length and CHF demonstrated that the CHF decreases significantly when the length increases to 0.2–0.3 m. However, the influence of length on CHF was accounted for in different ways [317, 322]. Savina [319] did not observe any correlation between the CHF and channel length at all.

The influence of the annular gap on CHF was studied in experiments [322]. An increase in CHF was observed when annular gap was increased from 1.5 to 2.0 mm. Heat input conditions (single-side or two-side heating) also influenced CHF. Hence, heat fluxes on the internal surface exceed external heating at the subcooled liquid flow. Analysis of the contents of Table 6.2 demonstrated that the CHF empirical correlations suggested in Refs. [315–325] are valid for restricted ranges of regime parameters' variations.

6.3.3 EXPERIMENTAL INVESTIGATION OF THE CHF AT FORCED FLOW IN CAPILLARIES: MODIFIED HYDRODYNAMIC THEORY OF BOILING HEAT TRANSFER CRISIS

Description of the experimental setup and methods were presented in several studies [294–297], and specified by Kozhelupenko [140]. It is known that the basis of experimental research on CHF is associated with selection of the crisis fixation procedure. Kozhelupenko [140] justified both the possibility of CHF fixation at the moment of burnout and the procedure of attaining the CHF. It was shown that variation between CHF values determined by the "burnout" method and data on CHF obtained by conventional "electric bridge" approach (the crisis moment can be set via the abrupt increase in electric resistance of the working element) did not exceed 30%. Parallel experiments and statistical treatment of experimental data were performed for each combination of independent parameters (subcooling, pressure, diameter, mass flow velocity, etc.). Special statistical analysis of data on crisis locations was performed, because it significantly influences the calculations of flow parameters (enthalpy, pressure, temperature) in the CHF cross section.

CHF dependency, $q_{CR} = f(w)$, based on experimental data [140] at 0.1 MPa outlet pressure and given inlet subcooling for capillaries with diameters of 0.5 and 1.0 mm is presented in Figure 6.3. CHF values were selected with respect to the corresponding approximately equal inlet subcooling values. It is obvious that CHF increases monotonously with increasing velocity. It shows agreement with numerous experimental data, existing physical explanations, and basic statements on the boiling heat transfer crisis hydrodynamic theory. Increase in static pressure via the special system allowed operators to obtain the CHF experimental data at pressures of 0.1, 0.6, and 1.1 MPa (Figure 6.4).

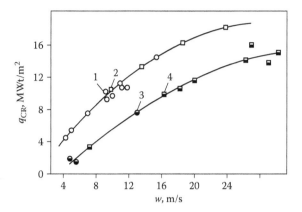

FIGURE 6.3 Influence of liquid velocity and subcooling on q_{CR} at water flow in capillary channels ($p = 0.1$ MPa) [140]. Channel diameter (d): 1 and 3, 0.5 mm; 2 and 4, 1.0 mm; inlet subcooling (ΔT): 1 and 2, 90 K; 3 and 4, 60 K.

Considerable uncertainty in the calculation of subcooling within the crisis cross section was caused by both the CHF value and distance between the channel's inlet and crisis cross section. On the other hand, a saturation temperature in the crisis cross section, $T_{S;CR}$, depends on the pressure drop at the channel outlet. In experimental data treatment, values $T_{S;CR}$ were determined by the saturation pressure in the crisis cross section, $p_{S;CR}$, accounting for liquid phase hydraulic losses from the outlet chamber to the crisis cross section.

Dependency between the CHF and subcooling at different velocities of 66% ethylenglycol–water solution is presented in Figure 6.5 [140]. A direct proportionality between q_{CR} and temperature drop $\Delta T = T_{S;CR} - T_{in}$ was observed.

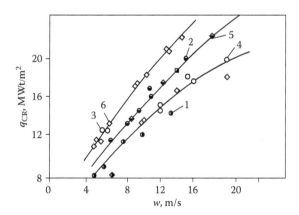

FIGURE 6.4 Influence of pressure on q_{CR} at water flow in capillary channels [140]. $\Delta T_{in} = 100$ K. Channel diameter (d): 1–3, 0.5 mm; 4–6, 1.3 mm. Pressure (p): 1 and 4, 0.1 MPa; 2 and 5, 0.6 MPa; 3 and 6, 1.1 MPa.

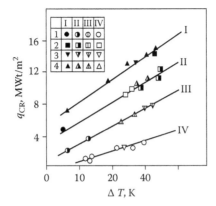

FIGURE 6.5 Influence of basic regime parameters on q_{CR} at 66% ethylenglycol–water solution flow in capillary channels [140]. Channel diameter (d): 1, 0.5 mm; 2, 1.0 mm; 3, 1.3 mm; 4, 2.0 mm. Heat carrier velocity (w): I, 20 m/sec; II, 14 m/sec; III, 8 m/sec; IV, 4 m/sec.

As observed in the figure, the influence of channel diameter and length on CHF values is absent. Similar dependences were observed for other heat carriers. Influence of the solution content on CHF is shown in Figure 6.6 [140]. Increase in ethylenglycol concentration led to a decrease in CHF values because of the change in the thermophysical properties of the solution [309].

Therefore, it could be supposed that in the considered conditions the dependency $q_{CR} = f(w, \Delta T)$ qualitatively concurs with the basics of the hydrodynamic theory of the boiling heat transfer crisis [268] both for pure liquids and for binary mixtures. Attaining data on two-phase flow internal parameters near the crisis cross section has significance in establishing justified physical explanations of crisis phenomena, when forced flow with high velocities occurs in capillaries. However, visualization of processes inside

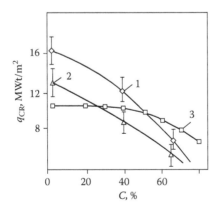

FIGURE 6.6 Influence of concentration of ethylenglycol–water solution on CHF at $\Delta T_{in} = 50$ K, $w = 10$ m/sec [140]. Channel diameter (d): 1, 1.0 mm; 2, 1.3 mm; 3, 4.0 mm.

the channel appears extremely complicated due to the small sizes of channels, external heating, and excess pressure in the loop. Therefore, in the majority of experiments local static pressures were measured under the conditions preceding the crisis.

Restricted empirical correlations (Figures 6.3–6.6) and flow conditions in capillaries assumed a possibility to generalize experimental data obtained at the subcooling liquid flow with high velocities inside the capillary channels, basing on the crisis hydrodynamic theory.

However, corresponding attempts to treat experimental data by using a subcooling in the crisis cross section as the determining parameter failed (Figure 6.7). Such drawbacks of the conventional approach to the treatment of experimental data on CHF were also observed in other studies (e.g., [334]).

Application of subcooling in the crisis cross section as the independent variable requires accounting for the fact that q_{CR} increases as the ΔT_{CR} value decreases. Inaccuracy in determining the value of ΔT_{CR} affects the accuracy of q_{CR} and, in particular, the entire correlation, $q_{CR} \approx f(\Delta T_{SUB})$. Considering such circumstances, Ornatsky et al. [334] suggested using independent parameters in CHF generalization that are actually independent in the experiments.

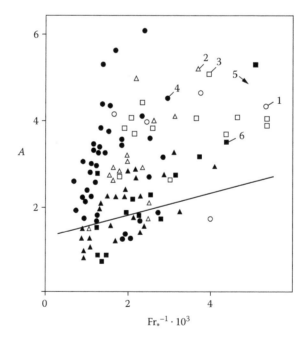

FIGURE 6.7 Treatment of experimental data on CHF at liquid forced flow boiling inside capillary channels [268]. $A = q_{CR} \times 10^3 \bigg/ \left\{ rw_0' \sqrt{\rho'\rho''} \left(1 + K_3 Pr^{-0.6} \frac{\Delta i_{CR}}{r} \sqrt{\frac{\rho'}{\rho''}} \right) \right\}$. $Fr^* = \rho \otimes W \otimes_0^2 l \bigg/ \sqrt{g\sigma(\rho' - \rho'')}$. $l/d = 80$, p_s: 1, 0.1 MPa; 2, 0.5 MPa; 3, 1.1 MPa. $l/d = 40$, p_s: 4, 0.1 MPa; 5, 0.6 MPa; 6, 1.1 MPa.

However, generalization of the CHF experimental data in the low-pressure region requires answering the following questions: What are regularities for the CHF? Do they correspond to the known hydrodynamic theory? Let us consider the approach to answering these questions.

The CHF hydrodynamic theory correlation [268] yields

$$q_{CR} = rw_0' \sqrt{\rho'\rho''} \left(1 + K_3 Pr^{-0.6} \frac{\Delta i_{CR}}{r} \sqrt{\frac{\rho'}{\rho''}}\right) \left[K_2 + \frac{K_1 \sqrt[4]{g\sigma(\rho'-\rho'')}}{w_0' \sqrt{\rho'}}\right] \quad (6.14)$$

Accounting for the straightforward correlation between dependent variables Δi_{CR}, q_{CR} and independent variable Δi_{in} allows transformation of Equation (6.14).

Through the correlation $\Delta i_{CR} = \Delta i + q\Pi l/G$, it is possible to obtain the dependency of q_{CR} on parameters Δi_{in} and $p_{S,out}$. Simple transformations yield

$$\frac{\left(1 + K_3 \sqrt{\frac{\rho'}{\rho''}} \frac{\Delta i_{in}}{r} Pr^{-0.6}\right) r \sqrt{\rho'\rho''} w_0' \cdot 10^{-3}}{q_{CR}} = \frac{1}{K_2 + K_1(Fr^*)^{-1}} + 6.4(Pr')^{-0.6} \frac{l}{d} \cdot 10^{-3}$$

$$(6.15)$$

where K_1 and K_2 are the stability criteria of liquid pool boiling and forced boiling near the heating surface, respectively, representing known empirical constants $K_1 = 140$ and $K_2 = 1.3$ defined by the treatment of numerous experimental data; $K_3 = 1.6$ is a coefficient accounting for the subcooling effect.

Kozhelupenko [140] performed treatment of experimental data basing on the transformed hydrodynamic theory (Figure 6.8). In addition, experimental data [303] obtained for different heat carriers including water, ethanol, and mono-isopropyl-diphenyl are presented in the same figure.

Comparison of Figures 6.7 and 6.8 with respect to the so-called "natural scattering" of experimental data in parallel experiments leads the following conclusions:

- CHF at flow boiling inside capillary channels can be predicted by the crisis hydrodynamic theory predictions.
- Generalization of CHF experimental data based on transformed fundamental correlation of the crisis hydrodynamic theory treating ΔT_{in} and $p_{S;in}$ (inlet subcooling and inlet saturation pressure, respectively) as independent variables provides satisfactory results.

Initial attempt to treat experimental data on CHF at forced binary mixtures (water–ethylenglycol) flow boiling inside capillary channels was based on the straightforward use of correlation (6.15) by assuming a water–ethylenglycol mixture as a homogeneous system, in which thermophysical properties can be defined simply by values of concentration and saturation temperature, $T_{S;M}$. As a result, significant inconsistency between the experimental data and calculations based on the modified correlation (6.15) of the crisis hydrodynamic theory was attained.

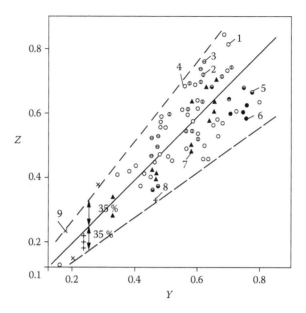

FIGURE 6.8 Treatment of experimental data on CHF based on Equation (6.15) at liquid forced flow boiling inside capillary channels. Water, $l/d = 80$, p: 1, 0.1 MPa; 2, 0.6 MPa; 3, 1.1 MPa. $l/d = 40$, p: 4, 0.1 MPa; 5, 0.6 MPa; 6, 1.1 MPa. 7, [303], $p = 1.12$ MPa; 8, ethanol [335]; 9, mono-isopropyl-diphenyl [335]. $Z = 10^{-3} r w_0' \sqrt{\rho' \rho''} \left(1 + K_3 Pr^{-0.6} \dfrac{\Delta i_{CR}}{r} \sqrt{\dfrac{\rho'}{\rho''}}\right) / q_{CR}$.

$$Y = \frac{1}{K_2 + K_1 (Fr^*)^{-1}} + 6.4 (Pr')^{-0.6} \frac{l}{d} 10^{-3}.$$

The binary water–ethylenglycol mixture is characterized by substantial distinction of saturation temperatures of components. It is known that the vapor phase properties of such mixtures are secured to the volatile component (water) properties in a wide range of ethylenglycol concentrations (0–90%). Phase equilibrium diagram shows that even at 66% concentration of ethylenglycol in the liquid phase, a water concentration in the vapor phase is about 94–98%.

As discovered in experiments on highly subcooled liquid flow in short channels, it is possible to assume that the vapor phase appearing near the wall condenses on the border of thermal boundary layer, and the average liquid phase content near the wall remains unchanged. Both are based on the hydrodynamic model of the heat transfer crisis, and supposing an analogy between boiling and barbotage of binary mixture presents an opportunity for modeling the vaporization of subcooling liquid flow near the wall as water vapor injection into the liquid binary mixture flow of given content.

Such an approach allows the use of the modified correlation of the Kutateladze hydrodynamic theory of boiling crisis [268] for generalization of experimental data on CHF at binary mixture flow boiling inside capillary channels. Consequently, it should be assumed that vapor phase at binary mixture boiling has properties of the water vapor at the given saturation pressure, whereas the liquid phase properties are determined by the initial mixture content. Subcooling is the difference between the saturation temperature of mixture and inlet channel temperature.

Crisis cross section was determined in each experiment. It allowed the evaluation of a saturation pressure in such a cross section and saturation temperature at liquid phase equilibrium state. Outlet hydraulic losses significantly influence the saturation temperature in the crisis cross section, especially at high values of inlet velocity. The effect of accounting for hydraulic losses on the treatment of CHF experimental data by Equation (6.15) is shown in Figure 6.9. As shown in the figure, accounting for these losses leads to systematic deviation of 25–30% from the generalized correlation (6.15).

Therefore, according to experimental data on binary solutions flow boiling, the CHF dependency on concentration demonstrates a nonmonotonous behavior. However, it is not a sequence of the change in heat transfer crisis physical mechanism. It was caused by the modification of thermophysical properties of the vapor–liquid mixture within the two-phase boundary layer, which can be accounted for in the first approximation via the equilibrium liquid–vapor interface.

Treatment of CHF experimental data based on the modified hydrodynamic theory correlation (6.15) is presented in Figure 6.10. The data of Tolubinsky and Motorin [309] on pure liquids and binary mixtures water–ethylenglycol at flow boiling in small channels are also included. As shown, experimental data coincide with calculations based on the modified hydrodynamic theory correlation (6.15). For this reason, the approach to CHF experimental data generalization at forced flow boiling of subcooled binary mixtures with considerable differences between the thermal properties of components can be used when such differences

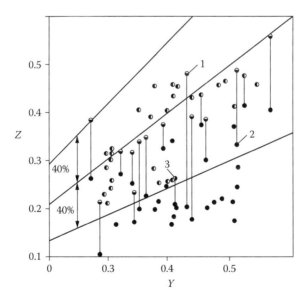

FIGURE 6.9 Effect of accounting for hydraulic losses on treatment of CHF experimental data by Equation (6.15) at water–glycol mixture flow boiling in capillary channels: 1, 39% solution; 2, 66% solution (neglecting hydraulic losses); 3, 66% solution.

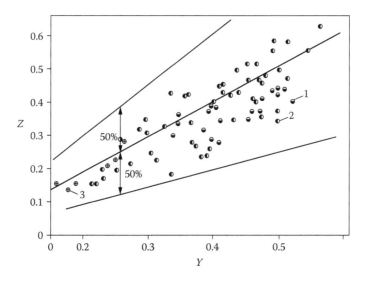

FIGURE 6.10 Generalization of experimental data on CHF water–ethylenglycol mixture flow boiling in capillaries based on correlation (6.15). 1, 39% water–ethylenglycol mixture; 2, 66% water–ethylenglycol mixture; 3, data from Tolubinsky and Motorin [309]; 4, calculations by (6.15).

are not present by correctly accounting for the existence of both components in the vapor phase.

Consequently, as the first approximation, the modified correlation (6.15) can be applied when liquid phase properties are to be determined by the mixture parameters, and the vapor phase properties are also to be based on the mixture characteristics but determined from thermodynamic equilibrium conditions at saturation pressure in the crisis cross section.

6.3.4 CHF in Narrow Annular Channels (Experimental Data and Semiempirical Models)

Results of previous crisis phenomena observations in "slug" regimes at forced flows in annular channels were presented in Chapter 4. Particular difficulties were connected with the treatment of experimental data on CHF in annular channels. Visual observations and video capture demonstrated that dissimilar physical natures of crisis phenomena creation correspond to different two-phase flow regimes. Thus, according to experimental data, conditions of CHF generation were classified into three groups based on different two-phase flow regimes:

* "Bubble" and "emulsion" flow regimes (experimental data and visual observations demonstrated both qualitative and quantitative agreement between the CHF regularities and the crisis hydrodynamic theory; treatment of corresponding experimental data [140] and calculations [319, 324, 336] based on correlation (6.15) are presented in Figure 6.11

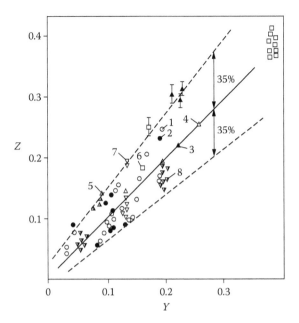

FIGURE 6.11 Comparison of CHF experimental data at "emulsion" and "bubble" flow regimes inside annular channels with calculations by Equation (6.15). 1, "Bubble" flow; 2, "emulsion" flow. Annular channel gap (s): 3, 2.87 mm; 4, 1.2 mm; 5, 1.2 mm [324]; 6, 3.0 mm; 7, 3.0 mm; 8, 5.0 mm [319].

- "Slug" flow regime
- "Stratified" flow regime

In the last two regimes mentioned, experimental data demonstrated essential peculiarities of crisis generation and development. The video capture results yielded several important physical explanations of the crisis at flow boiling inside short annular channels (Figure 6.12):

- "Dry spots" emerge on the heated wall at slug existence into the channel (Figures 4.34 and 4.35).
- "Dry spot" dimension depends on the quantity of heat transferred to liquid film and increases over time until the heating surface is completely dried (Figure 4.37).
- Liquid film return to the heating surface follows a complete vapor slug evacuation from the internal volume of the annular channel.
- Crisis is generated when the time of vapor slug existence in the channel exceeds the time needed for complete drying and overheating of the wall to the temperature surpassed corresponding to the metastable liquid state (spinodal), that is,

$$\tau_{SL} > \tau_{DS} + \tau_{OH} \qquad (6.16)$$

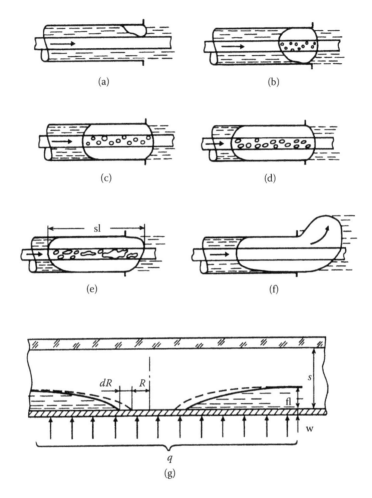

FIGURE 6.12 Physical model of the heat transfer crisis at "slug" two-phase flow. (a) Slug formation; (b, c, d, e) slug development; (c, d, e, f) "dry spots" appearance and growth; (f) slug exit; (g) schematic of single dry spot growth. s, gap; δ_{fl} and δ_w, thicknesses of liquid film and wall, respectively.

Consider a "dry spot" development on the heating surface (Figure 6.12). Because pressure changes slightly along the vapor slug, a liquid film remains steady on the wall.

Analysis of heat transfer conditions at slug growth is associated with the complex problem of unsteady heat conduction, when heat is spreading from the spot center toward its edge. Therefore, the approximated model of the CHF development was chosen:

1. "Dry spots" formation regularities are identical to boiling centers nucleation in thin liquid films (such an assumption is based on video capture of "dry spots" growth; Figures 4.34 and 4.35).

2. Intensity of liquid film evaporation was caused by heat input conditions in movable boundary zone of "dry spot."
3. Wall drying time, τ_{DS}, in the vapor slug foot is determined by assumptions (1) and (2).
4. Wall overheating time, τ_{OH}, to the temperature surpassed corresponding to the metastable liquid state depends on the heat flux and the heat accumulating properties of the wall.
5. Lifetime of a vapor slug was determined by the hydrodynamics of two-phase flow with respect to the influence of subcooling, vapor quality, thermophysical properties of two-phase mixture, and channel geometry.

Combining previously mentioned assumptions, physical explanations, and condition (6.16), the semiempirical correlation for slug flow in annular channels was obtained in two studies [297, 140] as:

$$q_{CR} = \frac{r\rho' w_0' s}{L_{CH}} \left(C_2 \sqrt{\frac{\rho''}{\rho'}} + C_1 c_p' \frac{\Delta T_{in}}{r} \right) \tag{6.17}$$

Generalization of experimental data based on Equation (6.17) is shown in Figure 6.13.

Thus, available experimental data on CHF at forced flow in capillaries and annular channels at low vapor qualities corresponded to the regularities predicted by the boiling crisis hydrodynamic theory, if the modified Equation (6.15) is used as generalizing correlation. Inlet flow parameters (subcooling, velocity, pressure, dimensionless numbers Pr and Fr) are used as independent variables.

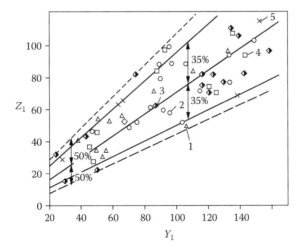

FIGURE 6.13 Generalization of experimental data on CHF at slug flow inside annular channels [140]. s: 1, 0.6 mm; 2, 1.0 mm; 3, 1.2 mm; 4, 1.5 mm; 5, 1.4 mm. $z_1 = ql/[w\rho s(1 + s/d)]$; $Y_1 = c_p \Delta T_{out}/r$.

6.4   EXPERIMENTAL RESEARCH ON THE CHF AT POOL BOILING IN SLITS, CAPILLARIES, AND CORRUGATED CAPILLARY CHANNELS

Experimental data on boiling heat transfer and CHF inside horizontal and vertical slits submerged in liquid pool were analyzed in Chapter 4. In addition, typical experimental dependencies and experimental methods, including schematics of test setups and data on boiling internal characteristics, were presented.

Consider generalization of experimental data on CHF at boiling in horizontal and vertical slits submerged in liquid pool performed by Koba [109]. Corresponding generalized semiempirical correlations were obtained by using the modified crisis hydrodynamic theory, and its modified "energy" form was treated as the initial correlation

$$E \geq k_1^2 \Pi g - k_2 \Pi_{FR}$$

where E and $k_1^2 \Pi_g$ are determined from the Kutateladze theory [268], and the term $k_2 \Pi_{FR} = k_2 \Delta p_{FR}$ accounts for the potential energy decrease in stable vapor film caused by the influence of the restricted spaces (Δp_{FR} denotes hydraulic losses at two-phase flow filtration in slit channel).

Using known correlations of hydraulics of the vapor–liquid flows, Koba [109] obtained the following correlation relating to horizontal slits with width, b and gap s:

$$\Delta p_{TP} = \xi \frac{\rho'}{2} \left(\frac{q}{r\rho''} \right)^2 \left(\frac{b}{s} \right)^3 f_1(\varphi) + \Sigma \zeta \frac{\rho'}{2} \left(\frac{q}{r\rho''} \right)^2 \left(\frac{b}{s} \right)^3 f_2(\varphi), \quad (6.18)$$

where ξ and $\Sigma \zeta$ are friction factor over the length and local hydraulic resistance, respectively; $f_1(\varphi)$ and $f_2(\varphi)$ are factors accounting for the increase in two-phase flow hydraulic resistance in comparison with the corresponding single-phase flow.

Complex character of correlations for $f_1(\varphi)$ can be further approximated by a simple constant-index power law as

$$f_1(\varphi) = (1 - \varphi)^{-m}$$

Initially, Koba [109] supposed that if $q \to q_{CR}$, then vapor quality $\varphi \to \varphi_{CR}$, and φ_{CR} is a constant that is autonomous with respect to the main independent variables of the process. However, experimental data treatment demonstrated significant deviations of the experimental points by the saturation pressures. Therefore, φ_{CR} increases with the increase in pressure, that is, with increase in phase densities ratio (ρ''/ρ'), and such dependence can be prescribed as

$$\varphi_{CR} \sim 1 - const(\rho'/\rho'')^p \quad \text{or} \quad \frac{1}{(1-\varphi)^m} \approx (\rho'/\rho'')^{-k} \quad (6.19)$$

At $m = 3$ and $p = 0.42$, the following generalized correlation was obtained

$$\frac{q_{CR}}{q_{CR0}} = \frac{1}{\sqrt{1 + (\rho''/\rho')^{0.25}\ [const_1\,(b/s)^3 + const_2\,(b/s)^2]}} \tag{6.20}$$

where q_{CR0} is a CHF at pool boiling.

Treatment of experimental data by Equation (6.20) is presented in Figure 6.14. Also included are the data of Katto et al. [337, 338] on CHF at water boiling in horizontal slit formed between butts of two cylinders with a diameter of 10 and 11 mm, and those of Klimov et al. [110, 116] on CHF at refrigerant R-113 boiling in plain horizontal slit [b (width) = 20 mm and gap (s) = 0.5 mm at saturation pressure (p) = 1 bar]. Accounting for the statistical nature of boiling crisis, generalization of the experimental data by Equation (6.20) can be considered satisfactory.

A similar approach was used by Bezrodny et al. [124] and Sosnovsky [125] in the treatment of experimental data on CHF at refrigerants R-11 and R-113 boiling inside short vertical slits submerged in the liquid pool.

In such a case, conditions of the CHF development yield the following correlation

$$E \geq k_1^2 \Pi_g - k_2 \Pi_T + k_3 \Pi_H \tag{6.21}$$

where Π_H is a potential energy of liquid column of a height H, which is equal to the channel length.

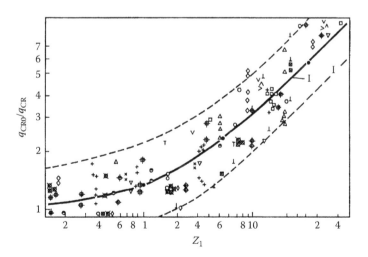

FIGURE 6.14 Generalization of CHF experimental data at boiling in horizontal slits [109, 110, 116, 337, 338]; $Z_1 = (\rho''/\rho')^{0.25}(b/s)^2(0.08 + 4 \times 10^{-3}(b/s))$; 1, by Equation (6.20). Heat carriers: water, ethanol, diethyl ether, refrigerant R-113; heating surface material: copper, nichrome; $15 \leq b \leq 30$ mm; $s = 0.5$ mm; $0.036 \leq p_s \leq 5$ bar.

Value Π_H is proportional to the lifting force developed in liquid due to vapor phase appearance, that is,

$$\Pi_H \sim \Delta p_H \qquad (6.22)$$

where Δp_H is the hydrostatic pressure defined by channel height and vapor quality in the two-phase flow:

$$\Delta p_H = \left\{ \rho' - \left[\left(1 - \bar{\varphi}_{CR}\right)\rho' + \bar{\varphi}_{CR}\rho'' \right] \right\} gH \sin \alpha \qquad (6.23)$$

where α is the angle of inclination between the channel and the horizontal line. Neglecting the small terms gives a following simplification

$$\Delta p \cong \rho' \bar{\varphi}_{CR} gH \sin \alpha \qquad (6.24)$$

Subsequently, accounting for correlations (6.18)–(6.24) modifies formula (6.21) to

$$\left(\frac{q_{CR}}{r\rho''}\right)^2 \rho'' \geq \left(\frac{q_{CR0}}{r\rho''}\right)^2 \rho'' - \left(const_1 + const_2 \frac{b}{s}\right)\left(\frac{q_{CR}}{r\rho'}\right)^2 \left(\frac{b}{s}\right)^2 \rho' + k_3 \rho' \bar{\varphi}_{CR} gH \sin \alpha \qquad (6.25)$$

Finally, the correlation for generalization of experimental data appears as follows:

$$\frac{q_{CR}}{q_{CR0}} = \sqrt{\frac{1 + const_3 \dfrac{\rho' gH\rho'' r^2}{q_{CR0}^2} \sin \alpha}{1 + \left(\dfrac{\rho''}{\rho'}\right)^{0.25} \left(\dfrac{b}{s}\right)^2 \left(const_1 + const_2 \dfrac{b}{s}\right)}}; \qquad const_3 = 6 \cdot 10^{-3} \quad (6.26)$$

At the horizontal location of slit channel ($\alpha = 0$ and $\sin \alpha = 0$), Equation (6.26) transforms into (6.20). For this reason, values $const_1$ and $const_2$ were determined from experiments of horizontal slits, and $const_3$ from experiments on vertical or inclined slits.

Treatment of CHF experimental data by Equation (6.26) presented in several studies [109, 110, 124, 337, 338] at boiling in horizontal and vertical plain slits is given in Figure 6.15.

The majority of experimental points were generalized by Equation (6.26) with scattering of ±35% when values of empirical constants were as follows: $const_1 = 0.08$, $const_2 = 4 \times 10^{-3}$, $const_3 = 6 \times 10^{-3}$. It justifies the current approach and allows using Equation (6.26) for calculations of CHF at boiling of different liquids in horizontal and vertical slits, when $0.2 \leq s \leq 4$ mm, $15 \leq b \leq 50$ mm, $0.036 \leq p_s \leq 5.0$ var, $H_{max} = 50$ mm.

Experimental data on heat transfer and CHF at boiling inside corrugated channels [215, 217, 218] demonstrate similar regularities with vaporization heat transfer in horizontal slits submerged into liquid.

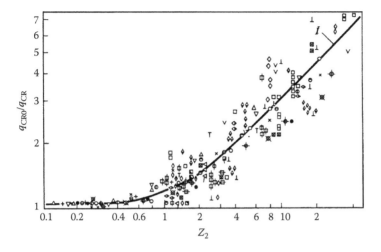

FIGURE 6.15 Generalization of CHF experimental data at boiling in horizontal and vertical slits. $Z_2 = \dfrac{1+0.08(\rho''/\rho')^{0.25}(b/s)^2[1+0.05(b/s)]}{1+0.006r^2\rho'\rho''gH\sin\alpha / q_{CR0}^2}$; data from Refs. [109, 110, 124, 337, 338]; 1, by Equation (6.26). Heat carriers: water, ethanol, refrigerant R-113, refrigerant 4, pentane; heating surface material: copper, nichrome, stainless steel; $15 \le b \le 50$ mm; $0.036 \le p_s \le 5.0$ bar.

Experiments [218] were performed to determine the CHF both for surfaces covered with corrugated capillary structures and heat pipes with corrugated wicks at horizontal, inclined, and vertical arrangements.

The time of occurrence of CHF was fixed by abrupt growth of surface temperature corresponding to heat transfer reduction. The CHF experimental data presented in the Table 6.3 and in Figure 6.16 were obtained for different values of modules and width of the corrugated strips at zero heat carrier depth on the heating surface. Based on data analysis of Table 6.3, increase in corrugated strips width causes decrease in CHF. Moreover, CHF decreases with the lessening of the characteristic size of the corrugations.

TABLE 6.3
Values of CHF in Experiments with Heating Surfaces Covered by Corrugated Structure

Module of Corrugation, m	Effective Diameter $(d_{Ef} \times 10^3)$, m	$q_m \times 10^{-4}$, W/m²			
		$b = 0.0055$ m	$b = 0.0106$ m	$b = 0.0207$ m	$b = 0.04$ m
0.5	0.95	46	39, 44	32, 33.2	24
0.3					
	0.49	–	36, 39.5	25, 28.5, 29.6, 31, 31.4	
0.2	0.4	–	32	24	

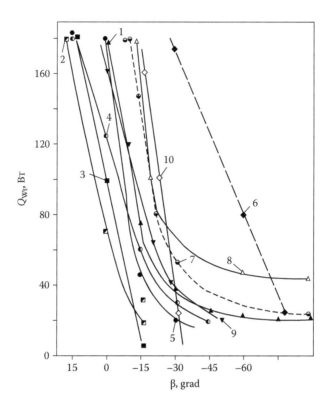

FIGURE 6.16 Dependency of CHF on heat pipe inclination angle, β [218]. Thermal wedge: 1, sample #1 at $T_{in} = 25°C$; 2, the same, #2; 3, the same #2 at $T_{in} = 50°C$; 4, the same #3 at $T_{in} = 50°C$. 5, $T_{in} = 50°C$; $m = 0.4$. Heat pipes: 6, heat pipe with $m = 0.3$, liquid filing volume $V_F = 2.13$ sm³; 7, the same, $m = 0.3$, $V_F = 3.7$ sm³; 8, the same, $m = 0.2$, $V_F = 5.8$ sm³; 9, the same, $m = 0.2$, $V_F = 2.17$ sm³; 10, the same, $m = 0.5$, $V_F = 2.6$ sm³.

It can be assumed that CHF depends on two factors:

- Two-phase filtration flow friction loss inside capillaries formed by corrugations.
- Local hydraulic resistance to vapor exit developing in contact zones between neighboring corrugated strips.

Actually, both the increase in corrugated strip length and decrease in corrugation module lead to a reduction in the number of holes and decrease in the cross section area of vapor output.

Heat carrier depth change also influences the CHF values. For example, for structures with module $m = 0.5$ m and corrugated strips width $b = 0.02$ m, values of $Q_m = 25 \times 10^4$, 33×10^4, and 46×10^4 W/m² corresponded to heat carrier depth of $h = -7.0$, 0, and 4.0 mm, respectively.

In experiments with manufacturing samples of heat pipes, maximum values of CHF were achieved at $\beta = 0°$: for sample #3, $Q_m \geq 180$ W; sample #2, $Q_m = 70$ W; sample #1, $Q_m = 180$ W.

The wick of the heat pipe #1 was manufactured from corrugations with module $m = 0.3$ and strip width $b = 0.02$ m. It is provided with higher values of heat transfer intensity and capillary tube potential $-\sigma/d_{Ef}$ compared to sample #3, which has a variable module value $m = 0.3, 0.5$. From one point of view, corrugated strips with $m = 0.5$ sustain higher values of CHF than structures with $m = 0.3$. Indeed, the crisis was not reached in the sample #3 at $\beta = 0°$ and heat pipe operated stably and transferred a thermal power $Q_m > 180$ W. At the same time, CHF was generated in heat pipe #1 at $\beta = 0°$ and $Q_m = 178$ W.

On the other hand, even at small inclinations ($\beta = -10°$), the capillary potential of heat pipe #3 did not provide sufficient liquid feed to the evaporation zone. Thus, in accordance with Figure 6.16, at negative angle $\beta = -15°$, CHF values were $Q_m = 44$ W (sample #3) and $Q_m = 76$ W (sample #1).

The CHF in heat pipe #2 with $m = 0.3$ and $b = 0.04$ m was equal to 100 W. Such decrease in maximum thermal power with respect to sample #1 can be explained by the increase in strip width and length of liquid transport zone ($L_{HP} = 0.44$ m). Consequently, it caused both the increase of two-phase flow friction loss inside heat pipe and local hydraulic resistances to vapor exit. Hence, such complex physical nature of CHF generation at boiling in corrugated structures require the development of appropriate physical explanations and mathematical models.

Modeling of CHF at boiling in corrugated coverings of heat pipe evaporative zone is based on the known hydrodynamic equation for low-temperature heat pipes:

$$\Delta p_{CP} \geq \Delta p' + \Delta p_{TW} + \Delta p_L \tag{6.27}$$

Pressure drops caused by capillary and volumetric forces were estimated as follows:

$$\Delta p_{CP} \sim \sigma/d_E; \quad \Delta p_L = \Delta \rho' g L_E \sin \beta \tag{6.28}$$

Quantitative estimations demonstrated that friction losses of liquid flow in corrugated channels within the condensation and transport zones $\Delta p \otimes$ is 2 or 3 orders of magnitude lower than the pressure drops in the evaporation zone, Δp_{TW}, representing a sum of two-phase flow friction loss and local hydraulic resistance of corrugated strip contacts. Characteristic velocity was assumed as proportional to evaporation velocity:

$$W = \frac{Q_m}{r\rho' S_{out}}$$

Two-phase flow friction loss was estimated as follows:

$$\Delta p_{TW1} = c_1 \frac{L_{EV}}{d_E} \rho' W^2 = \frac{32}{(1-\varphi)^k} \frac{v' L_{EV}}{d_E^2} \frac{Q_m}{r S_{out}} \tag{6.29}$$

Local hydraulic resistances of corrugated strip contacts were treated as vapor flow filtration loss, and characteristic velocity was assumed as equal to $(Q/r\rho'^{TM} S_{out})(b/L_{EV})$.

In addition, it was considered that outlet vapor quality is determined by vapor volumetric concentration, φ. Thus, the second term in Δp_{TW} can be determined as

$$\Delta p_{\text{TW2}} = const_2 \left(\frac{bQ_m}{L_{\text{EV1}} r S_{\text{out}}} \right)^2 \frac{1}{\rho'' \varphi^2} \qquad (6.30)$$

where S_{out} is a total outlet flow cross section constrained by the channel wall from one side and by corrugated structure on the other side, that is, S_{out} is the entire cross section of internal channels of the corrugated structure. Then, correlation (6.27) can be presented as follows:

$$\frac{4\sigma}{d_E} + \rho' g L_E \sin\beta \geq const_1 \frac{1}{(1-\varphi)^k} \frac{v' L_{\text{EV}}}{d_E^2} \frac{Q_m}{r S_{\text{out}}}$$

$$+ const_2 \frac{1}{\varphi^2 \rho''} \left(\frac{bQ_m}{L_{\text{EV}} r S_{\text{out}}} \right)^2 ; \quad 1.5 \leq k \leq 3 \qquad (6.31)$$

where $const_1$ and $const_2$ account for the specifics of local hydraulic resistance. Introduction of the parameter $X_0 = Q_m/(r S_{\text{out}})$ yields the following correlation:

$$X_0 = C_2 v' L_{\text{EV}}^3 \frac{\rho''}{d_E^2 b^2} \left[\sqrt{1 + C_3 \frac{d_E^4 b^2}{v'^2 \rho'' L_{\text{EV}}^4} \left(\frac{4\sigma}{d_E} + \rho' g L_E \sin\beta \right)} - 1 \right] \qquad (6.32)$$

Equation (6.32) yields the following generalization of experimental data:

$$X_0 \frac{d_E^2 b^2}{v' \rho'' L_{\text{EV}}^3} = C_2 \left[\sqrt{1 + C_3 \frac{d_E^4 b^2}{v'^2 \rho'' L_{\text{EV}}^4} \left(\frac{4\sigma}{d_E} + \rho' g L_E \sin\beta \right)} - 1 \right] \qquad (6.33)$$

Using the following equation assigned to the dimensionless parameters

$$X = X_0 \left(\frac{d_E^2 b^2}{v' \rho'' L_{\text{EV}}^3} \right) \qquad Y = \frac{d_E^4 b^2}{v'^2 \rho'' L_{\text{EV}}^4} \left(\frac{4\sigma}{d_E} + \rho' g L_E \sin\beta \right) \qquad (6.34)$$

yields the final correlation

$$X = C_2 \left(\sqrt{1 + C_3 Y} - 1 \right) \qquad (6.35)$$

Equation (6.35) shows a dependency between the maximum transferred power Q_m and both the geometric and regime parameters.

Hence, Q_m increases with the enlargement of the outlet cross section and evaporation zone length L_{EV}, and with the lessening of d_E and width of corrugated element, b. Equation (6.35) matches basic local regularities of the CHF for heat pipes' operational conditions. Analysis of the term $((4\sigma/d_E) + \rho' g L_E \sin\beta)$ demonstrated that Q_m increases when the angle of inclination of heat pipe adjusts from 0° to 90°, and it reduces abruptly when the angle changes from 0° to −30°.

Heat pipe hydrodynamic limit occurred when the angles were less than $-30°$. When this happens, a capillary potential becomes lower than $\Delta P_L = \rho' g L_E \sin \beta$ and consequently capillary forces cannot transport liquid to the evaporation zone.

Treatment of experimental data on CHF (Table 6.3) and testing results for heat pipes #1, #2, and #3 are presented in Figure 6.17. Equation (6.35) generalizes experimental data with scattering of 50%, when $C_2 = 8$ and $C_3 = 0.017$.

However, Equation (6.35) does not correspond with existing experimental data when the inclination angles of heat pipes with corrugated wicks are less than $-30°$. Experimental data confirmed that heat pipes with corrugated wicks operated reliably regardless of the angle of inclination (from less than $-30°$ to $-90°$). It can be explained by the fact that condensate flow in evaporation zone occupied only a small part of the cross section (by the so-called "conditional channel") formed by corrugation profile, wall of heat pipe, and interface meniscus (Figure 6.18). In such cases, balance of forces for the liquid element inside the conditional channel can be presented as

$$-\frac{\sigma}{R^2} dR = \frac{64}{d_{E0}^2} \frac{\mu' Q_m}{N_G r \rho' s_L} dZ + \rho' g \sin \varphi dZ \qquad (6.36)$$

Integrating the equation at $x/R \ll 1$ yields the following correlation for the CHF:

$$Q_m = C_4 t_G N_G \frac{\sigma r}{b} (\sin \beta) v' \qquad (6.37)$$

where t_G is the corrugation pitch.

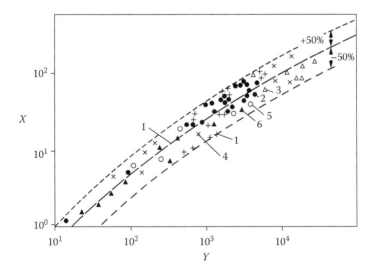

FIGURE 6.17 Treatment of experimental data on CHF in corrugated heat pipes, (by Equation [6.35]). Inclination angles: $-30°$ to $+90°$; m: 1, 0.2; 2, 0.3; 3, 0.5; heat pipe samples: 4, #1; 5, #2; 6, #3. I, $X = 8(\sqrt{1+0.017Y} -1)$.

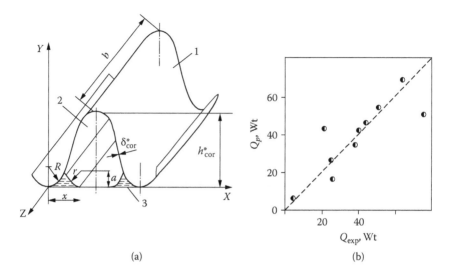

(a)   (b)

FIGURE 6.18 Maximum thermal power of corrugated heat pipes. Inclination angle (β), $-30°$ to $-90°$. (a) Schematic diagram of corrugated heat pipe operation. Evaporation zone above the condensation zone: 1. a corrugation profile; 2. internal channel of the corrugated structure; 3. heat carrier inside conditional channels of corrugated wick. (b) Comparison between experimental data and calculations of heat pipe maximum thermal power.

Equation (6.37) allows us to determine the maximum thermal power of corrugated heat pipes when inclination angle β changes in the range of $-30°$ to $-90°$ and $C_4 = 0.0479$. Comparison of experimental data and calculations is shown in Figure 6.18.

6.5 EXPERIMENTAL RESEARCH ON HEAT TRANSFER CRISIS AT BOILING ON SURFACES WITH POROUS COVERINGS

Experiments with screen wicks [79] and metal–fiber structures [176] belong to a small number of systematic experimental studies on CHF at pool boiling on surfaces covered by capillary-porous structures.

Experimental data on CHF at boiling on surfaces covered by sintered or gas-sprayed porous structures [36, 196, 197] presented specific results but with a lack of parallel experiments. For example, comparison between calculations and experimental data on CHF presented by Kovalev and Soloviev [36] was sourced from five publications consisting only of 16 experimental points. Each of them had its own set of independent variables (porosity, average pore radius, structure thickness, structure material, type of heat carrier, etc.). Data on reproducibility of coverings structure properties, technology and reliability of thermal contact between wall and structure, etc., are not presented in the cited papers.

Experimental values of CHF were determined for different screen wicks at vaporization of water, ethanol, and refrigerant R-113 [79].

Typical experimental data for uniform capillary-porous structures are presented in Figure 6.19. This figure shows CHF dependency on structure thickness and screen cell size when the liquid level above the heating surface was changed from 0 to 50 mm.

As shown, CHF decreases when the number of screen layers in the structure increases. CHF values are higher when vaporization takes place in the flooded structure than in the case of capillary feed. However, both cases are characterized by identical CHF values when the thickness of the wick structure becomes large.

As shown in Figure 6.19, variation in screen cell size has a complex influence on CHF due to the sequence of different tendencies:

- Decrease in cell size causes increase in the capillary potential, that is, increase in the ability to supply and to maintain liquid in the evaporation zone.
- On the other hand, decrease in cell size causes significant increase in hydraulic resistance at vapor removal from the near-wall region. In addition, it should be noted that increase in cell size, as a rule, is accompanied by increase in structure thickness.

A significant constraint limiting the CHF values exists in the liquid level related to the heating surface.

Experiments were also performed on nonuniform capillary structures formed from screens in such a manner that a layer with the minimum cell size was adjoined to the heating surface. At the absence of capillary feed, the CHF at vaporization

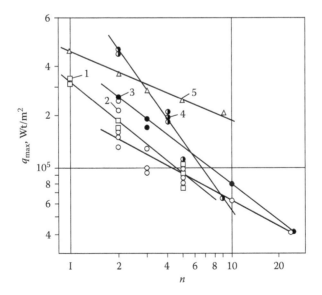

FIGURE 6.19 CHF dependency on cell size and thickness of screen structure (water at $p_s = 0.1$ MPa) [75–79]. 1, Copper screen, $a = 0.45$ μm; stainless steel screen: 2, $a = 60$ μm; 4, $a = 125$ μm; 5, $a = 200$ μm; 3, stainless steel screen in submerged conditions, $a = 60$ μm; n, number of layers.

of water under the saturation pressure of 0.1 MPa was unachievable, because these critical values exceeded the capability of the experimental setup ($q \leq 10^6$ W/m²).

However, for vacuum pressures, CHF rates at boiling on surfaces covered by screen wicks were obtained and their values exceeded 1.2–1.6 times the corresponding values for pool boiling.

Specific data on CHF are presented in Figure 6.20. As shown, in some cases the CHF were increased with the decrease in saturation pressure, which contradicted the known data on CHF at pool boiling. Increase of vaporization velocity in a vacuum corresponded to the increase of hydraulic resistance of the vapor phase removal, which is especially noteworthy for the so-called "thick" wicks ($n \geq 3$). Actually, increase in CHF was observed in nonuniform and thin-layer ($n < 3$) uniform screen structures with cell size $a \geq 200$ µm when vapor removal from the near-wall region was enhanced. As noted earlier, such structures also reveal a similar character of change in heat transfer intensity at high heat fluxes. Hence, it could be supposed that an identical heat transfer mechanism determines a similar crisis mechanism.

High CHF values within nonuniform screen structures were attained due to the essential decrease in vapor and liquid hydraulic resistance in the presence of high capillary potential in the near-wall zone. Thus, availability of "surplus" capillary

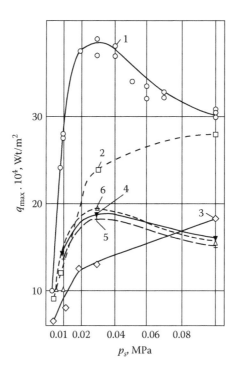

FIGURE 6.20 Dependency between q_{CR} and saturation pressure at ethanol vaporization inside the screen structures [75–79]. Two-layer stainless steel screens: 1, $a = 450$ µm; 2, $a = 125$ µm; 3, $a = 60$ µm; four-layer stainless steel screens: 4, $a_1 = 60$ µm; $a_2 = 200$ µm; $a_{3,4} = 125$ µm; three-layer stainless steel screens: 5, $a_1 = 60$ µm; $a_{2,3} = 200$ µm; two-layer brass screen: 6, $a = 160$ µm.

potential in the nonuniform capillary structures, in comparison to uniform structures, causes the increase of CHF values.

In addition, some manufacturing-induced defects significantly influence the CHF values.

For example, the appearance of gap formation between the structure and the wall causes a significant decrease in CHF values. By the same reason, the fall in CHF values is observed in screen structures with small rigidity manufactured from copper, brass, etc. Disjoining of such screens could be caused by dynamic pressure of the vapor occurring near the wall and can be avoided by high pressurizing screens to the wall.

Insufficient wetting of capillary structure can also lower CHF values. For example, the CHF of three-layer skimmed and not annealed screen with cell size $a = 60$ μm was decreased by almost two times (6.2 kW/m² instead of 11.4) in comparison to the same, but carefully treated, capillary structure.

Experimental data [176] on CHF in flooded submerged surfaces of the metal-fibrous capillary-porous structures (MFCPS) are presented in Figure 6.21. Shapoval [176] noted that because of capillary forces effect, capillary-porous covering opposes the vapor blanket formation typical for pool boiling on smooth surfaces. Application of high-porosity MFCPS manufactured from high-conductivity metals allows significant increase of CHF in comparison to a smooth surface (Figure 6.21). Optimum covering thickness estimated by Shapoval [176] was equal to $\delta_{CS} = 0.6$ mm.

Generalization of the CHF experimental data at water pool boiling on the surfaces covered by the MFCPS yields the following empirical formula:

$$q_{CR} = C q_{CR0} \delta_{CS}^n \lambda_{CS}^{0.1} \left(1 - \frac{\varepsilon_{max}}{1-\varepsilon} \right)^{0.1} D_E^{0.2} \qquad (6.38)$$

where q_{CR0} corresponds to CHF calculation at water boiling on smooth surfaces [268] and D_E is the effective pore diameter determined by the capillary head:

- At 0.1 mm $< \delta_{CS} < 0.6$ mm, $C = 105$, $n = 0.25$.
- At 0.6 mm $< \delta_{CS} < 10$ mm, $C = 0.28$, $n = -0.55$.

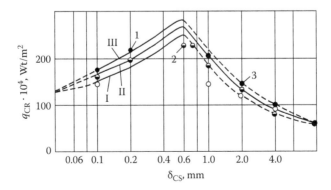

FIGURE 6.21 Dependency between q_{CR} and parameters of the metal-fibrous capillary-porous structure (MFCPS) at water boiling [176]. Copper MFCPS: 1, porosity (ε) = 40%; 2, $\varepsilon = 60\%$; 3, $\varepsilon = 80\%$. I–III, Calculations by (6.38).

Analysis of vaporization and CHF conditions at boiling on surfaces covered by screen wicks [75–79] demonstrated that, as a rule, experiments were performed with thin-layer coverings ($n < 3$), and it was only in severe cases that the number of layers was equal to 10 or reached 24 (Figure 6.19). In experiments [176] at characteristic pore sizes in the range of 20–40 μm depending on the porosity value, typical capillary structure thicknesses were about 0.6–1.0 mm, which corresponded to the conditional layer number $15 \leq n \leq 50$.

Thus, hydrodynamic phenomena at two-phase mixture filtration through screen coverings could be considered to have appeared due to flow through of local hydraulic resistances consisting of a small number of screen layers.

However, it is helpful to treat hydrodynamic phenomena at vapor–liquid mixture filtration in the multilayered MFCPS and other similar capillary structures as the countercurrent vapor and liquid flows through the porous media, when vapor flows from the wall inside vapor-generating channels, whereas liquid moves in the liquid pores and partially in the liquid film covering the skeleton of the porous structure.

In both cases, the capillary potential is forming at the points of contact between the wall and the capillary structure. It maintains a liquid presence near the wall and promotes increase in CHF values. Afanasiev [79] and Shapoval [176] provided direct experimental proof of the present approach by reporting a significant increase in CHF in comparison to a similar pool boiling case but without capillary structure on the heating surface.

These conditions were accounted for in the generalization of experimental data on CHF at pool boiling on surfaces covered by screen wicks based on the modified crisis hydrodynamic theory (see Section 6.2). Basics of the theory and generalized results are presented in several papers [79, 339–341].

Using the energy form of stability loss in two-phase boundary layer yields the following inequality:

$$E \geq k_1^2 \Pi_g - k_2 \Pi_{FR} + C_3 \Pi_\sigma - E_{BR} \qquad (6.39)$$

where E, Π_g, and Π_{FR} are values introduced in Section 6.2 and considered in cases when vaporization occurs inside the slit channels (described in Section 6.4). Analysis of these parameters obtained for vaporization on surfaces covered by screen capillary-porous structures is presented below.

The term $C_3 \Pi_\sigma$ accounts for the increase in two-phase boundary layer potential energy caused by the capillary forces acting in the region of contact between the wall and the structure. The term E_{BR} accounts for the increase in kinetic energy of the vapor insufflations due to throwing of liquid droplets out of the structure.

The Kutateladze theory [268] yields the following correlations:

$$k_1^2 \Pi_g = k_1^2 \left(\frac{q_{CR0}}{r\rho''} \right)^2 \rho'' \qquad (6.40)$$

$$E = \left(\frac{q_{CR0}}{r\rho''}\right)^2 \rho'' \tag{6.41}$$

Parameter Π_{FR} represents total hydraulic resistance consisting of the two following terms: Δp_1 is the hydraulic resistance of liquid flow to the vapor generation sites and Δp_2 is the hydraulic resistance of vapor filtration. Assigning characteristic velocity of liquid filtration and length as w' and L_{EV}, respectively, the following correlation is obtained:

$$\Delta p_1 = \frac{\mu' w' L_{EV}}{\kappa_0}$$

The value Δp_2 is treated as a local hydraulic resistance

$$\Delta p_2 \sim \zeta \rho'' \left(w_0''\right)^2 \sim 1.7n(1-\varepsilon)\frac{2.3-\varepsilon}{\varepsilon}\left(\frac{q_{CR}}{r\rho''}\right)\rho'' \tag{6.42}$$

where $\xi = 1.7n(1 - \varepsilon)(2.3 - \varepsilon)/\varepsilon$; n denotes the number of screens and ε is the fraction of crack-type pores.

The assumption that momentum source is proportional to capillary potential σ/a (a is cell size) allows the estimation of thrown liquid droplets out of a structure. Thus, the mass flow rate of such a throw-out is

$$g_{BR} \sim \sqrt{\frac{\sigma}{a}\rho'}\zeta_0$$

where ζ_0 is a factor accounting for resistance to throwing liquid droplets out of the porous structure.

Since, according to the present approach, the source of surplus capillary potential is forming near the wall, it is natural that increase in capillary structure thickness (screens number) and lowering of permeability (porosity and/or cell size) caused a decrease in the intensity of liquid throw-out, that is,

$$\zeta_0 = \zeta_0(n, K) \quad \text{or} \quad E_{BR} \approx \frac{q_{CR}}{r\rho''}\sqrt{\frac{\rho''\sigma}{a}}f(n, K) \tag{6.43}$$

Note that the term E_{BR} should be accounted for only in cases of capillary feeding of liquid, low height of the liquid suction, and when using thin-layer screen wicks.

Substituting (6.40)–(6.43) into the initial equation and accounting for the liquid filtration average velocity inside the screen wick as

$$w' = \frac{Q}{r\rho'\delta\Pi_{EV}n}$$

where $Q = q_{CR}F_{EV}$, yields the following correlation:

$$Z = B_1\left(\sqrt{1+YB_2} - 1\right) \tag{6.44}$$

where

$$Z = \frac{q_{CR}K^*\delta\Pi_{EV}}{r\rho''v'F_{EV}L_{EV}}\left(1+nC_2 1.7(1-\varepsilon)\frac{2.3-\varepsilon}{\varepsilon}\right),$$

$$Y = \left(\frac{\delta\Pi_{EV}nK^*}{F_{EV}L_{EV}}\right)^2 \frac{1}{\rho''v'^2}\left[\left(\frac{4\sigma}{a} - \rho'gh\sin\beta\right) + \frac{q_{CR0}^2}{r^2\rho''}\left(1+nC_2 1.7(1-\varepsilon)\frac{2.3-\varepsilon}{\varepsilon}\right)\right], \tag{6.45}$$

$$K^* = K_0\left(1+\frac{\delta\Pi_{EV}n}{F_{EV}L_{EV}v'\rho''}\sqrt{\frac{\sigma}{a}}f(n, K)\right)^{-1} \tag{6.46}$$

where K^* is the conditional liquid permeability accounting for the presence of the vapor phase inside the porous media; F_{EV}, L_{EV}, and Π_{EV} are the surface area, length, and perimeter of the evaporative zone, respectively; δ is the thickness of the single-layer screen covering; and h is the liquid capillary head.

Estimations done under the condition of heating surface flooding by the liquid and for the case of capillary feed, when the difference between the evaporation surface and liquid level was very small (corresponding to conditions of experiments [75–79] with screen structures), demonstrate that

$$\frac{4\sigma}{a} \gg \rho'gh\sin\varphi \tag{6.47}$$

or

$$\frac{4\sigma}{a} \gg C_1\left(\frac{q_{CR0}}{\rho''r}\right)^2\rho'' \tag{6.48}$$

Then

$$Y = \left(\frac{K^*\Pi_{EV}}{F_{EV}L_{EV}v'}\right)^2 \frac{4\sigma}{a\rho''}\left(1+1.7n(1-\varepsilon)\frac{2.3-\varepsilon}{\varepsilon}\right) \tag{6.49}$$

The coefficient C_2 was assumed as equal to 1. The artery permeability K_0 was calculated by the empirical correlation for woven-type wicks:

$$K_0 = (0.03-0.05)a^{-2.09} \tag{6.50}$$

A decrease in vapor content was observed in the direction from the near-wall layer to the external boundary of the structure.

The entire experimental data for water and ethanol obtained in several studies [75–79] were generalized by correlations (6.44)–(6.49) with maximum scattering of ±50% when constants B_1 and B_2 were equal to 1.9×10^{-5} and 2.21×10^5, respectively. However, a similar approach to generalization of experimental data for refrigerant R-113 failed. Further studies [110–113, 115, 116] revealed significant distinctions both in integral regularities of heat transfer and in microlayer evaporation mechanisms, when boiling of refrigerants occurs in restricted spaces. These primary peculiarities determine the specific generalized correlations on CHF [264].

The generalization of systematic experimental data based on Equations (6.44)–(6.49) are presented in Figure 6.22.

Kovalev and Soloviev [36] classified the conditions of the vapor film formation within the porous layer and on the external surface. Many researchers recognize vapor film formation in the porous structure as an "inside layer crisis" usually accompanied by a significant increase in wall superheat. The formation of vapor film on the external surface of porous layer covered by the liquid column is analogous to heat transfer crisis during transition from bubble boiling to film boiling regime. Kovalev and Soloviev [36] determined such a phenomenon as the heat transfer crisis at boiling on surfaces with porous coatings. In addition, it was suggested to define the heat flux related to reaching a crisis moment as q_{max} or $q_{CR,max}$ and to classify conditions of the CHF reaching for the sintered porous coatings by means of the porosity values as follows:

$$At\ \varepsilon > 0.48,\ 1 - \varphi = 0.25;$$
$$At\ 0.26 \leq \varepsilon \leq 0.48,\ 1 - \varphi = 0.65\varepsilon$$

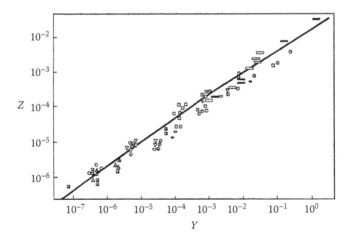

FIGURE 6.22 Experimental data on CHF at pool boiling on surfaces covered by screen wicks [75–79] and generalization by correlations (6.44)–(6.49). Heat carriers: water, ethanol; wick materials: copper, brass, stainless steel; layers number: 2–24; cell sizes: 40–450 μm.

The CHF at $0.26 \leq \varepsilon \leq 0.48$ was determined in one study [36] as

$$q_{CR\,max} = 0.52 r\varepsilon \sqrt{\sigma \frac{\rho'\rho''}{\rho'+\rho''}} R_{CL} \qquad (6.51)$$

where R_{CL} is the pore clamping radius.

The use of Equation (6.51) is based on the condition $\phi_S(R\delta) = 1$. In case of small pore-size distribution, Kovalev and Soloviev [36] recommended the following correlation:

$$q_{CR} = r\varepsilon \sqrt{2\sigma\rho'' \frac{R_{CL} - R_{min}}{R_{CL} R_{min}}} \qquad (6.52)$$

Comparison between calculations and a limited amount of experimental data was performed by Kovalev and Soloviev [36].

The most significant drawbacks of the present approach [36] in determining CHF at vaporization on surfaces with porous coatings are as follows:

1. It is not appropriate for vaporization regimes, when the structure of two-phase flow removed from the vapor-generating surface is different from the "vapor column," including regimes with capillary feeding, which are typical for operational conditions of heat pipes as well as with complex structures. Moreover, the present approach is not valid for boiling in restricted spaces, that is, in slits and capillaries, including a case of subcooling liquid flow to the vapor-generating surface.
2. The approach [36] excludes limit transition options, when both decrease in porous structure thickness and increase in pore size should lead to known correlations for CHF.
3. Large amount of experimental data on "inside layer crisis" were excluded from the generalization.
4. Extensive experimental information on thin-layer capillary structures, including screen wicks, was also eliminated from the analysis.

The present approach [36] allows determination of CHF in cases when abrupt falling of heat transfer intensity is caused by vapor film formation on the external surface of the porous structure that is most related to low-porosity and high-conductivity structures. It is especially effective and justified for conditions when the capillary structure is covered by the liquid layer, and the existing heat load is capable of maintaining the vapor film on the external surface.

In such restricted cases, using the approach [36], including direct application of Equations (6.51) and (6.52) simultaneously with thermohydrodynamic models accounting for both processes inside the porous structure in the contact zone between the porous structure and the wall as well as hydrodynamic phenomena caused by capillary and joint liquid transport, may provide positive results in the theoretical analysis of CHF at pool boiling on surfaces covered by capillary-porous structures.

Thus, combined consideration of the approach [36] and the thermohydrodynamic model of vaporization on surfaces covered by porous coatings presented in Chapter 5 eliminates the drawbacks of the approach [36], and offers an opportunity to analyze these critical phenomena from the common positions.

Comparison between the calculations based on the generalized model that accounts for Equation (6.51) and experimental data on CHF at ammonia and acetone boiling on surfaces with high-conductivity ($\lambda_E = 20$ W/(mK)) porous aluminum coverings (manufactured by Thermacore Inc.), supported the previous statement. Results of the comparison are presented by Smirnov [49] (see Figure 5.49).

Hence, common consideration of thermohydrodynamic models of internal vaporization processes and the model of boiling crisis on external surfaces allows correct description of these complex processes.

The problem of accurate accounting for various interacting thermal and hydrodynamic phenomena is especially important for studies on CHF developing inside the evaporation zones of heat pipes.

6.6   MAXIMUM HEAT FLUXES INSIDE HEAT PIPES

The problem of interaction between the known capillary limit of the heat pipe operation and the boiling heat transfer crisis on surfaces with porous coverings is of special interest and of particular complexity. It could be effectively solved via a common "hydrodynamic approach" (see Chapter 5).

Attention is particularly focused on the problem relating to design and influence of technological parameters on heat transfer crisis:

- Parameters of thermal contact zones involving the artery and the near-wall structure
- Design and technological factors in the near-wall and the arterial structure
- Presence or absence of noncondensable gas inside the liquid phase, including measures on its removal
- Influence of corrosion, inhibitor (if any), purity, and fill volume of working fluid, etc.

Thus far, hundreds or even thousands of types and designs of capillary-porous structures and heat pipes have been proposed. However, the present analysis is limited to the most common and the most attractive from the practical viewpoint of solutions.

These structures can be classified into screen wicks (without arteries) and artery-screen wicks (artery is fabricated from the metal screen).

There are both uniform screen wick structures manufactured from a single screen with regular cell size at constant porosity, as well as nonuniform wick structures, having different screen sizes and parameters.

Nonuniform screen capillary structures could have pore size increasing from the wall to the wick internal surface and vice versa. Structures of the first type are promising for achieving steady boiling on surfaces covered by the screen wick, whereas structures with decreasing pore size facilitate sustaining the enhanced evaporation mode in cases of heat pipe operation, when the boiling mode is restricted.

Heterogeneities are used to achieve a positive effect. Several heterogeneities appear in the sequence of technology, design, or thermal regime, such as

- Technological gaps between the heat pipe heating wall and the screen wick
- Nonuniformity of heat load caused by properties of the external thermal source, including the capability of spreading heat through the heat pipe wall
- Nonuniformity of the internal heat removal due to variation in wick thickness or wick thermophysical properties along the heat pipe perimeter

Known nonuniformities influence the heat pipe thermal mode, as follows:

1. Appearance of the gap between the wall and the screen structure can cause dry spots to form on the internal surface of the heat pipe wall and corresponding local superheats, that is, heat transfer deterioration modes. Under some conditions, the existence of gaps can stimulate premature dry-out of capillary structure and development of heat transfer crisis. This problem was considered in several studies [75–79].

2. The nonuniform external heat load causes heat load concentration in specific places and its de-concentration in others—hence, the significance of local heat transfer crises and their influence on the CHF maintaining reliable thermal mode increase in the maximum heat flux zones. Thermal influence of the wall is estimated by the characteristic scale, L_0:

$$mL_0 \leq 2; \quad m = \sqrt{\frac{\alpha u}{\lambda f}}; \quad L_0 \leq \frac{2}{\sqrt{\alpha u / (\lambda f)}} \tag{6.53}$$

where u and f are the perimeter and cross section area of the elementary cell, respectively.

3. The nonuniformity of the internal heat removal influences heat pipe thermal modes similar to the external nonuniformity. Temperature nonuniformities in the heat load zones can initiate atypical thermohydrodynamic phenomena. Hence, the creation of high heat fluxes in the zones covered by the thin-layer capillary structures (screen layer number, ≤ 3) can cause intensive throwing of liquid [75–79], and the throwing liquid flow rate can exceed a few times the flow rate required for vaporization (see Chapter 4).

Maintaining both reliable hydrodynamic contact between the artery-screen and the near-wall structure and efficient conditions for vapor removal is required to prevent a vapor filling of capillary structure in these places and to prevent a premature reaching of heat pipe capillary limit. This problem could be resolved by using a nonuniform capillary structure with the pore size increasing from the wall to the vapor channel [75–79] (see Chapter 4).

Pulsating thermohydrodynamic regimes can precede stable crisis modes with partial or complete dry-out of heat load zones in both artery-screen and artery-groove heat pipes. Temperature pulsations in such regimes were observed by various researchers.

Hydrodynamic regimes of liquid flow in artery-screen and artery-groove structures of heat pipes are dissimilar for artery and for near-wall structure. Typical regimes have sufficient hydrodynamic potential to maintain liquid flow inside the artery, but it is not enough to support flow near the wall. It causes dry-out of the near-wall structure accompanied by both interface deepening and decreasing of liquid transport length in this zone.

When the drying front is approaching the artery, then capillary potential, increasing due to the deepening of meniscus, could significantly exceed the liquid hydraulic resistance in the near-wall zone, which decreases because of the reduction in liquid transport length. This frequently happens when the drying front appears so close to the artery that surplus capillary potential causes throwing out of liquid from the artery into the near-wall zone, followed by its flooding and the consequent decrease in capillary potential. The liquid supply decreases abruptly and becomes less than the vaporization flow rate. Consequently, the dry-out is accomplished again, that is, the cycle is completed. Depending on the heat flux, wall superheat, geometric, and capillary properties of the near-wall zone and artery, the driven pressure scales can be different, and at high heat fluxes the stable dry-out takes place in the near-wall zone, while the liquid remains in the artery and a zone nearby. In such cases, evaporation occurs only in part of the near-wall zone, which is adjoined to the artery, that is, partial failure of thermal regime ensues. Consequently, it increases the wall superheat in the near-wall zone, developing with the heat flux growth. However, the entire liquid supply remains stable. Hence, this regime was classified as the local thermohydrodynamic crisis.

Further development of the heat flux can be followed by the consequent extension of the dry-out zone and decrease in the artery wetted part, until the appearance of vapor inside the artery, artery blockage, and irreversible growth of the wall temperature. Such a phenomenon represents general or integral crisis of heat transfer at vaporization on surfaces covered by the artery-capillary structures.

The CHF related to the local or integral crisis are essentially dependent on the orientation when the condenser section is disposed at a location below the evaporator section in the gravity field. In the opposing case, the influence of orientation on CHF is not important.

These physical explanations associated with the crisis phenomena in the artery-screen and artery-groove heat pipes were considered in several studies [342–344] (see Chapter 5).

Hence, treatment of thermal regimes in arterial, artery-screen, and artery-groove structures requires accounting for such options as appearance of local or integral crisis, existence of the pulsating modes, throwing liquid out the near-wall structure, presence of structure, and thermal nonuniformities (involving their influence on the heat pipe operational conditions).

The local crisis is accompanied by increase of the wall superheat and wall temperature pulsations. The dependency between the average wall superheat and the average temperature drop is consistent. The integral crisis is caused by the collapse of the liquid flow in the artery, leading to its partial or complete dry-out and the irreversible growth of the wall temperature.

Unfortunately, a reliable analytical model of the local crisis is currently unattainable because of the absence of adequate data on structure and hydraulic parameters of the near-wall zone, such as porosity, permeability, effective thermal conductivity, capillary potential, etc. Determination of these data requires special experimental modeling and cannot guarantee a high effectiveness because of the small values of linear scales (pore size, effective hydraulic diameter, flow cross section area, etc.).

However, the CHF related to the integral crisis can be reliably determined using the correlation suggested in two reports [170, 342]. The maximum heat fluxes when the condenser section disposed at a location below the evaporator section in the gravity field and at the heat pipe horizontal position are calculated using the capillary potential that accounts for the hydrostatic term,

$$\frac{4\sigma}{a_0} + \rho' gL \sin \varphi$$

where L is the length of heat pipe.

When the evaporator section is located below the condenser section in the gravity field, calculations of the integral CHF are based on the capillary potential of the near-wall zone of the capillary-porous structure. Calculations of the CHF for the artery-screen wicks with the single-layer structure in the near-wall zone or with the nonuniform structure as well as for the basic designs of heat pipes with screen wicks were performed using the following correlations:

$$Z = B_1 \left(\sqrt{1 + B_2 Y} - 1 \right), \quad \text{where} \quad B_1 = 1.9 \cdot 10^{-5}, \quad B_2 = 2.21 \cdot 10^5; \quad (6.54)$$

$$Y = \frac{1}{\rho'' F_{EV}^2 v'^2} \left[1 + 1.7(1-\varepsilon)\frac{2,3-\varepsilon}{\varepsilon} \right] \left(\frac{4\sigma}{a_0} + \rho' gL \sin \varphi \right); \quad (6.55)$$

$$Z = \frac{q_{CR} \dfrac{1}{r\rho'' F_{EV} v'} \left[1 + 1.7(1-\varepsilon)\dfrac{2.3-\varepsilon}{\varepsilon} \right]}{\dfrac{L_{1-2} + 2L_{2-3} + \quad + nL_{n-k}}{SK_0} + \dfrac{L_{EV}}{2\kappa^* \delta_{CS} \Pi}} \quad (6.56)$$

where F_{EV} and L_{EV} are the surface area and length of evaporator section located under each heat generating element, respectively; L is the length of the heat pipe; subscripts 1, 2,..., n denote the order of heat supply sections, starting with the most distant from the condenser; L_{n-k} is the distance between the nearest to the condenser "turned on" heat load section and half part of the condenser; $L_{1-2}, L_{2-k},..., L_{(n-1)(n-2)}$ are the distances between the neighboring heat load sections; ε is the porosity; δ_{CS} is the thickness of the near-wall capillary structure; a_0 is the cell size of the near-wall capillary structure layer; Π is the perimeter of heat load zone; S is the artery cross section area; K_0 and K^* are the permeabilities accounting for vaporization inside the artery and near-wall layer [170].

The present approach to calculation of CHF was experimentally justified for artery-screen heat pipe structures via generalization of experimental data obtained both for continuous and discrete heat load in artery-screen heat pipes [342, 344]. Such treatment of experimental data by Equation (6.54) is shown in Figure 6.23. One can suppose that this approach can also be qualitatively valid for the artery-groove heat pipes when the permeability will be substituted by the corresponding value proportional to the square of the groove characteristic size, the multiplier $[1+ 1.7(1 - \varepsilon)$ $(2.3 - \varepsilon)/\varepsilon]$ will be set as equal to 1, and the corresponding values of S, δ_{CS}, L_i, etc., will be applied. However, reliable calculations based on the above-mentioned suggestion require experimental validation.

To estimate the moment of reaching CHF, the simplified procedure could be used when $q_0 \leq q_{CR}$. In such a case, the failure of low-temperature heat pipe thermal regime is determined by the known heat pipe capillary limit.

Deteriorating thermal regimes of operation of heat pipe heat load sections appearing in zones between the local and integral crises can be estimated by assuming the heat transfer as heat conduction through the wall and heat pipe capillary structure to the artery, that is, $\Delta T \approx ql_w^2/(2\lambda_E\delta_w)$. Here, δ_w is the conditional thickness of the heat transferring wall from the metal container and capillary structure, and λ_E is the effective thermal conductivity.

It is known that in the general case the maximum (limit or critical) heat flux has a particular physical nature that determines the appropriate heat transport limit (discussed below).

The *capillary heat transport limit* appears when the sum of liquid and vapor flow pressure drops ($\Delta p'$ and $\Delta p''$) exceeds the capillary pumping head of the heat pipe

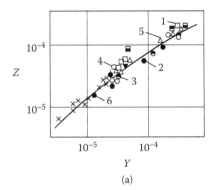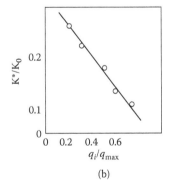

(a)   (b)

FIGURE 6.23 Generalization of experimental data on q_{CR} in heat pipes with the artery-screen structures for continuous and discrete heat load (by Equation 6.55). Discrete heat load: 1 (two layers), $a_1 = 40$ μm, $a_2 = 140$ μm; 2 (three layers), $a_1 = 40$ μm and $a_2 = a_3 = 70$ μm; 4 (single layer), $a_1 = 71$ μm; 5 (two layers), $a_1 = 40$ μm and $a_2 = 70$ μm. Continuous heat load: 3 (single layer), $a_1 = 71$ μm; 6 (two layers), $a_1 = 40$ μm, $a_2 = 70$ μm.

wick structure Δp_0, determined as the sum of the capillary potential Δp_σ and the hydrostatic pressure, that is,

$$\Delta p_0 = \Delta p_\sigma + \Delta p_g \leq \Delta p' + \Delta p'' \qquad (6.57)$$

The *sonic heat transport limit* occurs once the vapor flow reaches sonic velocity when exiting the evaporator section, resulting in a constant heat pipe transport power and large temperature gradients. Its value is given by the following correlation:

$$Q_{\max a} = \rho'' S'' a'' r \qquad (6.58)$$

where a'' is the vapor sonic velocity and S'' is the cross section area of the vapor channel.

The *boiling heat transport limit* takes place when high radial heat fluxes cause film boiling, resulting in heat pipe dry-out and large thermal resistances.

The *combined heat transport limit* is related to the interaction of capillary and boiling limits.

Development of reliable models accounting for capillary heat transport limit and crisis phenomena at boiling inside porous structures, including their interaction, requires extensive experimental data. Such data for the heat pipes with fiber–metal porous coverings were presented Semena et al. [19]; their work contained complete information about geometric parameters (shape, entire heat pipe length, lengths of zones, inclination angle, diameter, etc.), thermophysical, and structure properties of porous coverings, such as pore effective diameter, pore distribution, permeability, porosity, effective thermal conductivity, etc.

The following semiempirical correlation of the maximum heat flux in the heat pipes with fiber–metal structures was suggested [19] for zero-gravity conditions:

$$Q_{\max} = 2,2 \frac{r\sigma}{v'} \frac{S_f F_{CS}}{L_{ef}} \qquad (6.59)$$

where L_{ef} is the heat pipe effective length and F_{CS} is the liquid flow cross section area.

Calculating for the value of L_{ef} is very easy when the change in liquid viscosity along the heat pipe can be neglected because of small temperature drops and weak temperature dependency of viscosity, V. Otherwise, according to Semena et al. [19], it can be determined using the following correlation

$$L_{ef} = 0,5\left(L_{EV} + \frac{v'_C}{v'_{EV}} L_C\right) + \frac{v'_{TR}}{v'_C} L_{TR} \qquad (6.60)$$

Here, L_{EV}, L_C, L_{TR}, and v_{EV}', v_C', v_{TR}' are the lengths and liquid kinematic viscosities of the evaporator, condenser, and transport sections of heat pipes, respectively.

When the heat pipe is operated in the gravity field, partial dry-out of porous wick is accounted for by the following correlation [19]

$$F_{CS}(z) = F_0 \exp\left(-\frac{m\Delta p_z}{\Delta p_0}\right); \quad K_{CS}(z) = K_0 \exp\left(-\frac{n\Delta p_z}{\Delta p_0}\right)$$

where F_{CS}, K_{CS} and F_0, K_0 are the actual and initial values of the cross section area and permeability, respectively; Δp_z and Δp_0 are the liquid hydrostatic pressure in the heat pipe cross section with coordinate z and capillary pumping head; and m and n are empirical constants. Consequently, Semena et al. [19] suggested calculation of Q_{max} with respect to heat pipe working against gravity conditions as

$$Q_{max} = \frac{4r\sigma}{v'} \frac{K_{CS}}{D_E} \frac{F_{CS}}{L^*}\left(1 - \frac{\rho'g(L-L_1)\sin\varphi}{\Delta p_0}\right) \tag{6.61}$$

In zero gravity conditions, L^* is equal to L_{ef}; otherwise, it depends on L_{EV}, L_C, L_1, and the parameters S and B_1 determined as

$$S = 0,631g\left(20,25\frac{D_E}{D_0 - 1}\right) \tag{6.62}$$

where D_E is the effective pore diameter.

$$B_1 = \frac{\rho'g\sin\varphi}{\Delta p_0} \tag{6.63}$$

Here, $D_0 = 45 \times 10^{-6}$ m. Correlation (6.62) is valid when $50 \leq D_E \leq 200$ μm.

Thus, the analysis [19] allows determining of the heat pipe heat transport capillary limit. However, Semena et al. [19] noted the existence of boiling regimes in warm zones of the considered heat pipes. At the same time, satisfactory concurrence between experimental data and calculations with scattering of ±45% was demonstrated, explaining the empirical nature of initial correlations for calculation of such values as S, D_E, L_1; both significant influence of the capillary limit and application of fiber-metal porous structures have high thermal conductivities.

Fundamentals of the generalized model of heat transfer and crisis phenomena at vaporization and boiling on the coated surfaces with respect to the conditions of liquid capillary feed and in case of flooded structure (the third level model) were discussed in Section 5.4. Comparison between calculations based on the generalized model, and experimental data is shown in Figure 5.48. The lack of information on the correlation between actual filling conditions and the ideal case affects the reliability of Q_{max} calculations. Therefore, in case of horizontal heat pipe operation, two different values of Q_{max} were calculated. The first Q_{max} value was obtained using the assumption of the absence of the liquid "bottom" layer; the second one accounted for the significant decrease in liquid hydraulic resistance due to heat pipe overfilling with working fluid.

Maximum heat flux values corresponding to heat transfer crisis at vaporization on the porous structure external surface flooded by liquid are shown in Figure 5.49. Calculations were performed using Equation (6.51) [36]. As shown, combined consideration of this approach and the generalized heat transfer crisis model at vaporization inside porous structures [49] demonstrated a satisfactory concurrence with experimental data.

Hence, analysis of boiling heat transfer regularities and crisis phenomena in the "thick" capillary-porous structures can be based on the physical models, assuming that hydrodynamic phenomena at the liquid and vapor flow are similar to filtration inside extended porous structures.

Liquid and vapor filtration inside the flooded porous structures represents countercurrent flow normal to the heating surface. In case of capillary liquid feed, the vapor filtration is also normal to the wall, whereas liquid filtration is directed mainly along the heating surface. Supplementing these models with appropriate equations, thereby revealing the crisis processes on the external surface of porous structure, creates the theoretical basis for analysis of the entire set of feasible heat transfer crisis phenomena at boiling on surfaces covered by porous structures.

References

1. Petukhov, B.S., *Heat Transfer and Hydraulic Resistance of Laminar Liquid Flow in Tubes*, Energy, Moscow, 1967, 441 pp.
2. Kalinin, E.K., Dreitzer, G.A., and Yarcho, S.A., *Enhancement of Heat Transfer at Forced Convection in channels*, Mechanical Engineering, Moscow, 1981.
3. Petukhov, B.S., and Roizen, L.I., Heat transfer generalized correlations for gas turbulent flow in annular cross section tubes, *High-Temp. Thermophys.*, 2(1), 78–81, 1964.
4. Peng, X.F., and Wang, D.X., Liquid flow and heat transfer in microchannels with/without phase change, *Proceedings of the Tenth International Heat Transfer Conference*, Brighton, UK, Vol. 1, 159–177, 1994.
5. Wu, P.Y., and Little, W.A., Measurement of friction factor for the flow of gases in very fine channels using for microminiature Joule–Thomson refrigerators, *Cryogenics*, 23(5), 273–277, 1983.
6. Wu, P.Y., and Little W.A., Measurement of the heat transfer characteristics of gas flow in fine channels heat exchangers used for micro-miniature refrigerators, *Cryogenics*, 24(8), 415–420, 1984.
7. Tuckermann, D.B., and Pease, R.F., High performance heat sinking for VLSI, *IEEE Electro. Device Lett.*, EDL-2(5), 126–129, 1981.
8. Tuckermann, D.B., and Pease, R.F., Optimized convective cooling using micromachined structure, *Electrochem. Soc.*, 129(3), 98 pp., 1982.
9. Choi, S.B., Barron, R.F., and Warrington, R.O., Fluid flow and heat transfer in micro-tubes, micromechanical sensors, actuators and systems, *ASME DSC*, 32, 123–134, 1991.
10. Harley, I., et al., Fluid flow in micron and submicron channels, *Proc. of the IEEE*, Vol. 89-THO249-3, pp. 25–28, 1989.
11. Harlow, I.N., et al., Transport processes in micron and submicron channels, *Convection Heat Transfer and Transport Processes*, ASME, HTD, 116, 1–5, 1994.
12. Zhang, P.I., and Xin, M.O., Liquid and convective heat transfer in microtubes, *Proc. of China National Heat Transfer Conf.*, Beijing, 1992, pp. III89–III94.
13. Koshkin, V.K., et al., *Unsteady Heat Transfer*, Mechanical Engineering, Moscow, 327 pp., 1973.
14. Burundukov, A.P., Thermal and hydrodynamic non-steady processes in single-phase and two-phase media, Preprint N209, Novosibirsk, 1989.
15. Dulnev, G.N., and Yu.P., *Mixtures and Composite Materials Heat Conduction*, Energy, Leningrad, 1974.
16. Isatchenko, V.P., Osipova, V.A., and Sukomel, A.S., *Heat Transfer Fundamentals*, Energy, Moscow, 1981.
17. Cheiphez, L.I., and Naimark, A.V., *Multi-Phase Processes in Porous Media*, Chemistry, Moscow, 1982.
18. Luikov, A.V., *Heat and Mass Transfer*, Energy, Moscow, 1972.
19. Semena, M.G., Gershuni, A.N., and Zaripov, O.K., *Heat Pipes with Metal-Fiber Capillary Structures*, High School, Kiev, 1984.
20. Burdo, O.G., and Smirnova, Zh.B., Investigations of grooved structures thermal resistances, *Eng.-Phys. J.*, 10(3), 535–536, 1983.
21. Burdo, O.G., and Smirnova, Zh.B., Calculations of heat pipes grooved structures thermal resistances, *Radioelectron. Thermal Problems*, 2, 117–118, 1982.
22. Ratiani, G.V., Mestverishvili, I.A., and Shekriladze, I.G., Thin liquid laminar films evaporation from a surface, two cases analysis, *Inf. Georgian Acad. Sci.*, 55(3), 645–649, 1969.

23. Shekriladze, I.G., and Rusishvili, D.G., Evaporation and condensation on grooved capillary surfaces, *Proceedings of the Sixth International Heat Pipe Conference*, 1987, pp. 173–175.

24. Shekriladze, I.G., Evaporation and condensation on capillary surfaces, achievements and unsolved problems, *Proceedings of the Tenth International Heat Pipe Conference*, September 1997, pp. 1–14.

25. Shekriladze, I.G., et al., *Heat Pipes of Thermal Control Systems*, Energoatomizdat, Moscow, 175 pp., 1991.

26. Mistchenko, L.N., Hydrodynamic and heat transfer in low-temperature heat pipes, PhD thesis, OSAR, Odessa, 1975.

27. Aerov, M.I., and Todes, O.M., Hydrodynamic and thermal fundamentals of fluidized bed apparatuses in steady and unsteady state, *Chemistry*, 1968.

28. Belov, S.V., *Porous Materials in Mechanical Engineering*, Mechanical Engineering, Moscow, 1981.

29. Mayorov, V.A., Flow and heat transfer of single-phase cooling liquid in porous metal–ceramic media, *Thermal Engineering*, 1978, N1.

30. Polyaev, V.M., Mayorov, V.A., and Vasiliev, L.L., *Hydrodynamic and Heat Transfer in Porous Parts of Flying Apparatus Design*, Mechanical Engineering, Moscow, 1988, 168 pp.

31. Mori, Y., and Miyazaki, H., Heat transfer and mass loss characteristics of matrices for regenerative heat exchangers, *Tokyo Journal*, Sept., (Vol. 1), 79–87, 1967.

32. Baumann, K.H., and Blab, E., Schmidt-pathman geometrische, stromungs techisehe und termische untersuchungen an drahtggweben, *Inst. Fur Kerntechnik der Technischen Untersibat*, Berlin, 1966, pp. 34–72.

33. Pritula, V.V., and Zablotskaya, N.S., *Experimental Investigations of Matrix Heat Exchangers Aerodynamic Characteristics*, Deposit manuscript in the USSR Patent and Publishing House USSR, 1987, N3925-78.

34. Pritula, V.V., Screen metal-polymeric heat exchangers design and investigations, PhD thesis, OSAR, Odessa, 1983.

35. Kastornov, A.G., Permeable metal-fiber materials, Tech., Kiev, 1983.

36. Kovalev, S.A., and Soloviev, S.L., *Evaporation and Condensation in Heat Pipes*, Science, Moscow, 1989.

37. Luikov, A.V., *Transition Phenomena Inside Capillary-Porous Media*, Gostehizdat, Moscow, 1954.

38. Luikov, A.V., *Heat and Mass Transfer in Drying Processes*, Gostehizdat, Moscow, 1956.

39. Luikov, A.V., *Theory of Drying*, Energy, Moscow, 1968.

40. Charny, I.A., *Underground Hydro-Gas Dynamic*, Gostehizdat, Moscow, 1963.

41. Mayorov, V.A., Structure and hydraulic resistance of two-phase evaporative flow inside porous materials, *USSR Acad. Sci. News Energy Transp.*, 5, 126–132, 1980.

42. Kovalev, S.A., and Ovodkov, O.A., Hydraulic resistances at liquid and gas counter-current flow inside porous layer, *HTT*, 25(5), 1032–1035, 1987.

43. Kovalev, S.A., and Ovodkov, O.A., Vaporization inside porous layer mechanism investigations, Abstracts of papers of Minsk Int. Forum of HMT, Vol. 4, pp. 108–110, 1988.

44. Ovodkov, O.A., Liquid and vapor counter-current flows at bubble boiling on surfaces with porous coatings, Heat Transfer and Filtration, Dissertation thesis, Moscow Inst. of HTT of USSR Academy of Science, 1989.

45. Kovalev, S.A., Ovodkov, O.A., et al., Heat transfer at boiling on porous coatings of surfaces, *Proc. of USSR Conf. of HMT*, Minsk, Vol. 4(4), 10–15, 1984.

46. Smirnov, H.F., and Vinogradova, E.P., Regularities of heat transfer and thermal regimes at vaporization inside capillary-porous structures, *USSR Acad. Sci. News Energy Transp.*, 4, 128–135, 1985.

47. Smirnov, H.F., The generalized model of boiling heat transfer, *Advances in Heat Pipe Science and Techn.*, Proc. of 8th IHPC, Beijing, pp. 60–65, 1992.

48. Smirnov, H.F., Afanasiev, B.A., and Poniewski M., Boiling in capillary-porous structures, *Proc. of Int. Conf. on Heat Transfer with Change of Phase*, Part 11, Kielce, pp. 197–220, 1996.

49. Smirnov, H.F., Thermodynamic irreversibility minimization for determination of two-phase structure parameters, *Proc. of Int. Symp. on Physics of Heat Transfer in Boiling and Condensation*, Moscow, pp. 411–417, May 1997.

50. Kovalev, S.A., and Soloviev, S.L., Heat transfer and hydrodynamics in the inverted meniscus evaporator of a heat pipe, *Abstracts of 6th IHPC*, France, pp. 61–67, 1987.

51. Kovalev, S.A., and Soloviev, S.L., Heat transfer in an inverted meniscus evaporator investigations, *HTT*, 24(1), 196–198, 1986.

52. Yatscenko, E.S., High-heat devices cooling enhancement, Dissertation thesis, Moscow Inst. of Power Eng., 1989.

53. Demidov, A.S., Sasin, V.Ya., and Yatscenko, E.S., Heat and mass transfer in heat pipes evaporation zone, Moscow, *Pow. Eng. Inst. Works*, 173, 44–48, 1988.

54. Styrikovich, M.A., Malyshenko, S.P., et al., Boiling on surfaces with non-heat conductive coatings peculiarities, *Papers Acad. Sci. USSR*, 241(2), 345–348, 1978.

55. Malyshenko, S.P., Andrianov, A.B., et al., Hysteresis and transient phenomenon by boiling on surfaces with porous coatings, *Papers Acad. Sci. USSR*, 256(3), 591–595, 1981.

56. Malyshenko, S.P., et al., Transient processes and boiling on surfaces with porous coatings curves peculiarities, *Papers Acad. Sci. USSR*, 273(4), 866 pp., 1983.

57. Styrikovich, M.A., Malyshenko, S.P., et al., Investigations of boiling processes on porous surfaces, *Proc. USSR. Conf. HMT*, Moscow, 4(4), 10–15, 1984.

58. Mayorov, V.A., and Vasiliev, L.L., Two-phase system of porous cooling properties, *Processes of Substance and Energy Transfer at Low Temperatures in Vacuum*, Minsk, 1975, pp. 55–75.

59. Mayorov, V.A., and Vasiliev, L.L., Determination of two-phase porous cooling stable system parameters, *Processes of Substance and Energy Transfer at Low Temperatures in Vacuum*, Minsk, 1975, pp. 76–93.

60. Vasiliev, L.L., and Mayorov, V.A., Analytical investigation of two-phase porous reversible cooling system stability, *Heat Mass Transfer Probl.*, Minsk, 219–231, 1976.

61. Luikov, A.V., Vasiliev, L.L., and Mayorov, V.A., Theoretical analysis of two-phase cooling by evaporation inside porous structures, *Int. J. HMT*, 18(7/8), 863–885, 1975.

62. Mayorov, V.A., and Vasiliev, L.L., Two-phase cooling of porous heat generating element heat transfer and hydraulic resistance analytical investigation, *Heat Mass Transfer Probl., Minsk*, 232–258, 1976.

63. Mayorov, V.A., Heat Transfer in cooler evaporation region inside heated porous wall, *USSR Acad. Sci. News, Energy Transp.*, 6, 141–145, 1979.

64. Mayorov, V.A., and Vasiliev, L.L., Porous wall with external heat input evaporative cooling system temperature state, *Heat Transfer in Cryogenic Devices*, Minsk, pp. 97–108, 1979.

65. Mayorov, V.A., and Vasiliev, L.L., Temperature state of porous heat exchanger with account for heat carrier thermal conductivity, *Heat Transfer in Cryogenic Devices*, Minsk, pp. 108–119, 1979.

66. Mayorov, V.A., Vasiliev, L.L., and Polyaev, V.M., Evaporation inside porous materials flow structure, heat transfer and hydraulic resistance, *Proc. USSR Conf. of HMT*, Minsk, 4(4), 9–13, 1984.

67. Nakoryakov, V.E., and Petrick, P.T., Heat exchange during film boiling and condensation in a granular lager, *Two-Phase Flow Modeling and Experimentation, Proc. of the First Intern. Symp. on Two-Phase Modeling and Experimentation*, Rome, pp. 253–260, 1995.

68. Nakoryakov, V.E., Petrick, P.T., et al., Film boiling in granular bed, *Russ. J. Eng. Thermophys.*, 2(4), 200–310, 1992.

69. Bau, H.H., and Torrance, K.E., Boiling in low-permeability porous materials, *Int. J. Heat Mass Transfer*, 25(1), 45–55, 1982.

70. Bau, H.H., and Torrance, K.E., Thermal convection and boiling in a porous media, *Lett. Heat Mass Transfer*, 9, 431–441, 1982.

71. Smirnov, H.F., Boiling on surfaces covered by capillary-porous structures, heat transfer approximated theory, *Thermal Eng.*, 9, 40–43, 1977.

72. Khvostov, V.I., and Marinichenko, S.I., Stability of evaporation from porous structure at liquid transport by capillary forces, *EPJ*, 10(2), 135–139, 1979.

73. Ogniewez, Y., and Tien, C.L., Analysis of heat pipe phenomenon in porous media, *AIAA Paper*, 1–6, 79–109, 1979.

74. Wang, C.Y., and Groll, M., Dry-out heat fluxes in composite porous layers heated from below, *Advances in Heat Pipe Science and Technology*, Proc. of 8th IHPC, 1992, Beijing, pp. 65–68.

75. Afanasiev, B.A., and Smirnov, H.F., Experimental investigations of heat transfer and critical heat fluxes at boiling in capillary-porous structures, *Thermal Eng.*, 5, 65–67, 1979.

76. Afanasiev, B.A., Zrodnikov, V.V., Koba, A.L., et al., Boiling in capillaries, slits and capillary-porous structures heat transfer, *Heat Transfer and Hydrodynamics at Boiling and Condensation*, Collected works, pp. 75–82, Nov. 1979.

77. Smirnov, H.F., Koba, A.L., Afanasiev, B.A., and Zrodnikov, V.V., Boiling in capillaries, slits and other tighten conditions heat transfer, *Heat and Mass Transfer-V, Sci. Technol.*, Minsk, 3, 1, 193–197, 1976.

78. Smirnov, H.F., and Afanasiev, B.A., Experimental investigations of boiling inside screen structures of heat pipes, *RETP*, 2, 22–27, 1979.

79. Afanasiev, B.A., Vaporization inside screen capillary-porous structures investigations (heat transfer, critical heat fluxes, mechanism), Dissertation thesis, Odessa, 1981.

80. Moss, A.R., and Kelly, I.A., Neutron radiographic study of limiting planar heat pipe performance, *Int. J. Heat Mass Transfer*, 14, 491–502, 1980.

81. Vershinin, S.V., Fershtater, Yu.G., and Maidanik, Yu.F., Contact thermal resistance influence on evaporation from small-porosity capillary structures heat transfer, *HTT* 30(4), 811–817, 1992.

82. Borodkin, A.A., Liquid evaporation from capillary open surface grooved channels heat transfer intensity calculation method working out for HP structures optimization, Dissertation thesis, Moscow Inst. of Power Eng., 1988.

83. Borodkin, A.A., and Sasin, V.Ya., Liquid evaporation from capillary open surface grooved channels heat transfer intensity, *Moscow Power University Scientific Works, Industrial Energy Application*, Moscow, Vol. 28, 94–100, 1984.

84. Gorbis, Z.R., Smirnov, H.F., and Savchenkov, G.A., Liquid boiling in granular bed heat transfer investigations, *(EPJ)*, 26(3), 544–545, 1974.

85. Gorbis, Z.R., and Berman, M.I., Experimental investigations of boiling heat transfer in thermal fluidization particles layer conditions, *(EPJ)*, 27(4), 389–396, 1974.

86. Berman, M.I., and Gorbis, Z.R., Distillate water boiling in dispersed granular bed heat transfer, *Thermal Eng.*, 11, 86–88, 1973.

87. Gorbis, Z.R., and Berman, M.I., Liquid boiling inside dispersed particles layer heat transfer, *HMT-V, Sci. Technol.*, Minsk, 3, 1, 56–60, 1976.

88. Mankovsky, O.N, Fridgant, L.G., et al., Boiling on submerged with porous coatings surfaces mechanism discussion, *(EPJ)*, 30(2), 310–316, 1976.

89. Fridgant, L.G., Boiling heat transfer enhancement with porous coating help in heat exchangers of low-temperature gas separation apparatuses, Dissertation thesis, 1985.

90. Chernobulsky, I.I., and Tananyako, Yu.M., Liquid boiling in annular slit heat transfer, *J. Tech. Phys.*, 26(10), 2316–2322, 1956.

91. Ishibashi, T., and Nishikawa, H., Saturated boiling heat transfer in narrow spaces, *J. Heat Mass Transfer*, 12(8), 863–893, 1969.

92. Azarskov, V.M., Refrigerants boiling in slit channels of plate evaporators experimental investigations, Dissertation thesis, *Leningrad Food Tech. Refrig. Inst. Publ.*, 1973.

93. Gogolin, A.A., Danilova, G.N., Azarskov, V.M., and Mednikova, N.M., Heat transfer enhancement in refrigeration machines evaporators, *Food Ind.*, 287 pp., 1982.

94. Malyugin, G.I., Ice generator and sublimator internal heat transfer investigation and enhancement, Dissertation thesis, *Leningrad Food Tech. and Refrig. Inst. Publ.*, 1976.

95. Zemskov, B.B., Refrigerants boiling in vertical complex form channels heat transfer and hydrodynamic investigations, Dissertation thesis, Leningrad Food Techn and Refrig. Inst. Publ., 1978.

96. Grigoriev, V.A., Krokhin, Yu.I., and Kulikov, A.S., Vertical slit channels boiling heat transfer, Moscow, *Power University Works, Industrial Energy Application*, Vol. 141, 58–68, 1972.

97. Grigoriev, V.A., and Krokhin, Yu.I., Real phase velocity and vapor quality by boiling in slit channels, *HTT*, 2(1), 145–148, 1972.

98. Antipov, V.I., Grigoriev, V.A., and Illarionov, A.G., Liquid nitrogen boiling in small diameter tubes (capillaries) experimental investigations, Moscow, Final papers, pp. 146–151, 1968–1969.

99. Labuntsov, D.A., et al., Boiling inside small diameter tubes heat transfer coefficients experimental investigations, *HEIN, Energy and Transport Mechanical Engineering*, N5, 1971.

100. Labuntsov, D.A., et al., Boiling inside small diameter tubes heat transfer coefficients experimental and analytical investigations, *HEIN, ETME*, 7, 68–73, 102–105, 1970.

101. Leontiev, A.I., et al., One vaporization center in slit channel heat transfer experimental investigation, Moscow High Tech. School Work, 195, 34–41, 1975.

102. Leontiev, A.I., et al., Ethanol water solution boiling in slit channel heat transfer experimental investigation, *HEIN, ETME*, 3, 85–87, 1977.

103. Leontiev, A.I., et al., Microfilm by liquid boiling in weak gravity condition role, nonlinear wave, *Proc. in Two-Phase Media*, pp. 266–275, Nov. 1977.

104. Leontiev, A.I., Mironov, B.M., and Korneev, S.D., Ethanol water solution boiling in slit plate-parallel channel heat transfer characteristics investigations, *Heat Transfer Problems* (Moscow Wood Tech. Ins. W.), 101, 98–108, 1977.

105. Koba, A.L., and Smirnov, H.F., Boiling in submerged into liquid pool horizontal slit heat transfer and critical heat fluxes, *EPJ*, 27(4), 738–739, 1974.

106. Koba, A.L., and Smirnov, H.F., Boiling in submerged into liquid pool inverted plate heat transfer and critical heat fluxes, *RETP*, 2, 70–74, 1974.

107. Koba, A.L., and Smirnov, H.F., Boiling in submerged into liquid pool horizontal plate slit heat trans., and critical heat flux, *RETP*, 3, 126–134, 1977.

108. Koba, A.L., Afanasiev, B.A., et al., Heat transfer investigations at boiling on horizontal plate slit, Works of En. Inst. Krig. Dep. in All-Un. In. P., N1029-78 DSP, 1978.

109. Koba, A.L., Boiling on horizontal plate channels submerged into liquid pool (heat transfer, mechanism, internal characteristics, heat transfer crisis), Dissertation thesis, Odessa., Acad. of Refrigeration, 1979.

110. Klimov, S.M., Bologa, M.K., and Smirnov, H.F., Boiling in pool horizontal plate slit heat transfer under electric field action investigations, *ETM*, 5, 48–52, 1978.

111. Klimov, S.M., Bologa, M.K., and Smirnov, H.F., Electric field action influence on boiling in pool narrow horizontal plate slits heat transfer and critical heat fluxes, *Electron. Treatment Mater.*, 5, 45–49, 1979.

112. Klimov, S.M., Bologa, M.K., and Smirnov, H.F., Mechanism, internal characteristics of boiling in pool horizontal plate slit heat transfer under electric field action investigations, *Electron. Treatment Mater.*, 5, 46–50, 1980.

113. Klimov, S.M., Bologa, M.K., and Smirnov, H.F., Mechanism of formation and pressing of dry spots by boiling in liquid pool submerged slit under electric field action, *Electron. Treatment Mater.*, 5, 56–62, 1980.

114. Mironov, B.M., Lobanova, L.S., and Shadrin A., Liquid boiling in vertical plate par. slit chan local heat transfer crisis, *HEIN, Energy*, 7, 85–88, 1978.

115. Klimov, S.M., Bologa, M.K., and Smirnov, H.F., Boiling in pool horizontal plate slit heat transfer under electric field action, *Heat Transfer, Minsk*, 4(part 1), 22–27, 1980.

116. Klimov, S.M., Boiling in pool horizontal plate slit heat transfer under electric field action, Dissertation thesis, Odessa, Acad. of Refrigeration, 1982.

117. Danilova, G.N., and Dyundin, V.A., Refrigerants R-12 and R-22 boiling on finned tubes bundle heat transfer, *Refrigeration*, 7, 40–42, 1971.

118. Dyundin, V.A., Refrigerants boiling on finned surfaces heat transfer, Dissertation thesis, 1971.

119. Tiktin, S.A., *Vapotron Technology*, Technika, Kiev, 151 pp., 1971.

120. Kovalev, S.A., Traditional and dissociating liquids pool boiling on isothermal and non-isothermal (finned) surfaces heat transfer and stability mechanism investigations, Dissertation thesis, Moscow, 1977.

121. Petukhov, B.S., Kovalev, S.A., and Geshele, V.D., Boiling on finned surface, *Heat Transfer and Hydrodynamics*, Science, Moscow, pp. 5–14, 1977.

122. Petukhov, B.S., and Geshele, V.D., Liquid boiling in slit experimental methods and results, *HTT*, 13(1), 321–329, 1975.

123. Bezrodny, M.K., Butuzov, A.I., and Sosnovsky, V.I., Value of gap in finned spaces of finned tube influence on liquid pool boiling heat transfer, *RETP*, 2, 19–28, 1975.

124. Bezrodny, M.K., Sosnovsky, V.I., and Alekseenko, D.V., Refrigerant R-11 pool boiling in slit channels heat transfer and critical heat fluxes experimental investigations, *Thermal Processes in Electronic High-Wave Devices*, Ser. HWD. Electronics, Vol. 3(43), 1975.

125. Sosnovsky, V.I., Boiling in slit channels and on finned surfaces refrigerants heat transfer and critical heat fluxes experimental investigations, Dissertation thesis, Kiev National Polytechnic University, 1977.

126. Yagov, V.V., Liquid pool boiling in low pressures field investigations, Dissertation thesis, Moscow Inst. of Power Eng., 1971.

127. Belsky, V.K., Refrigerant R-12 boiling on tubes bundle and on single wrapped tubes heat transfer investigations, *Refrigeration*, 2, 41–44, 1970.

128. Kupriyanova, A.V., Ammonia boiling on tubes heat transfer investigation, Dissertation thesis, Leningrad, 1970.

129. Vishnev, I.P., et al., Helium bubbles boiling thermal parameters, *Heat Mass Transfer*, Inst. of HMT of Acad. Sci. BSSR, 2(part 1), 263–270, 1972.

130. Vishnev, I.P., et al., Helium-1 bubble pool boiling in vertical channels pressure influence, *Acad. of Sci. of USSR*, 206(5), 1090–1092, 1970.

131. Grigoriev, V.A., Pavlov, and Ametistov, E.D., *Cryogenic Liquids Pool Boiling*, Energy, Moscow, 288 pp., 1983

132. Gorokhov, V.V., Helium boil. by natural circulation in heat exchangers investigation, Dissertation thesis, Moscow Inst. of Chem. Eng., 1975.

133. Petukhov, B.S., Zukov, V.M., and Shildkret, V.M., Helium in small diameter channels boiling heat transfer investigation, *Cryo-Electro-Technology and Energy*, Part III, Kiev, pp. 104–109, 1977.

134. Shildkret, V.M., Helium two-phase flow in channels heat transfer and hydraulic resistance, Dissertation thesis, Moscow, 1981.

135. Kutateladze, S.S., and Nakoryakov, V.E., *Heat and Mass Transfer and Waves in Gas–Liquid Systems*, Science, Novosibirsk, 1984, 300 pp.

136. Kutepov, A.L., Sterman, L.S., and Styushin, N.G., *Hydrodynamic and Heat Transfer by Vaporization*, Moscow High School, 484 pp., 1983.

137. Lippert, L., and Pitts, A., *Heat Transfer Problems*, Atomizdat, Moscow, pp. 142–198, 1967.

138. Smirnov, H.F., Initial micro-layer under bubble by boiling thickness elementary calculation method, *EPJ*, 28(3), 503–508, 1975.

139. Kozhelupenko, Yu.D., Smirnov, H.F., and Koba, A.L., Subcooled liquid boiling heat transfer crisis in narrow annular channels by small liquid velocities, Abstracts of paper. *Boiling and Condensation Processes Thermo-Physics and Hydrodynamics*, Riga, Vol. 1, pp. 154–155, 1982.

140. Kozhelupenko, Yu.D., Water and water solution boiling in capillary and annular channels heat transfer and critical heat fluxes by cooling systems of radio-electronic equipment (REE) working conditions, Dissertation thesis, Odessa, 1984.

141. Baranenko, V.I., and Chichkan, L.A., Liquid microfilm thickness measurement method and some results by boiling inside slit channel, *Boiling and Condensation*, Riga, Vol. 2, pp. 3–12, 1978.

142. Chichkan, L.A., Liquid boiling heat transfer local characteristics and mechanism investigation on laser diffraction interferometer application base, Dissertation thesis, Kiev National Polytechnic University, 1982.

143. Kutateladze, S.S., *Heat Transfer and Hydraulic Resistance*, *Reference Manual*, Energoatomizdat, Moscow, 366 pp., 1990.

144. Diev, M.D., A study of vapor bubble departure during boiling in narrow slit channel, the physics of heat transfer in boiling and condensation, *Proc. of Int. Symp.*, May 21–24, pp. 345–350, 1997.

145. Labuntsov, D.A., Developed bubble boiling heat transfer approximated theory, *USSR Acad. Sci. News Energy Transp.*, 1, 58–71, 1963.

146. Ferell, I.K., Alexander, E.G., and Piver, W.T., Vaporization heat transfer in heat pipe wick materials, *AIAA Paper*, 256, 1–12, 1972.

147. Davis, W.R., and Ferell, I.K., Evaporative heat transfer of liquid potassium in porous media, *Heat Transfer Ther. Contr. Appl.*, 39, 198–199, 1975.

148. Ferell, I.K., and Alexander, E.G., Vaporization heat transfer in capillary wick structures, *Chem. Eng. Symp. Ser.*, 66, DD2, 1970.

149. Ferell, I.K., and Johnson, D.E. The mechanism of heat transfer in the evaporative zone of a heat pipe, ASME Heat 70-HT/sptn, June 1971.

150. Abhat, A., and Seban, R.A., Water, acetone and ethanol boiling and evaporation in heat pipe wicks, *Heat Transfer*, 3, 74–83, 1988.

151. Seban, R.A., and Abhat, A., Steady and maximum evaporative from screen wicks, *Paper ASME*, WA/HT-12, 1971.

152. Corman, I.C., and Walmet, W.L., Vaporization from capillary wick structures, *Paper ASME NHT-35*, 1–8, 1971.

153. Marto, P.I., Mosteller, W.L., Effect of nucleate boiling on the operation of low-temperature heat pipes, *ASME Paper,* No-69-HT-24, pp. 1–8, 1969.

154. Sasin, V.Ya., Fedorov, V.N., and Sorokin, A.Ya., Heat pipe with low-boiled heat carrier experimental investigation, Moscow, *Pow. Eng. Inst. Wor. Coll. Connected with Sci.-Techn. Conf.*, pp. 79–F84, 1969.

155. Voronin, V.G., Revyakin, A.V., Sasin, V.Ya., and Tarasov, V.S., Low-temperature heat pipes for flying apparatuses, Mech. Engin., Moscow, 200 pp., 1976.

156. Asakyavichus, I.P., Zhukauskasm A.A., et al., Heat pipes working by moderate temperatures heat transfer mechanism*, Lithonia SSR Works, Ser. B*, 5(102), 81–88, 1977.

157. Asakyavichus, I.P., Zhukauskas, A.A., et al., Refrigerant-113, ethanol and water boiling in screen wicks heat transfer, *Lithonia SSR Works, Ser. B.*, 1(104), 87–93, 1978.

158. Asakyavichus, I.P., and Eva, V.K., Heat transfer inside low-temperatures heat pipes problems, *Heat Transfer*, Soviet investigations, pp. 458–461, 1978.

159. Antonenko, V.A., Boiling and Evaporation in thin liquid films heat transfer investigations, Dissertation thesis, Kiev Ins. Tech. Therm., 1978.

160. Tolubinsky, V.I., Antonenko, V.A., and Ostrovsky, Yu.N., Evaporative zone of heat pipe vaporization heat transfer regularities and mechanism, *USSR Acad. Sci. News Energy Transp.*, 1, 141–148, 1979.

161. Tolubinsky, V.I., et al., Liquid evaporation in capillary wicks of low-temperature heat pipes limiting heat fluxes, *HTT*, 18(12), 367–373, 1980.
162. Tolubinsky, V.I., et al., Drop carry over phenomenon in liquid evaporative from capillary structures, *Lett. Heat Mass Transfer*, 5(6), 339–347, 1978.
163. Tolubinsky, V.I., et al., Vaporization regimes inside low-temperature heat pipes wicks, *Proc. of USSR Conf. of HMT, Minsk*, 6(1), 26–31, 1984.
164. Kudritsky, G.R., Evaporation on small sizes surfaces, coated capillary-porous structures (by heated surface local heat source using), heat transfer enhancement, Dissertation thesis, Kiev Institute of Technical Thermophysics, Kiev, 1985.
165. Tolubinsky, V.I., et al., Boiling on surfaces, covered by thin screen wick, heat transfer, *Ind. Appl. Thermal Eng.*, 11(4), 3–7, 1989.
166. Kudritsky, G.R., Ostrovsky, Yu, N., and Antonenko, V.A., Vaporization mechanism inside screen wicks of low-temperature heat pipes investigation, *Ind. Appl. Thermal Eng.*, 12(4), 61–65, 1990.
167. Antonenko, V.A., and Kudritsky, G.R., Vaporization process inside heat pipes wick mechanism investigation results, *EPJ*, 90(1), 12–19, 1991.
168. Kudritsky, G.R., and Kolomiez, E.A., Surface geometry parameters influence on heat transfer and initial boil, *Moment, IATE*, 16(1), 18–21, 1994.
169. Kudritsky, G.R., Boiling in defined geometry characteristics of heating surface conditions influence on heat transfer, Dissertation thesis, Kiev Institute of Technical Thermophysics, Kiev, 1995.
170. Smirnov, H.F., Afanasiev, B.A., et al., Heat transfer and limiting heat fluxes at vaporization in construction capillary-porous structures of heat pipes, *Prepr. Proceeding of 7th IHPC*, Minsk, USSR, paper A-16, 1990.
171. Semena, M.G., et al., Transient from evaporation up to boiling regimes in low-temperature heat pipes with fiber-metal wicks, *Heat and Mass Transfer-7, Minsk*, 6, 117–123, 1980.
172. Hasegawa, S., Echigo, R., and Ikie S., Boiling characteristics and burnout phenomena on heating surface covered with woven screens, *J. Nucl. Sci. Technol.*, 12(11), 722–724, 1975.
173. Semena, M.G., Zaripov, O.K., and Gershuni, A.N., Heat Transfer in heat pipes heating zone with metal-fiber capillary structures, *HTT*, 24(2), 67–73, 1982.
174. Ornatsky, A.P., Semena, M.G., and Timofeev, V.I., Maximum heat fluxes by boiling on plane metal-fiber wicks in conditions, typical for heat pipes investigation, *EPJ*, 35(1), 212–219, 1978.
175. Shapoval, A.A., Zaripov, O.K., and Semena, M.G., Water boiling on surfaces with metal-fiber porous coatings heat transfer intensity investigation, *Thermal Eng.*, 12, 63–65, 1983.
176. Shapoval, A.A., Water and acetone boiling on surfaces, covered metal-fiber capillary-porous coatings, Dissertation thesis, Kiev Nat. Polyt. Univ., 1978.
177. Semena, M.G., et al., Water boiling on porous surfaces heat transfer intensity pressure influence peculiarity, *EPJ*, 62(1), 779–782, 1992.
178. Semena, M.G., et al., Water boiling on porous surfaces heat transfer intensity pressure influence investigation, *Proc. Int. For. Minsk., Heat and Mass Transfer*, Vol. 4, Part 1, pp. 59–62, 1992.
179. Fridrikhson, Yu.V., Kravetz, V.Yu., and Semena, M.G., Vaporization acting centers density calculation by liquid boiling on metal-fiber capillary-porous structures, *EPJ*, 66(5), 600–605, 1994.
180. Fridrikhson, Yu.V., Liquid pool boiling on metal-fiber coating heat transfer, Dissertation thesis, Kiev Nat. Polyt. Univ., 1995.
181. Labuntsov, D.A., Bubble boiling heat transfer problems, *Thermal Eng.*, 9, 14–19, 1972.
182. Maidanik, Yu.F., Working out and investigation of low-temperature heat pipes, which can be working by different positions in gravity field, Dissertation thesis, Sverdlovsk Polyt. Univ., 1977.

183. Dolgirev, Yu.E., Heat pipes, working by different position in volumetric forces field, investigation, calculation and optimization, Dissertation thesis, Sverdlovsk Polyt. Univ., 1977.

184. Maidanik, Yu.F., Fershtater, Yu.G., and Pastukhov, V.G., Loop heat pipes: working out; investigations; engineering calculations elements, *Acad. Sci. USSR Sci. Pap., Ural Part.* Sverdlovsk, 1989.

185. Kiseev, V.M., et al., Plane heat pipe with separate vapor and liquid channels evaporation zone optimization, Sverdlovsk, N4162-84DEP, 1984.

186. Kiseev, V.M., et al., Antigravity heat pipe with flat evaporator wick closing wall thickness optimization, Sverdlovsk, N4160-84DEP, 1984.

187. Maidanik, Yu.F., Loop heat pipes and two-phase loops with capillary pumping, Dissertation thesis, Moscow Inst. of Power Eng., 1993.

188. Dyundin, V.A., and Borishanskaya, A.V., Liquid boiling heat transfer surface conditions influence on, HMT M., 2, 177–179, 1972.

189. Danilova, G.N., Guigo, E.I., et al., Low-temperature liquids boiling at low heat fluxes heat transfer enhancement, *Proc. of Int. For in Minsk, HMT* HMT-5 M., 3(Part 1), 22–31, 1976.

190. Dyundin, V.A., Danilova, G.N., and Borishanskaya, A.V., Refrigerants boiling on surfaces with coated coverings, *Boiling and Condensation*, Riga, Vol. 2, pp. 7–19, 1977.

191. Dyundin, V.A., et al., Refrigerants boiling on tubes surfaces with metallic coverings experimental investigations, *Refrigeration and Cryogenic Technology*, pp. 121–127, 1975.

192. Bukin, V.G., Danilova, G.N., and Dyundin, V.A., Refrigerants moving film washing horizontal tubes bundle coated porous covering by boiling heat transfer, *Heat Mass Transfer-5*, 4(part 1), 16–21, 1980.

193. Poznyak, V.E., Orlov, V.K., and Saviliev, V.N., Coating porous covering on boiling surface of condenser–evaporator heat transfer enhancement, *Chem. Oil Mech. Eng.*, 3, 43–47, 1980.

194. Orlov, V.K., and Saveliev, V.N., Cryogenic liquids boiling by lower atmospheric pressure heat transfer enhancement, *Thermal Eng.*, 4, 62–64, 1980.

195. Orlov, V.K., et al., Cryogenic liquids boiling heat transfer enhancement, Heat Transfer and Hydrodynamic by Boiling and Condensation, *21th Sibirsk Thermophysical Seminar Proc.*, October, pp. 83–89, 1978.

196. Royezin, L.I., et al., Nitrogen and refrigerant-113 boiling on metal porous coatings heat transfer, *HTT*, 20(2), 304–310, 1982.

197. Tunik, A.T., Bolshakov, A.A., and Techwer, Ya.Ch., Liquid dielectric boiling on porous coated surfaces heat transfer intensity influence on, *High School News, Acad. Sci. USSR Ser. Phys. Math.*, 27(3), 364–369, 1978.

198. Ito, T., Fudzita, K., and Nisikawa, K., Surfaces with high boiling heat transfer coefficient, *Sci. Mech.*, 31(1), 1979.

199. Sirotin, A.G., Dvoiris, A.D., et al., Refrigeration machines efficiency increasing for account of application of tubes with porous coatings in heat exchangers-evaporators, *Gas Ind.*, 12, 28–32, 1976.

200. Dvoiris, A.D., Sirotin, A.G., and Sakharova, G.P., Natural gas liquefaction processes enhancement, *USSR Acad. Sci. News Eng. Transp.*, 4, 145–151, 1980.

201. Sirotin, A.G., Liquefied hydrocarbon gas boiling on porous surfaces heat transfer, New equipment and technology of preparation and treatment gas and condensate, 1, 104–116, 1982.

202. Nakayama, W., Daykoku, T., Kuwahara, H., and Nakajima, T., Boiling on porous surface heat transfer enhancement dynamic model. Part 1, Experimental Investigation; Part 2. Analytical Model, *ASME Works*, 102(3), 62–69, 69–74, 1980.

203. Bergles, A., Pool boiling on metal porous coatings heat transfer characteristics, *ASME Works*, 104(2), 56–64, 1982.

204. Nakayama, W., Daikoku, T. and T. Nakajima, Pores parameters and system pressure influence on boiling heat transfer in saturation regime, *ASME Works*, 104(2), 65–72, 1982.

205. Martin, P.J. and Lepert, V. Dielectric liquids pool boiling on structural surface heat transfer, *ASME Works*, 104(2), 72–80, 1982.

206. Styrikovich, M.A., et al., Mass Transfer by boiling in capillary-porous structures investigations modern state. Review, *HTT*, 18, 625–633, 1980.

207. Picone, L.F., Whyte, D.D., and Taylor, G.R., Radiotracer studies of hideout at high temperature and pressure, *Wes. Elect. Corp. Atom. Pow. Div*, WCAP-3731, June 1963.

208. Polonsky, V.S., et al., Boiling in capillary-porous structures accompanied by concentration model, *Paper Acad. Sci. USSR*, 241(3), 579–583, 1978.

209. Styrikovich, M.A., et al., Iron oxides deposits influence on mass transfer intensity in vapor generating channels, *Moscow Pow. Eng. Inst. Wor. Coll.*, 128, 54–62, 1972.

210. Styrikovich, M.A., et al., Mass transfer intensity in vapor generating channels experimental investigations, *HTT*, 15(2), 225–233, 1977.

211. Styrikovich, M.A., et al., Mass Transfer conditions in vapor generating channels experimental investigations with heat-generating in cosine law, *HTT*, 16(3), 325–333, 1978.

212. Dvoryaninov A.V., et al., Corrugated screen wick heat pipe characteristics experimental investigation, *EPJ*, 37(1), 13–19, 1979.

213. Aptekar, B.F., and Baum, Ya.M., Heat carrier mass inside low-temperature heat pipes with corrugated screen wicks influence on limiting heat fluxes, *HTT*, 20(1), 150–154, 1982.

214. Zigalov, V.G., et al., Heat pipes with corrugated capillary structures experimental investigation, *RETP*, 3, 117–118, 1983.

215. Afanasiev, B.A., et al., Condensation and evaporation inside heat pipes with corrugated wicks, *7th All-Union HMT Conf.,* Minsk, 1, Part 2, 8–14, 1984.

216. Silinsky, A.L., Heat pipes with corrugated wicks produced from smooth metal foil for power transistors cooling, Dissertation thesis, L. In. of Fd. Tech, 1986.

217. Afanasiev, B.A., Vinogradova, E.P., and Smirnov, H.F., Heat pipes with corrugated wicks produced from smooth metal foil heat transfer and it maximum thermal power, *EPJ*, 50(1), 58–65, 1986.

218. Vinogradova, E.P., Heat Transfer and maximum heat fluxes of heat pipes with corrugated capillary structures, Dissertation thesis, Odes, Acad. of Refrigeration, 1988.

219. Styrikovich, M.A., Leontiev, A.I., Polonsky, V.S., et al., Mass transfer intensity in vapor generating channels experimental investigations, Int. Sem. Mom. Heat and Mass Transfer in Two-Phase Energy and Chemical Systems, Dubrovnic, Jugoslavia, September, pp. 4–9, 1978.

220. Cohen P., Concentration by solution boiling theoretical model, *Am. Inst. Chem. Eng. Symp.*, 138, 70–71, 1974.

221. Malyshenko, S.P., and Andrianov, A.B., Initial boiling on surfaces with porous coatings curve part and boil hysteresis, *HTT*, 25(3), 563–572, 1987.

222. Andrianov, A.B., and Malyshenko, S.P., Porous coatings influence on boiling heat transfer intensity, *USSR Acad. Sci. News Energy Transp.*, 1, 139–149, 1989.

223. Malyshenko, S.P., and Andrianov, A.B., Irreversible phase transients by boiling on surfaces with porous coatings, Preprint of Inst. High Temp. of Acad. of Sci. of USSR, N1-293, 1990.

224. Andrianov, A.B., et al., Composite super-conductivity materials cryostat stabilization improvement with porous coating help, Preprint of Inst. High Temp. of Acad. of Sci. of USSR, N4-227, 1987.

225. Techwer, Ya.Ch., Syi, Ch.N., and Temkina, V.S., Liquid boil on porous surface and heat transfer hysteresis, Preprint of Ac. of Est. SSR. Tallin, 1989.

226. Malyshenko, S.P., Percolation transition at boiling on porous coated surfaces, *Physics of Heat Transfer in Boiling and Condensation*, Proceedings of Int. Symp., Moscow, pp. 319–333, 1997.

227. Zuev, A.v., and Malyshenko, S.P., Dry-out phenomena at two-phase in tubes with porous coating, *10th Int. Heat Trans. Conf.,* Brighton, UK, Vol. 6, pp. 147–150, 1994.

228. Kovalev, S.A., and Soloviev, S.L., Heat Transfer in an inverted meniscus evaporator investigations, *HTT*, 24(1), 196–198, 1986.

229. Poniewski, M.E., Afanasiev, B.A., and Wojcik, T.M., Hysteresis of boiling heat transfer on porous covering, *Proceedings of the 30th National Heat Transfer Conf.*, Poland, Vol. 7, pp. 197–220, 1996.

230. Styrikovich, M.A., Leontiev, A.I., and Malyshenko, S.P., Non-volatile impurities transfer mechanism by boiling on surfaces covered by porous structures, *HTT*, 14(1), 998–1006, 1976.

231. Makbeth, R.V., Boiling on surface overlaid with a porous deposit, heat transfer rates obtainable by capillary actions. AEEW-R-711, Winfrith, 1971.

232. Pichlak, U.L., and Techwer, Ya.Ch., Porous coating influence on boiling heat transfer, *USSR Acad. Sci. News Energy Transp.*, 3, 13–17, 1983.

233. Techwer, Ya.Ch., and Tunik, A.G., Boiling on porous coating surface, *USSR Acad. Sci. News, Ser. Phys. Math.*, 28(1), 68–27, 1979.

234. Krokhin, Yu.I., and Kulikov, A.S., Vaporization process in capillary-porous structures approximated hydrodynamic theory, *HTT*, 21(5), 952–958, 1983.

235. Kulikov, A.S., Liquid flow inside pores and limiting heat fluxes definition by vaporization on porous coated surfaces problem solution, Dissertation thesis, Moscow Inst. of Power Eng., 1984.

236. Borishanskaya, A.V., Gotovsky, M.A., and Dyundin, V.A., Boiling on surfaces with capillary-porous coverings heat transfer calculation, *Boiling and Condensation*, Riga, Vol. 2, pp. 29–36, 1978.

237. Aleshin, A.N., et al., Vaporization on porous coated surfaces heat transfer investigation, *HTT*, 18(2), 1098–1101, 1980.

238. Soloviev, S.L., and Kovalev, S.A., Liquid evaporation from capillary-porous body heat transfer, *7th All-Union HMT Conf.*, Minsk, Vol. 6, pp. 22–25, 1984.

239. Kovalev, S.A., and Soloviev, S.L., Liquid evaporation and boiling on porous coated surfaces heat transfer, *Problems papers* of *7th All-Union HMT Conf., Minsk*, pp. 3–12, 1985.

240. Kovalev, S.A., and Soloviev, S.L., Liquid boiling on porous coated surfaces models, *Problems papers*, Heat Mass Transfer-8 Int. For. Minsk, pp. 3–18, 1988.

241. Webb, R.L., *Principles of Enhanced Heat Transfer*, Wiley-Interscience, New York, 1994.

242. Webb, R.L., Advances in Modeling Enhanced Heat Transfer Surface, *Proc. of the 10th Int. Heat Trans. Conf.*, Brighton, UK, Vol. 4, pp. 445–449, 1994.

243. Rosenfeld, I.H., Nucleate Boiling Heat Transfer in porous wick structures, ASME, *Heat Transfer Equipments Fund. Book*, NH00500, 108, 47–55, 1989.

244. Nishikava, K., Ito, T., and Tamaka K., Enhanced heat transfer by nucleate boiling on sintered metal layer, *Heat Transfer Jpn. Res.*, 8, 65–81, 1979.

245. O'Neil, P.S., Gottzman, G.F., and Turbot, T.W., Novel heat exchanger increases cascade cycle efficiency for national gas liquefaction, in *Advances in Cryogenic Engineering*, K.D. Fimmounhaus, ed., Plenum, NY, pp. 402–437, 1972.

246. Webb, R.L., Nucleate boiling on porous coated surface, *Heat Transfer Eng.*, 3(304), 71–82, 1983.

247. Rosenfeld, I.H., Modeling of heat transfer into a heat pipe for a localized heat input zone, *AICHE Symp. Ser. Heat Transfer*, Pittsburgh, 83, 71–76, 1987.

248. Shaubakh, R.M., Dussinger, P.M, and Bogart, J.E., Boiling in heat pipe evaporator wick structures, *7th IHPC Proceedings, Minsk*, pp. 1–15, 1990.

249. Singh, B.S., and Shaubakh, R.M., Boiling and two-phase flow in the capillary porous structure of a eat pipe, *ASME Fed.*, 60/91, 1988.

250. Gorbis, Z.R., Berman, M.I., and Beliy, L.M., Invention Certificate N449227 of USSR Heat Transfer Enhancement Method, Bulletin of Inv., N41, 1974.

251. Chulkin, O.A., Hydrodynamic heat transfer deposit prevention in evaporator with solid particles fluidized bed, Dissertation thesis, 1984.

252. Berman, M.I., Boiling process on heating surfaces submerged into solid particles dispersed layer investigation, Dissertation thesis, Odessa Academy of Refrigration, 1984.

253. Berman, M.I., Gorbis, Z.R., and Alexandrov, V.I., Boiling on horizontal tubes bundle heat transfer in developed fluidization conditions, *EPJ*, 36, 337–343, 1979.

254. Berman, M.I., and Gorbis, Z.R., Process analysis and boiling on heating surface, submerged into solid particles dispersed layer, heat transfer experimental data generalization, *EPJ*, 38, 5–15, 1980.

255. Berman, M.I., and Chulkin, O.A., Boiling into fluidized particles bed deposits prevention process experimental investigation and model working out, Minsk, N3140-83DEP, 1983.

256. Berman, M.I., and Chulkin, O.A., Boiling on horizontal tubes bundle in developed fluidized particles bed conditions heat transfer and hydrodynamic, *EPJ*, 43, 718–727, 1982.

257. Smirnov, H.F., Liquid pool boiling in dispersed material submerged bed heat transfer, *EPJ*, 34, 792–798, 1978.

258. Tuchnin, A.M., Vasanova, L.K., and Siromyatnikov, N.I., Sub-cooled boiling in liquid fluidized particles bed by cylinder cross flow heat transfer, *EPJ*, 36, 389–394, 1979.

259. Tuchnin, A.M., Heat transfer in liquid fluidized bed by high heat fluxes investigation, Dissertation thesis, Sverdlovsk Polyt. Univ., 1980.

260. Baranenko, V.I., Asmolov, V.G., and Kirov, V.S., *Thermodynamic and Heat Transfer in Nuclear Power Systems with Gas-Saturated Heat Carriers*, Energoatomizdat, Moscow, 1993.

261. Siromyatnikov, N.I., Vasanova, L.K., and Tuchnin, A.M., Heat transfer in liquid fluidized bed by sub-cooled boiling of liquefied media, *Boiling and Condensation*, Riga, Vol. 2, pp. 112–119, 1978.

262. Veeman, A.W., and Potze W., Wall-to-liquid heat transfer in a liquid-solid fluidized bed, *6th Heat Transfer Conf.*, Toronto, Vol. 4, pp. 123–128, 1978.

263. Ipfelkofer R., and Blenke, H., Beitray zum Warmeubergang in Flussigkeits, Friebbettern, *Waerme*, 84(1), 47–50, 1978.

264. Bologa, M.K., Smirnov, H.F., et al., *Heat Transfer by Boiling and Condensation under Electric Field Action*, Stiinza, Kishinev, 240 pp., 1987.

265. Smirnov, H.F., Closed heat transferred evaporative systems theory fundamentals, Dissertation thesis, Leningrad, 1979.

266. Kutateladze, S.S., Pool boiling heat transfer crisis hydrodynamic theory, *J. Tech. Phys.*, 20(11), 1389–1392, 1950.

267. Zuber, N., On the stability of boiling heat transfer, *Trans. ASME*, 80(3), 711–720, 1958.

268. Kutateladze, S.S., *Heat Transfer Theory Fundamentals*, 5th Issue, Atomizdat., Moscow, 415 pp., 1979.

269. Borishansky, V.M., and Gotovsky, M.A., Two-phase near-wall layer hydraulic stability destroying by pool and forced convection boiling theory, *Heat Mass Transfer, Minsk, Sci. Technol.*, 3, 125–130, 1965.

270. Linchard, J.Ch., and Dhir, V.N., Maximum and minimum heat fluxes by pool boiling hydrodynamic base of calculation, HMT M., 2(Part1), 273–298, 1972.

271. Kirichenko, Yu.A., and Chernyakov, I.S., First critical heat flux by boiling on flat heaters determination, *EPJ*, Vol. 20(6), 982–987, 1971.

272. Labuntsov, D.A., Boiling critical heat fluxes top border about, *HTT*, 10(6), 1337–1338, 1972.

273. Yagov, V.V., and Pusin, V.A., Critical heat flux by forced saturated liquid motion calculation approximated model, *Thermal Eng.*, 3, 2–5, 1985.

274. Kovalev, S.A., and Smirnova, L.F., Fin washed by boiled liquid, temperature field, *HTT*, 6(4), 698–701, 1968.

275. Kovalev, S.A., and Rubchinskaya, G.B., Boiling on non-isothermal surface heat transfer stability, *HTT*, 11(1), 117–122, 1973.

276. Isaev, S.I., et al., *Heat Transfer Theory*, A.I. Leontiev, ed., Moscow Tech. Univ, pp. 684, 1997.

277. Tolubinsky, V.I., *Boiling Heat Transfer*, Naukova Dumka, Kiev, 316 pp., 1980.

278. Yagov, V.V., and Zudin, Yn.V., Mechanistic model for nucleate boiling crisis at different gravity fields, *Heat Transfer. Proc. of the 10th Int. HTC*, Brighton, UK, Vol. 5, 189–194, 1994.

279. Haramura, Y., and Katto, Y., A new hydrodynamic model of critical heat flux, applicable widely to both pool and forced convection boiling on submerged bodies in saturated liquid, *J. Heat Mass Transfer*, 26(3), 389–399, 1983.

280. Katto, Y., A physical approach to critical heat flux of sub-cooled flow boiling in round tubes, *J. Heat Mass Transfer*, 33(4), 611–620, 1990.

281. Katto, Y., Prediction of critical heat flux of sub-cooled flow boiling in annular tubes, *J. Heat Mass Transfer*, 33(9), 1921–1928, 1990.

282. Yagov, V.V., Boiling in tubes with saturated and subcooled liquid heat transfer crisis mechanism, *Thermal Eng.*, 5, 16–22, 1992.

283. Celata, G.P., Critical heat flux in water sub-cooled flow boiling, experiments and modeling, *Proc. 3rd European Thermal Sciences and 14th UIT National Heat Transfer Conference*, May 29–31, Vol. 1, pp. 27–40, 1996.

284. Celata, G.P., Modeling of critical heat flux in subcooled flow boiling, *The Physics of Heat Transfer in Boiling and Condensation*, May, Moscow, pp. 324–332, 1997.

285. Tong, L.S., Boundary layer analysis of the flow boiling crisis, *Proc. of 3rd Int. Heat Trans. Conf. Hemisphere*, New York, Vol. 111, pp. 1–6, 1966.

286. Bergles, A.E., Burn out under conditions of subcooled boiling and forced convection, *Thermal Eng.*, 27(1), 48–50, 1980.

287. Smogalev, I.P., Calculation on of Critical heat fluxes with flow of subcooled water at low velocity, *Thermal Eng.*, 28(4), 208–211, 1981.

288. Weisman, I., and Ying, S.H., Theoretically based CHF prediction at low qualities and intermediate flows, *Trans. Am. Nucl. Soc.*, 45, 832–833, 1983.

289. Weisman, I., and Pei, B.S., Prediction of critical heat flux in flow boiling at low qualities, *Int. J. Heat Mass Transfer*, 26, 1463–1477, 1983.

290. Styrikovich, M.A., Nevstrueva, E.I., and Dvorina, G.M., The effect of two-phase flow pattern on the nature of heat transfer crisis in boiling, *Proc. 4th Int. Heat Transfer Conf.*, Hemisphere, New York, Vol. 9, pp. 360–362, 1970.

291. Celata, G.P., et al., Rationalization of existing mechanistic models for the prediction of water subcooled flow boiling critical heat flux, *Int. J. Heat Mass Transfer*, 37(Suppl 1), 347–360, 1994.

292. Lee, C.H., and Mudawar, L.A., Mechanistic critical heat flux model for subcooled flow conditions, *J. Multiphase Flow*, 14, 711–728, 1988.

293. Kafarov, V.V., *Heat and Mass Fundamentals*, High School, Moscow, 453 pp., 1979.

294. Koba, A.L., et al., Liquid by forced convection inside capillary channels boiling heat transfer critical heat fluxes, *Thernal Eng.*, 6, 73–74, 1980.

295. Kozhelupenko, Yu, D., et al., Binary liquids mixtures by forced convection inside capillary channels boiling heat transfer critical heat fluxes investigations, *RETP*, 3, 90–98, 1981.

296. Kozhelupenko, Yu, D., and Koba, A.L., Smirnov, H.F., Binary liquids mixtures by forced convection inside capillary channels heat transfer crisis, *Thermal Eng.*, 1, 66–68, 1983.

297. Kozhelupenko, Yu, D., Smirnov, H.F., and Koba, A.L., Subcooled liquids flow inside narrow annular channels by low motion velocity heat transfer crisis, *Thermo-Physics and Hydrodynamics by Boiling and Condensation*, Riga., Vol. 1, pp. 154–155, 1982.

298. Vasiliev, A.N., and Kirilov, P.L., Boiling in small diameters channels heat transfer crisis, connected with liquid abrupt evaporation investigation, *Nuclear Reactors Thermophysical Problems*, Atomizdat, Moscow, Vol. 2, pp. 49–57, 1970.

299. Gluschenko, L.F., Subcooled boiling critical heat fluxes experimental data generalization, *Thermo-Phys. Thermal Eng.*, 5, 125–129, 1969.
300. Grigoriev, V.A., Antipov, V.P., and Pavlov, Yu.M., Nitrogen and helium boiling in channels heat transfer experimental investigation, *Thermal Eng.*, 4, 11–14, 1977.
301. Ornatsky, A.P., Main process regime parameters and channel geometry sizes influence on critical heat flux of boiling by forced convection of sub-cooled liquid, *Convective Heat Transfer*, Kiev, pp. 19–28, 1968.
302. Ornatsky, A.P., and Vinyarsky, L.S., Sub-cooled water forced motion in tubes of small diameters heat transfer crisis, *HTT*, 3(3), 444–451, 1965.
303. Ornatsky, A.P., and Kichigin, A.M., Critical heat flux dependency on liquid mass velocity, sub-cooling and pressure investigation, *Thermal Eng.*, 2, 75–79, 1961.
304. Ornatsky, A.P., and Kichigin, A.M., Sub-cooled water boiling in tubes of small diameter in high pressure field critical heat fluxes, *Thermal Eng.*, 6, 44–47, 1962.
305. Labuntsov, D.A., Boiling critical heat fluxes by sub-cooled water forced motion, *Atomic Power Eng.*, 10(5), 523–525, 1961.
306. Povarnin, P.I., 96% sub-cooled ethyl alcohol forced flow boiling critical heat fluxes investigation, *Thermal Eng.*, 12, 57–60, 1962.
307. Povarnin, P.I., and Semenov, S.T., Sub-cooled water flow in small diameter tubes by high pressures boiling heat transfer crisis investigation, *Thermal Eng.*, 1, 79–85, 1960.
308. Ribin, R.A., Diameter tube influence on critical heat fluxes by water boiling investigation, *EPJ*, 6(2), 15–19, 1963.
309. Tolubinsky, V.I., and Motorin, A.S., Binary liquids mixtures by forced convection inside capillary channels heat transfer crisis, *HMT M*, 2(Part 1), 62–66, 1972.
310. Bernath, L., A theory of local boiling burnout and it's application to existing data, *Chem. Eng. Prog. Symp. Ser.*, 56(30), 95–116, 1960.
311. Giarratano, P.I., et al., Forced convection heat transfer to liquid helium, *J. Adv Cryogenic Eng.*, 19, 404–409, 1974.
312. Shtokolov, L.S., Ethyl alcohol forced flow boiling heat transfer crisis in high velocities field, *EPJ*, 12, 3–7, 1964.
313. Lowdermilk, W., Lanzo, C.D., and Siegel, B.L., Investigation of Boiling Burnout and Flow Stability for Water Flowing in Tubes, NASA-TW-4383, 1958.
314. Macbeth, R.V., Burnout analysis. Part 2. The basic burnout curve, U.K. Report AEEW, R-167, Winfrith, 1963.
315. Gomelauri, V.T., Magrakvelidze, T.Sh., and Dvaladze, S.V., Sub-cooling value influence on forced water flow heat transfer crisis by surface with two-dimensional roughness streamline experimental investigation, *Commun. Acad. Sci. Georgia*, Tbilisi, 85(3), 673–675, 1977.
316. Ornatsky, A.P., et al., Annular channel critical power value changing regularities by high inequality along by length investigation, *HMT-V, Minsk, Sci. Technol.*, 3(Part 2), 21–27, 1976.
317. Ornatsky, A.P., Chernoby, V.A., and Vasiliev, A.F., Heat transfer crisis regularities of flow in annular channels by different heat generation laws, *Thermophys. Thermal Eng.*, 31, 13–14, 1976.
318. Ornatsky, A.P., et al., Annular heat generating channel critical power value experimental definition, *Appl. Thermal Eng.*, 1, 85–88, 1979.
319. Savina, V.N., Water, glycol and antifreeze-65 forced flow in annular channels heat transfer crisis values and hydraulic resistance investigation, Dissertation thesis, Kiev Nat. Polyt. Univ., 1971.
320. Sterman, L.S., and Checheta, P.G., Organic liquids in coaxial annular channels critical heat fluxes, *HMT-4, M.*, 2, 315–321, 1972.
321. Tolubinsky, V.I., et al., Boiling heat transfer crisis in concentric and eccentric annular channels and slits, *HMT-2, M.*, 3(part 2), 315–321, 1967.

322. Tolubinsky, V.I., Litoshenko, A.I., and Shevzov, A.L., Critical heat fluxes in annular channels experimental data generalization, *Heat Mass Transfer, Minsk*, 2, 315–321, 1968.
323. Tolubinsky, V.I., Ornatsky, A.P., and Litoshenko, A.I., Sub-cooling in narrow annular channels heat transfer crisis, *Heat Transfer by Phase Change*, Kiev, 1966, pp. 17–24.
324. Chirkin, V.S., and Yukin, V.P., Flow boiled liquid in annular gap heat transfer (heat removal) crisis, *J. Techn. Phys.*, 26(7), 1542–1555, 1965.
325. Shah, M.K., A general correlation for critical heat flux in Annuly, *Int. J. HMT*, 23(2), 225–234, 1980.
326. Alekseev, G.V., Zenkevich, B.A., and Subbotin, V.I., Experimental Data of critical heat fluxes by boiling in annular channels, Boiling crisis and evaporative heating surfaces temperature regimes, Leningrad, *ZKTI*, 58, 91–98, 1965.
327. Doroshuk, V.E., and Lanzman, V.P., Channel diameter influence on critical heat fluxes, *Thermal Eng.*, 8, 1963.
328. Miropolsky, Z.L., and Shizman, M.E., Water boiling in channels critical heat fluxes, *Nucl. Power Eng.*, 2(6), 515–521, 1961.
329. Kronin, I.V., Pokhvalov, Yu.E., and Voskresensky, K.D., Water forced flow critical heat fluxes study, *Nuclear Reactors Thermo-Physics Problems*, Atomizdat, Moscow, 3, pp. 32–38, 1971.
330. Smolin, V.N., Polyakov, V.K., and Esikov, V.N., Heat transfer crisis experimental investigation, *Nucl. Power Eng.*, 5, 417–422, 1964.
331. Styricovich, M.A. (Ed), Water boiling by flow in cylindrical channels heat transfer crisis calculation recommendations Scientific Council of complex problem. *Thermophysics*, Ser. HMT., Preprint 1-57. High Tem. In. of Acad. of Sci. of USSR, 67 pp., 1980.
332. Ivashkevich, A.A., Liquid forced flow boiling heat transfer coefficients and critical heat fluxes, *Thermal Eng.*, 10, 74–78, 1961.
333. Chigareva, T.S., and Chigarev, N.V., Binary mixtures characteristic peculiarities and theoretical imaginations of their explanations, *Boiling Physics Investigation*, Stavropol, Iss. 5, pp. 21–35, 1979.
334. Ornatsky, A.P., et al., Heat transfer crisis by forced flow boiling in annular channels experimental data generalization experience with using only liquid inlet parameters, *Thermo-Phys. Thermal Eng.*, 26, 33–38, 1974.
335. Sterman, L.S., et al., Organic heat carriers forced flow in tubes and pool boiling critical heat fluxes, *HMT, Minsk*, P-E., 1–29, 1964.
336. Ornatsky, A.P., Chernoby, V.A., and Savina, V.N., Water, glycol and antifreeze-65 sub-cooled forced flow in annular channels heat transfer crisis values study, *RETP*, 2, 86–91, 1970.
337. Katto, Y., Ishkoya, S., and Teraoka, K., Bubble and transient boiling in narrow media between two parallel disks, *Nichon Kikay Gakkay Rombuncyu*, 42(361) 2854–2861.
338. Katto, Y., and Shoji, M., Microlayer and bubble growth in pool boiling, *J. Heat Mass Transfer*, 12(N8), 1969.
339. Afanasiev, B.A., and Smirnov, H.F., Heat Transfer crisis hydrodynamic theory, modified for vaporization on surfaces covered by capillary-porous structures, Russia Nation, 1st HMTC, Moscow, Vol. 4, pp. 13–18, 1994.
340. Afanasiev, B.A., and Smirnov, H.F., The boiling crisis phenomenon on capillary porous covering. MTD-Vob. 309. *National Heat Transfer Conference*, ASME, Vol. 7, pp. 17–21, 1995.
341. Afanasiev, B.A., Vinogradova, E.P., and Smirnov, H.F., Crisis phenomenon by vaporization in screen capillary-porous coatings and artery structures of heat pipes, *EPJ*, 48(4), 607–614, 1985.
342. Smirnov, H.F., et al., Limiting characteristics of low-temperature heat pipes by discontinuous heat input, *RETP*, 2, 8–15, 1987.

343. Bezrodny, M.K., Volkov, S.S., and Moklyak, V.F., *Two-Phase Thermo-Siphons in Industrial Thermal Engineering*, High School, Kiev, 76 pp., 1991.

344. Smirnov, H.F., and Afanasiev, B.A., The crisis phenomenon when vaporization take place on surfaces covered porous structure, 9th IHPC Preprints, Albuquerque, May 1995.

345. Petukhov, B.S., Zukov, V.M., and Shildkret, V.M., Helium forced flow boiling heat transfer crisis investigation, *Heat Transfer and Hydrodynamic by Boiling and Condensation*, Coll. of Works., Novos, pp. 220–226, 1979.

346. Krieth, F., and Bohn, S.M., *Principles of Heat Transfer*, 5th ed., 720 pp. + 74 Appendixes, 1993.

347. Arefiev, K.M., and Gnedina, E.A., Forced flow sub-cooled liquid critical heat fluxes experimental data generalization, *Boiling Crisis and Heating Evaporative Surfaces Temperature Regimes*, pp. 191–200 (Works of ZKTI; Iss. 58), 1965.

348. Dann, P., and Reay, D., *Heat Pipes*, Energy, Moscow, 272 pp., 1979.

349. Ivanovsky, M.N., Sorokin, V.I., and Yagodkin, I.V., *Heat Pipes Physical Fundamentals*, Energoatomizdat, Moscow, 252 pp., 1987.

350. Styrikovich, M.A., Polonsky, V.S., and Ziklauri, G.V., *Heat Mass Transfer and Hydrodynamic at Two-Phase Flows of Nuclear Power Stations*, Nauka, Moscow, 368 pp., 1982.

351. Krukov, A.P., Kinetic analysis of evaporation and condensing processes on the surface, Paper collect, "Kinetic Theory of Transfer Processes When Evaporation and Condensing," Paper of International Seminar, Minsk, 1991, Belarus Academy of Science, Institute of Heat-and-Mass Exchange, pp. 3–21.

352. Labuntzov, R.A., and Krukov, A.P., Intensive evaporation processes, *Teploenergetika*, 4, 8–11, 1977.

353. Labuntzov, R.A., and Krukov, A.P., Analysis of intensive evaporation and condensing, *Int. J. Heat Mass Transfer*, 22, 989–1002, 1979.

354. North, M.T., Shaubakh, R.M., and Rosenfeld, I.H., Liquid film evaporation from bidisperse capillary wicks in heat pipe evaporators, *Proc. 9th Int. Heat Pipe Conf.*, Albuquerque, NM, May 1–5, 1995.

355. Rosenfeld, I.H., Anderson, W.G., and North, M.T., Improved high heat flux loop heat pipes using bidisperse evaporators wicks, *10th IHPC*, September, Stuttgart Preprints session, A1, pp. 5–6, 1997.

356. Maidanik, Yu.F., Vershinin, S.V., and Fershtater Yu.G., Heat transfer enhancement in a loop heat pipe evaporator, *10th IHPC*, September, Stuttgart Preprints session, A1 Paper, A1, pp. 3–6, 1997.

357. Kiseev, V.M., and Pogorelov, N.P., A study of loop heat pipe thermal resistance, *10th IHPC*, September, Stuttgart Preprints session, A1, Paper A1, pp. 6–9, 1997.

358. Gerasimov, Yu.F., et al., Low temperature heat pipe with segregated vapor and liquid lines, *TVT*, XII(5), 1131–1134, 1974.

359. Gerasimov, Yu.F., et al., Low temperature heat pipe with segregated vapor and liquid lines, *Inst. Liquid Phys.*, XXVIII(6), 957–960, 1975.

360. Gerasimov, Yu.F., et al., Some investigation results obtained for low temperature heat pipes operating against gravity forces, *Inst. Liquid Phys.*, XXX(4), 581–586, 1976.

361. Demidov, A.S., and Yatsenko, Eu.S., Investigation of heat and mass transfer in the evaporation zone of a heat pipe operating by the "inverted meniscus" principle, *Int. J. Heat Mass Transfer*, 37, 2155–2163, 1994.

362. Faghri, A., *Heat Pipe Science and Technology*, Taylor & Francis, 874 pp., 1994

363. Silverstein, C., *Design and Technology of Heat Pipes for Cooling and Heat Exchange*, Taylor & Francis, Washington, D.C., 368 pp., 1992.

364. Peterson, J.P., *Introduction to Heat Pipes: Modeling, Testing and Application*, John Wiley and Sons, New York, 1994.

Index

A

Adiabatic index, 22
Ammonia boiling heat transfer, 94
Annular channels
  Blasius formula for, 12
  boiling heat transfer crisis at forced flow in, 313–315
  boiling heat transfer crisis at low-velocity flow inside, 318–319
  bubbles flow boiling regime into horizontal, 137
  CHF in narrow, 326–329
  emulsion and wavy stratified two-phase flow regimes, 140
  experimental data on CHF at slug flow inside, 329
  experimental studies, vaporization mechanism, 121–146
  gap sizes, 136
  nonsteady heat transfer in, 10
  Nusselt number, 7, 13
  slug pulsations inside, 139
  true heat transfer conditions, 7
Annular slits, boiling of water, ethanol and NaCl solution in, 94
Annular slug flow regime zone, 95
APIMAS investigations, 105
Archimedean force, and heat transfer for transient flow, 12
Archimedes number, 117
Artery heat pipe, test sample, 172
Artery-screen wicks, 347, 350
Average heat transfer coefficient, 6
Average wall temperature measurements, reliability issues, 14

B

Blasius formula, for annular tubes, 12
Boiling
  conventional model inside horizontal slit, 154
  elementary cell model at, 234
  in heating surfaces submerged in dispersed layer, 277
  in liquid film covering heating surface, 83
  inside coated heating surfaces, 84
  inside vacuum evaporator with dispersed layer at tube bundle, 276
  Malyshenko phenomenological theory of, 214–226
  vapor bubbles growth curves in horizontal slits, 121–122
  volumetric flow vapor quality measurement results, 123
  with deepened meniscus, 233
Boiling heat transfer
  CPS model, 246
  experimental setup, 101
  horizontal slit gap influence, 102, 104
  in slits, 93
  influence of FMPS thickness on, 187
  inside horizontal slits, 101
  on clothed coverings, 113
  on coated surfaces in subcritical thermal regimes, 163–167
  on finned surfaces, 108, 109
  on plate nickel capillaries system, 99
  on surfaces covered by screen wicks, 168–182
  on surfaces covered by sintered and gas-sprayed coatings, 201–214
  R-113, 107
Boiling heat transfer crisis. *See also* Heat transfer crisis
  at forced flow in annular channels, 313–315
  at low-velocity flow inside annular capillary channels, 318–319
  in restricted spaces, 303–305
  modified hydrodynamic theory of, 319–326
  on surfaces with porous coverings, 338–347
  physical explanations and semiempirical models, 299–303
Boiling heat transport limit, 352, 353
Boiling initiation, 39
Boiling temperature curves, horizontal plate slits, 106
Bond number, 42
Boundary conditions
  first- or second-type, 5
  porous media, 19
  third-type, 8
  time dependency of, 15
Boundary layer removal model, 302
Bubble cloud formation model, 302

Bubble flow regime, 326
  CHF at, 327
Bubble growth curves, 121
Bubbles flow boiling regime
  in horizontal annular slit, 137
  in refrigerant boiling, 95
Burnout value, 83
  vapor zones at heat load approaching, 62

C

Capillary channels
  analysis of transient processes in, 16
  and reduction in non-steady-state effect, 16
  boiling crisis at forced flow in, 306, 316–319
  CHF at forced flow in, 319–326
  CHF at pool boiling in, 330–338
  CHF inside, 306
  correlation for single-phase heat
    transfer in, 14
  defined, 1
  experimental studies, boiling heat transfer
    crisis in, 305–306, 307–312
  heat transfer at boiling in, 93–115
  heat carrier supply mechanisms at
    vaporization, 42
  hydrodynamic phenomena and vaporization
    in, 115–120
  hydrodynamics and heat transfer
  single-phase turbulent flow in, 12–16
  steady single-phase flow in, 1–9
  influence of pressure on CHF in, 320
  influence of roughness and curvature on heat
    transfer in, 8
  theoretical models for heat transfer at
    vaporization, 147–153, 153–162
  thermohydrodynamics at vaporization in, 93
  vaporization conditions, 41–45
Capillary corrugated structures, 191, 192. *See also*
    Corrugated surface structures
Capillary liquid supply, 41
  volumetric heat generation, 43
Capillary pressure, 50, 51
  and heat carrier flow rate, 69
  and saturation, 47
  equalizing with friction pressure losses, 69
  Leverett functions for, 48
Capillary pumped loops (CPL), 85
  heat transfer models inside, 291–298
  thermal regime mode at narrow
    slit boiling, 294
Capillary-porous coverings. *See also* Coated
    surfaces; Porous coatings
  heat and mass transfer at vaporization,
    surfaces with, 163
Capillary-porous structures (CPS)

and irreversibility minimum, 55–63
boiling in microfilm thermal regime, 233
dependency between two-phase property
    parameter an vapor quality, 49
double-layer structure, 46
evaporating meniscus model, 248
evaporation regime, 233
experimental setup, 51, 72
hydrodynamics at vaporization
    inside, 41, 45–54
isothermal liquid meniscus in cylindrical
    capillary, 248
liquid transfer in, 45
microheat pipe thermal regime, 233
model of elementary cell at boiling, 234
model of heat transfer at boiling, 246
nonuniform forms, 46
pulsation regimes at vaporization in, 74, 75
relative phase permeabilities and phase
    saturations, 47
thermal regime, boiling with deepened
    meniscus, 233
vaporization conditions in, 41–45
vaporization model inside, 233
wetting and drying functions, 48
Channel parameter measurements, reliability
    problems, 14
Characteristic pore size, in porous media, 18
Circulation schematic, heat transfer at
    vaporization, 147
Closed pore effect, 53
Clothed coverings, 112, 113
  boiling heat transfer, 113
Coated surfaces
  boiling heat transfer
  subcritical thermal regimes, 163–167
  classification of vaporization regimes in
    capillary and submerged
    conditions, 261
  ethanol and acetone boiling, 166
  experimental results, 165
  fiber-metal nickel and fiber-metal
    copper wicks, 167
  general model of vaporization
    processes on, 260–271
  heat transfer at liquid nitrogen and
    refrigerant R-113 boiling on, 209
  water boiling on tube wrapped by wick, 166
  working zone for experimental study, 164
Combined heat carrier flow, 41
Combined heat transport limit, 352
Condensate thin film, heat conduction
    through, 36
Condensation, heat transfer at, porous structure
    external surface, 33
Contact angle value, 39

Convective heat input, to external evaporative
    surface of capillary structure, 43
Correlations
    and non-steady flow regimes, 15
    for turbulent flow, 13
    single-phase heat transfer in capillary
        channels, 14
    single-phase turbulent flow, 14
Corrugated capillary channels, CHF at pool
    boiling in, 330–338
Corrugated heat pipes
    experimental CHF data in, 337
    maximum thermal power, 338
Corrugated surface structures
    CHF values with heating surfaces
        covered by, 333
    experimental investigations, vaporization
        heat transfer on, 190–194
    from metallic foil, 191
    heat transfer at water boiling on, 193
    parameters and dimensions, 191
    structure profiles inside heat pipe
        container, 192
    water boiling on, 192
Counter flows. *See* Gas counter flows; Liquid
    counter flows
Countercurrent liquid/vapor mass flow rates, 54
Coupled structures, 25
Critical heat fluxes (CHF), 199, 300
    and characteristic corrugation size, 333
    and corrugated strip width, 333
    and gap formation, 341
    and mass flow velocity, 317
    and MFCPS at water boiling, 341
    and saturation pressure inside screen
        structures, 340
    and subcooling, 320
    at boiling in horizontal and vertical slits, 333
    at high velocities of ethanol flow, 317
    at pool boiling in slits, capillaries, and
        corrugated capillary channels,
        330–338, 331
    at pool boiling on surfaces covered by screen
        wicks, 345
    at slug flow inside annular channels, 329
    boundary layer removal model, 302
    bubble cloud formation model, 302
    channel cross section parameters, 316
    comparison of correlations, 318
    dependence on cell size and thickness of
        screen structure, 339
    dependence on channel geometric
        parameters, 306
    dependence on heat pipe inclination
        angle, 334
    dependence on slit's gap value, 112

effect of hydraulic losses on treatment of
    experimental data, 325
experimental data, liquid forced flow boiling
    inside capillary channels, 322, 324
experimental data at emulsion and bubble
    flow regimes inside annular
    channels, 327
experimental data in corrugated
    heat pipes, 337
experimental investigation at forced flow in
    capillaries, 319–326
experimental setup, 101
for artery-screen wicks, 350
generalization of experimental data, 326
in narrow annular channels, 326–329
influence of basic regime parameters on, 321
influence of ethylenglycol-water solution
    concentration on, 321
influence of liquid velocity and
    subcooling on, 320
influence of pressure in capillary
    channels, 320
influence of slit gap on, 106
inside cylindrical capillary channels, 306
liquid flow-blocking model, 302
liquid layer limit overheating model, 302
liquid sublayer drying model, 302
screen cell size influence on, 339
slit influence on, 108
theoretical models, 302
values in experiments with heating surfaces
    covered by corrugated structures, 333
vapor removal restriction model, 302
Crosscurrent phase flow, 56
Curvature
    and heat transfer in capillary channels, 8
    and secondary flows, 9
Cylindrical surfaces, vaporization heat transfer,
    uniform screen structures, 170

D

Darcy's law, 17, 29, 45
Deepened meniscus, thermal regime
    of boiling with, 233
Disjoined porous structures, 90
Dispersed multilayer particles
    at tube bundle, 276
    boiling on heating surfaces submerged in
        dispersed layer, 277
    dependence of heat transfer on layer
        height, 274
    experimental data, pool boiling, 285
    flooded dispersed multilayer bed, 276
    heat transfer and heat flux inside vacuum
        evaporator, 276

inside vacuum evaporator, 276
movement classification at boiling,
274, 275
Droplet liquid throwing, 79
boiling inside coated heating surfaces
with, 84
energy characteristics, 84
Dry spots, 129, 132, 135, 301
and microlayer rupture, 133
and vapor slug formation, 137
due to microlayer rupture, 133
due to surface overheating, 138–139
dynamic development on heating
surface, 138
factors determining dimension, 327
formation, 328
formation in microlayer under
vapor slug, 133
growth velocity of, 134
in R-113 boiling inside horizontal
slit, 130–131
R-113 slit boiling, 135
Dry-out zone, and heat flux development, 349
Drying processes, and vaporization in porous
structures, 44

E

Economizer zone, 147, 149
Effective thermal conductivity
porous media, 20–28, 87
vapor-filled porous structure, 69
Electrochemical coverings, heat transfer
intensity at boiling, 204
Emulsion flow regimes, 326
CHF at, 327
Emulsion two-phase flow regime, 140
Energy conservation equation, 3
Entropy generation minimization
principle, 55
Equilibrium saturation temperature, 39
Evaporating meniscus model, 248
Evaporation, 83
contact conditions and heat transfer
intensity, 90
cooling conditions with excess skeleton
thermal conductivity, 67
heat balance equation, 67, 68
heat transfer at, 33
in flooded triangular grooves, 39
in porous structures, 28
one-dimensional energy equation, 50
stability analysis, 69
with deepened meniscus, 83
Evaporator design, 59
Evaporator layout, inverted meniscus
principle, 59

Experimental studies
boiling heat transfer crisis at forced liquid flow
in slits and capillary tubes, 305–306
boiling heat transfer on coated surfaces in
subcritical thermal regimes, 163–167
boiling heat transfer on surfaces covered by
screen wicks, 168–182
boiling heat transfer on surfaces covered by
sintered and gas-sprayed coatings,
201–214
boiling heat transfer, CHF, and internal
process characteristics, 101
CHF at forced flow in capillaries, 319–326
CHF at pool boiling in slits, capillaries,
and corrugated capillary channels,
330–338
experimental setup principal schematic
diagram, 136
heat and mass transfer inside grain
bed, 30
heat transfer at boiling in slits and capillary
channels, 93–115
heat transfer crisis at boiling on surfaces
with porous coverings, 338–347
heat transfer inside evaporators of loop heat
pipes, 194–201
internal convection heat transfer in porous
materials, 31
on thermohydrodynamic phenomena, 71–92
reliability issues, 15
vaporization heat transfer on corrugated
surface structures, 190
vaporization heat transfer on fiber-metal
surfaces, 182–190
vaporization heat transfer on surfaces with
porous coatings, 163–214
vaporization mechanism in plain slits and
annular channels, 121–146
working elements, 102

F

Fiber-metal copper wicks
experimental results, 165
maximum heat fluxes transferred inside heat
pipes with, 270
Fiber-metal nickel wicks, experimental results,
165
Fiber-metal structures (FMS)
dependence of thermal conductivities on
porosity, 27
experimental *vs.* calculation data
comparisons, 28
experiments on thermal conductivity, 25
heat transfer at vaporization inside, 183–184
heat transfer at water boiling, surfaces
covered by, 188

influence of thickness on water boiling heat transfer, 187
vaporization heat transfer on, 182–190
water and acetone boiling on surfaces coated by, 187
Filtration equations, porous media, 17
Finned surfaces
boiling heat transfer on, 108, 109
heat transfer intensity on, 204
Flooded trapezoidal grooves, heat transfer in, 38
Flooded triangular grooves, 38
evaporation in, 39
Flow conditions, in screen structures *vs.* dense tube bundles, 32
Flow rate measurements, reliability, 14
Flow velocity
dependence on two-phase flow structure, 97
through porous media, 17
Fluid energy equation, porous media, 18
Fo number, 147
Forced convection, in slits and capillaries, 116
Forced flow, experimental investigation of CHF at, 319–326
Forced heat carrier flow, 42
Forced liquid supply, 41, 119
experiments on boiling heat transfer crisis at, 305–306, 306, 316–319
volumetric heat generation, 43

G

Gap formation, and decrease in CHF values, 341
Gap size, 162
and departure vapor bubble diameter, 107
and vapor bubble departure size, 101
CHF decrease with, 108
Gas counter flows
in porous structure, 51
inside porous layer, 52
Gas flow
effect of nonisothermity in, 12
internal heat transfer, screen packing, 33
Gas thermal dust raising coating method, 208
Gas-sprayed coatings
boiling heat transfer on surfaces covered by, 201–214
covering parameters, experimental samples, 205
Geometry constancy, 14
Glass slits, microhills and microwaves formation by vapor bubble growth into, 141
Grain bed
experimental data comparisons, heat/mass transfer inside, 30
heat and mass transfer in, 28
temperature distribution, 35

Granular bed, 20
heat and mass transfer in, 29
Gravitational heat supply, 41
at vaporization, 43
Gravitational liquid supply
at vaporization, 44
volumetric heat generation at vaporization, 43
Gravitational mechanism, at vaporization, 43

H

Heat carrier flow rate, and capillary pressure, 69
Heat conduction, 34
through condensate thin film, 36
through wetted porous structure, 36
Heat flux
and heat transfer intensity, 162
and temperature drop, 77
and transient heat transfer, 15
and vaporization on external liquid surface, 80
and wall superheat in nonuniform heat pipe screen structures, 168
dependence of average heat transfer coefficient in LHP evaporators, 200
dependence of LHP thermal resistance on, 198
dependence of R-113 boiling heat transfer on, 107
in loop heat pipe evaporators, 197
vaporization in slit and capillary channels, 99–100
Heat flux measurements
maximum, 76
reliability issues, 14
Heat generating element (HGE), in two-phase porous cooling, 65
Heat input conditions
at vaporization, 43
two-phase flow and, 43
Heat load enlargement, and growth of vapor zones, 61
Heat losses, in vaporization, 63
Heat pipe evaporator, capillary structure superheat and heat flux relationships, 182
Heat pipe heat transport capillary limit, 353
Heat pipe hydrodynamic limit, 337
Heat pipe inclination angle, CHF and, 334
Heat pipe surfaces
dependencies, heat transfer coefficients on contact area, 87
design schematics, two-sided and single-side heat removal, 181
heat flux and wall superheat at vaporization, 168

Heat pipes
  generalization of CHF experimental data, 351
  influence of nonuniformities on, 348
  maximum heat fluxes inside, 347–354
  wick manufacture from corrugations, 335
Heat transfer, 80, 88
  and Malyshenko phenomenological theory
    of boiling, 214–226
  at boiling on coated surfaces, 92
  at evaporation, 50
  at pulsating flow with low-frequency
    oscillations, 11
  at vaporization inside capillary-porous
    structures, 43
  at vaporization inside fiber-metal
    surfaces, 183–184
  at vaporization inside flooded dispersed
    multilayer bed, 276
  at vaporization on surfaces covered by
    movable capillary structures,
    271–298
  at vaporization on surfaces with
    capillary-porous coverings, 163
  at vaporization on surfaces with sintered
    coatings, 214–226
  average and local, 6
  average coefficient, 6
  between single sphere and passing flow, 28
  between wall and single-phase flow, 33, 34
  by cryogenic liquid boiling inside small size
    channels, 113
  comparison of coefficients for low-finned
    and smooth tubes, 108
  dependence on effective height of coating
    layer, 274
  dependence on relative part of contact, pipe
    surfaces, 87
  due to changing of bed element
    interpositions, 35
  experimental studies, boiling in slits and
    capillary channels, 93–115
  external surface, wetted PS, 34
  flooded trapezoidal grooves, 38
  for transient flow, 12
  fundamentals, models, and equations of
    momentum, 17–20
  generalization of experimental data inside
    horizontal slits, 160
  generalization of experimental data inside
    short vertical slits, 161
  in nonsintered screen structures, 32
  in spherical granular bed at various porosity
    levels, 29
  influence of contact conditions on
    evaporation, 90
  influence of roughness and curvature, 8
  inside grain bed, 30
  inside loop heat pipe evaporators, 194–201
  inside microdispersed layer, 60
  inside screen wicks, 171
  internal, in porous media, 28–33
  internal liquid turbulent flow, 13
  laminar flow, 2
  liquefied gases boiling on surfaces covered
    by porous coatings, 211
  models in CPL, 291–298
  models in LHP evaporators, 291–298
  nitrogen and R-113 boiling on coated
    surfaces, 209
  nitrogen and R-113 boiling on gas thermal
    dust raising and sintered
    coatings, 208
  nonsteady single-phase laminar flow, 9–12
  on corrugated surface structures, 193
  on surfaces coated by FMS, 187, 188
  on surfaces covered by metallized porous
    coatings, 207
  on surfaces with porous coatings, 163–214
  porous structure external surface, 33–40
  single-phase flow through porous media, 17
  single-phase turbulent flow in capillaries and
    slits, 12–16
  steady single-phase flow in capillaries and
    slits, 1–9
  theoretical models at vaporization in
    horizontal and short slits and
    capillaries, 153–162
  theoretical models at vaporization in vertical
    slits and capillaries, 147–153
  thermal modes inside LHP evaporators, 293
Heat transfer crisis. *See also* Boiling heat
    transfer crisis
  and time of vapor slug existence, 327
  at forced flow in capillaries, 306, 316–319
  at slug two-phase flow, 328
  at vaporization inside slits, capillaries, and
    capillary-porous structures, 200
  experimental data, in capillary
    channels, 307–312
  experiments, forced liquid flow in slits and
    capillary tubes, 305–329
  initiation by loss of vapor jet stability, 299
  modified hydrodynamic theory in restricted
    spaces, 303–305
  on surfaces with porous coatings, 338–347
  physical explanations and semiempirical
    models, 299–303
Heat transfer regularities, 70
Heating time, unsteady, 124
Hertz formula, 23
Horizontal slits
  and heating surface changes, 132
  average volumetric flow vapor quality
    measurements, 123, 125, 128

boiling heat transfer, 100
boiling temperature curves for, 104, 106
bubbles flow boiling regime into, 137
CHF experimental data at boiling in, 331
conventional model for boiling inside, 154
emulsion regime, boiling inside slits, 130
experimental setup, 101
generalization of experimental data, heat
    transfer boiling inside, 160
influence of gap on CHF, 106
influence of gap on ethanol boiling heat
    transfer, 102, 104
R-113 boiling heat transfer in, 107, 134
slug flow boiling regime inside, 137
theoretical models, heat transfer at
    vaporization, 153–162
vapor bubble growth curves, boiling
    inside, 121
vapor bubble growth regularities, 124
water boiling high-speed video, 127
with one-single side heat input, 153
Horizontal tubes, boiling heat transfer
    in, 307, 308
Hydraulic losses, effect on treatment of CHF
    experimental data, 325
Hydraulic losses minimum, 55, 57
Hydraulic resistance, laminar flow, 2
Hydrodynamics
    and vaporization in slit and capillary
        channels, 115–120
    at vaporization in capillary-porous
        structures, 45–54
    liquid and gas counter flows in porous
        structure, 51
    nonsteady single-phase laminar flow, 9–12
    single-phase flow through porous media, 17
    single-phase turbulent flow in capillaries and
        slits, 12–16
    steady single-phase flow in capillaries and
        slits, 1–9
Hysteresis phenomena, 213
    with porous coatings, 218

I

Inclusions
    in porous media, 20
    Odolevsky formula for, 21
Inertial flow, in porous media, 29
Inertial heat carrier supply, 41
Inlet conditions, and relative channel length, 4
Inlet liquid temperature, 5
Inside layer crisis, 345
Internal heat transfer
    between wall and single-phase flow, 34
    in porous media, 28–33, 31
Internal layer crisis, 56

Inverted meniscus principle, 58
    evaporator layout based on, 59
Irreversibility minimum, in two-phase filtration
    modeling, 55–63
Isolated bubbles regime, 100
Isothermal flows, liquid and vapor, 1

K

Kozeny equation, 17
Kutateladze hydrodynamic theory of boiling
    crisis, 317, 324, 342

L

Laminar flow, 1
    heat transfer and hydraulic resistance in, 2
    in small-diameter cylindrical channels, 3
    nonsteady single-phase, 9–12
    steady-state isothermal, 2
    transient heat transfer for, 11
Large velocities and large subcooling, 136
Layer-wall contact, dynamic effect of
    vaporization on, 91
Liquid counter flows
    in porous structure, 51
    inside porous layer, 52
Liquid film thickness, direct tare
    determination, 126
Liquid flow
    countercurrent or crosscurrent with
        vapor flow, 51
    separation from vapor flow, 85
Liquid flow rate, for boiling in slits, 148
Liquid flow-blocking model, 302
Liquid layer limit overheating model, 302
Liquid microfilm surface
    decrease in microhill sizes, 142
    structure near phase border, 144
Liquid microfilm thickness, and phase border
    movement velocity, 144
Liquid microlayer, 141
    curling on slit surface video, 142, 143
    thickness at growing vapor bubble
        base, 118, 126
Liquid microlayer thickness, 128, 154, 155, 156
    and average volumetric flow vapor
        quality, 126
    oscillograms, 126
Liquid movement momentum equation, 157
Liquid sublayer drying model, 302
Liquid supply, in porous-coated structures, 261
Liquid throw flow rate, 58
    at vaporization, 80
Liquid transfer, in capillary-porous
    structures, 45
Liquid velocity, influence on CHF, 320

Loop heat pipe evaporators
  compensation chamber, 293
  cylindrical, 293
  dependence of thermal resistance on heat
      flux, 198
  design schematics, 293
  second-generation, 198
  experimental investigations, heat transfer
      inside, 194–201
  experimental setup, 199
  heat transfer coefficient dependence on
      heat flux, 200
  heat transfer models, 291–298
  schematics and evaporations zones, 196
  surface vaporization, 89
  temperature dependence on heat
      flux, 197
  thermal modes of heat transfer in, 293
  thermal regime mode at narrow slit
      boiling, 294
  vapor generation channels, 293
  with vapor removal pitch, 200
Loop heat pipes (LHP), 85
Low-finned tubes, average heat transfer
      coefficients, 108
Low-velocity flow, boiling heat transfer
      crisis at, 318–319

M

Malyshenko phenomenological theory of
      boiling, 214–226
Mass conservation equation, 2
Mass flow, 15
Mass flow velocity, and CHF increase, 317
Mass losses, in vaporization, 63
Mass transfer
  at vaporization on surfaces with
      capillary-porous coverings, 163
  in spherical granular bed, 29
  inside grain bed, 30
  inside microdispersed layer, 60
  path length in porous media, 32
Maximum heat flux, 77
  inside heat pipes, 347–354
Maxwell formula, 21
Mean free path length, vapor phase, 1
Mesh model, for porous media, 32
Metal-fibrous capillary-porous structure
      (MFCPS), CHF and parameters at
      water boiling, 341
Metallized coverings
  boiling on surface with, 303
  heat transfer at boiling on surfaces coated
      by, 207
Metastable state temperature, *vs.* skeleton
      temperature, 66

Micro-heat pipes, 81, 83
  thermal regime, 233
Microchannels, single-phase hydrodynamic
      flow in, 13
Microdispersed layer, heat and mass transfer
      inside, 60
Microfilm thermal regime, 233
Microhills
  formation by vapor bubble growth, 141
  size decreases in liquid microfilm
      surface, 142
Microlayer evaporation process, thermal
      resistance, 158
Microlayer existence time, 156
Microlayer rupture, due to single nucleation
      center, 133
Microwaves, formation by vapor bubble
      growth, 141
Momentum conservation equations, 2
  at uniform inlet velocity profile, 4
  single-phase flow through porous media,
      17–20
Movable capillary structures, heat transfer at
      vaporization on surfaces covered
      by, 271–298

N

Natural convection, 118
  and vaporization, 41
  at augmented external heat flux
      values, 72
  Raleigh number as determinant of, 75
  refrigerant boiling heat transfer coefficients
      in slit channels by, 97
Near-wall porous structure
  characteristic cell in, 86
  hydrodynamic and thermal influences, 88
Non-steady state
  correlations and, 15
  major effects of, 15
Nonannular channels, 5
Nonequilibrium systems, stable stationary
      states of, 56
Nonisothermity, effect in gas flow, 12
Nonsintered screen structures, heat transfer and
      hydraulic resistance in, 32
Nonstabilized flow, interpolation equations,
      initial section, 6
Nonsteady heat transfer, in annular tubes, 10
Nonsteady single-phase laminar flow, 2
  hydrodynamics and heat transfer, 9–12
Nonsteady single-phase turbulent flow,
      hydrodynamics and heat transfer, 15
Nonuniform screen structures, 347
  boiling heat transfer inside, 174
  influence on heat pipe thermal mode, 348

Nusselt number, 5, 151
    and channel shape, 7
    and heat transfer coefficient, 6
    annular channels, 7
    dependence on *Re* values in porous
            structures, 36
    dependence on *Re* values in transient flow, 12
    dependence on *Re* values in turbulent flow, 14
    for third-type boundary conditions, 8
    modifications along annular tube, 7

O

Obliteration effect, porous media, 18
Odolevsky formula, 21
One-dimensional filtration, 49
    gas-liquid two-phase flow through porous
            media, 46
One-dimensional steady-state heat transfer,
            porous media, 19
OSAR experiments, 100, 109

P

Particle bed coverings, vaporization on surfaces
            coated by, 237
Particle bed layer, 90
Permeability coefficient, 18
    single-phase flow through porous media, 17
Phase border movement velocity, and liquid
            microfilm thickness, 144
Phase countercurrent flow, 56
Phase permeability model, 54, 66, 76
    and conditions affecting maximum
            heat flux, 77
Phase saturation, in capillary-porous
            structures, 47
Plane slit channels, true heat transfer
            conditions, 7
Plate nickel capillaries system, boiling heat
            transfer in, 99
Plugging lid, 72
Pool boiling, 299
    CHF at, 330–338
    heat transfer crisis regularities, 300
    on submerged flooded dispersed layer, 281
    on submerged surface by flooded dispersed
            layer surfaces, 285
Pore distribution effect, 30
Pore distribution law, 258
Pore size, 22
Porosity, 17
    and heat/mass transfer in spherical
            granular bed, 29
    dependence of FMS skeleton on, 27
    dependence of quartz sand relative thermal
            conductivity on, 25

Porous coatings
    classification, 245
    experimental investigations
        boiling heat transfer on surfaces covered
            by screen wicks, 168–182
        boiling heat transfer on surfaces covered
            by sintered and gas-sprayed coatings,
            201–214
        heat transfer inside evaporators of loop
            heat pipes, 194–201
        vaporization heat transfer on corrugated
            surface structures, 190–194
        vaporization heat transfer on fiber-metal
            surfaces, 182–190
        vaporization heat transfer on surfaces
            with, 163–214
    generalization of experimental data, 259
    heat transfer at boiling on, 215
    heat transfer at liquefied gases boiling on
            surfaces covered by, 211
    heat transfer at R-12 boiling on surfaces
            with, 203
    heat transfer crisis on surfaces with, 338–347
    heat transfer on surfaces with stainless
            steel, 217
    horizontal cross-section, porous layer, 246
    metallized, 207
    models of vaporization on surfaces
            with, 226–260
    pressure changes and coating thickness, 246
    Thermacore experimental data, 270
    transient and hysteresis processes, 218
    vaporization physical models, 227
Porous layer, liquid and gas counter flows
            inside, 52
Porous m, permeability, 71
Porous media, measurement of pressing force, 88
    accommodation coefficient, 22
    characteristic pore size, 18
    connection to heating wall, 92
    contact angle value, 39
    contact spot size-fiber radius ratio, 26
    coupled structures, 25
    disjoined, 90
    effective thermal conductivity, 20–28, 69, 87
    energy equation, 18–19
    evaporation in, 28
    experimental conditions and research
            results, 31
    filtration mass flow rate in, 30
    flooded triangular grooves, 38
    fluid energy equation, 18
    function theta, 24
    fundamentals, models, and equations, 17–30
    granular bed structures, 20
    heat conduction through wetted, 36
    heat generation inside, 43

hydrodynamics and heat transfer,
      single-phase flow through, 17
inertial flow in, 29
influence of radiation, 22
internal heat transfer in, 28–33
meniscus shape influence, 37
mesh model, 32
movability and separation, 92
nucleation limit, 54
obliteration effect, 18
one-dimensional filtration of gas-liquid
      two-phase flow through, 46
one-dimensional steady-state heat
      transfer, 19
parallel resistance model, 20
pulsation regimes at vaporization in
      low-permeability, 75
rectangular grooves, 37
screen structures, 27
sintered, 30
skeleton thermal conductivity, 23
skeleton thermal resistance, 20
stability of two-phase flow through, 63
steel, 26
stratified model, 21
structures with interpenetrated
      components, 20
thermal conductivity comparisons, 26
thermal contact resistance, 23
thermal resistance
      pore volumes, 20, 37
      representative cell, 20
thin-layered structures, 37
transient flow in, 29
vapor channel formation inside, 54
vaporization investigations, internal heat
      generation, 63–71
viscosity in, 29
void concentration, second-order structure, 24
wall temperature jump, 22
with inclusions, 20, 21
Porous nonmetallic coverings, 204
Porous structure external surface
      cases, 34
      heat transfer on, 33–40
Porous structures, video capture of vaporization
      inside, 79
Pressure drop
      across liquid-vapor interface, 50
      and capillary capacity, 54
      and inertia forces, 155
      and phase permeability model, 66
      due to vaporization inside porous channels, 55
Pulsating flow, 9
      heat transfer with low-frequency
            oscillations, 11
      hydrodynamic parameters, 10

Pulsating heat carrier supply, 41
Pulsation period, dependence on heat flux, 74
Pulsation thermal regimes, at vaporization, 71

Q

Quasi-steady heat transfer coefficient, 15

R

R-11 boiling heat transfer, in vertical slit
      channels, 130
R-113 boiling heat transfer, 173
      by saturation pressure, 111
      dependence on heat flux, 107
      dry spots and emulsion regime, 130
      dry spots in, 135
      empirical constants values, 110, 111
      inside horizontal slit, 134
      on coated surfaces, 209
R-113 vaporization heat transfer, on cylindrical
      surfaces with uniform screen
      structures, 170
Radiation, influence in porous media, 22
Raleigh number, and natural convection
      development, 75
Rectangular grooves, 37
Refrigerant boiling heat transfer coefficients.
      See also R-11; R-113 boiling heat
      transfer
      in slit channels by natural convection, 97
Refrigerant vaporization heat transfer,
      cylindrical surfaces with uniform
      screen structures, 170
Refrigerants
      and ammonia boiling heat transfer, 94
      and dry spot formation, 133
      observed zones, 95
      vertical slit boiling heat transfer, 96
Relative channel length, and heat transfer for
      transient flow, 12
Relative phase permeability, 48
      dependence on saturation, 75
      in capillary-porous structures, 47
Reliability problems, 14
      experimental data, 14, 15
Reynolds numbers, 1, 12, 117, 151
Roughness
      and heat transfer in capillary channels, 8
      in porous media, 23

S

Saturation jump, at border between two
      layers, 78
Saturation pressure, heat transfer by, 111
Screen cell size, influence on CHF, 339

Screen packing, 33
    stratification effect on heat transfer
        intensity, 33
Screen structures, 27
    CHF and saturation pressure at vaporization
        inside, 340
    CHF dependence on cell size and thickness
        of, 339
    effect of turbulence on internal heat transfer
        in, 33
    flow conditions in, 32
    thermal conductivity comparisons, 28
    vaporization on surfaces coated by, 237
Screen wicks, 347
    boiling heat transfer, nonuniform screen
        structures, 174
    boiling heat transfer on surfaces covered by,
        168–182
    CHF at pool boiling on surfaces covered
        by, 345
    CHF conditions at boiling on surfaces
        covered by, 342
    CHF rates at boiling on surfaces covered
        by, 340
    comparison of vaporization inside uniform
        and nonuniform screen structures, 175
    experimental data on heat transfer
        inside, 171
    experimental setups, 172
    heat flux and wall superheat in nonuniform
        heat pipe screen structures, 168
    heat pipes design schematics, 181
    influence of wall-structure contact and gap
        formation conditions on heat transfer
        intensity, 176
    layer thickness influence on boiling heat
        transfer, 173
    liquid level influence on boiling heat
        transfer, 173
    R-113 vaporization heat transfer on
        cylindrical surfaces, uniform screen
        structures, 170
    thermal wedge covered by single-layer and
        multilayer, 172
    uniform multilayer screen structures, 173
    with different skeleton thermal
        conductivities, 173
    working element, external surface covered
        with single-layer screen, 178
Secondary flows, curvature and, 9
Short slits, 99, 120
    generalization of experimental data, heat
        transfer boiling inside, 161
    theoretical models for heat transfer at
        vaporization in, 153
Single-phase flow zone, in boiling
    refrigerants, 95

Single-phase hydrodynamic flow
    in Microchannels, 13
    through porous media, 17
Single-phase laminar flow, nonsteady, 9–12
Single-phase turbulent flow
    correlation, 14
    hydrodynamics and heat transfer in
        capillaries and slits, 12–16
Single-scale porous coatings, 245
Single-side heat removal, heat pipes, 181
Sintered coatings
    boiling heat transfer on surfaces covered by,
        201–214, 203
    covering parameters, experimental
        samples, 205
    covering thickness, 216
    heat transfer at nitrogen and R-113 boiling
        on, 208
    heat transfer at vaporization on surfaces
        with, 214–226
    particle diameter, 216
    porosity, 216
Skeleton temperature *vs.* metastable state
    temperature, 66
Skeleton thermal resistance, 24
    porous media, 20
Slit channels
    boiling character and pool boiling
        condition, 124
    boiling flow regimes inside, 115
    boiling of water in annular, 94
    CHF at pool boiling in, 330–338
    closed, opened, trapezoidal types, 110
    defined, 1
    experimental studies on heat transfer at
        boiling, 93–115
    experimental studies on vaporization
        mechanism, 121–146
    experiments on boiling heat transfer crisis
        in, 305–306
    gap value and CHF, 112
    horizontal, 101
    hydrodynamic phenomena and vaporization
        in, 115–120
    hydrodynamics and heat transfer
        single-phase turbulent flow in, 12–16
        steady single-phase flow in, 1–9
    influence of gap on CHF, 106
    liquid microlayer curling, 143
    local heat transfer coefficient changes over
        height, 96
    refrigerant boiling heat transfer coefficients, 97
    thermohydrodynamics at vaporization, 93
    vertical, 96
    water and ethanol boiling in vertical, 98
    water boiling with transparent walls, 99
    working elements, channel design, 110

Slit flooding height, 152
Slug existence time, 139
Slug flow, 10, 327
  experimental data on CHF at, 329
Slug flow boiling regime, inside horizontal
      annular slit, 137
Slug formation, 328
Slug outlet time, 139
Slug pulsations, in annular channels, 139
Slug regime, 129
Slug two-phase flow, physical model, heat
      transfer crisis at, 328
Slug waiting time, 139
Small velocities/large subcooling regime, 136
Small velocities/small subcooling regime, 136
Small-diameter channels, 1
  laminar flow in, 3
Smooth tubes
  average heat transfer coeffiients, 108
  comparison with wicks, 166
Sonic heat transport limit, 352
Stainless steel porous coatings, 217
Steady single-phase flow, hydrodynamics and
      heat transfer, 1–9
Steady-state isothermal laminar flow, 2, 3
  in curved channels, 2
Stefan-Boltzmann constant, 22
Stratified flow regime, 327
Subcooling boiling, 317, 320
  in crisis cross section, 322
  in refrigerants, 95
  influence on CHF, 320
Subcritical thermal regimes, boiling heat transfer
      on coated surfaces in, 163–167
Submerged porous structures
  boiling on heating surfaces submerged in
      dispersed layer, 277
  heat transfer at vaporization, 272
  pool boiling at surface, 281
  vaporization inside, 261
Surfaces
  covered by movable capillary structures,
      271–298
  models of vaporization on porous-coated,
      226–260
  pool boiling on, 281
  vaporization heat transfer on porous-coated,
      163–214
  with sintered coatings, 214–226
Surplus capillary potential, in nonuniform
      capillary structures, 341

T

Temperature distribution, grain bed, 35
Theoretical models, heat transfer at
      vaporization

  horizontal and short slits and capillaries,
      153–162
  vertical slits and capillaries, 147–153
Thermacore, experimental data, 270
Thermal boiling regimes, 94
Thermal conductivity
  comparisons, coupled porous structures, 26
  dependence on porosity, 25
  experimental vs. calculated data, screen and
      metal-fiber structures, 28
  porous media, 20–28
  screen structure comparisons, 28
Thermal edge schematic, 107
Thermal oscillatory instability phenomena, 71
Thermal pseudo-fluidization regime, 44
Thermal pulsating regimes, 10
Thermal resistance
  in loop heat pipe evaporators, 198
  pore volumes, 20
Thermodynamic similarity form base, 159
Thermohydrodynamic phenomena
  at vaporization in slit and capillary
      channels, 93
  experimental investigations, internal
      characteristics and mechanisms, 71–92
Thermohydrodynamics, at vaporization inside
      capillary-porous structures, 41, 45–54
Thin layer porous structures, evaporation in, 50
Thin liquid microlayer, 128
Third stage model, 260–271
  comparison with maximum heat fluxes,
      fiber-metal wicks, 270
Three-layer screen structure,
      vaporization at, 82, 83
Total heat flux maximization, 58
Transient flow, 1, 150
  heat transfer for, 12
  in porous media, 29
  inside capillaries, 2
Tubes, velocity profile inside, 9
Turbulence, effect on internal heat transfer in
      screen structures, 33
Turbulent flow
  generalized correlation for, 13
  inside capillaries, 2
  internal liquid heat transfer, 13
Turbulent flow transient heat transfer coefficient,
      and quasi-steady heat transfer
      coefficient, 15
Two-layer screen structure, vaporization at, 81
Two-phase boundary layer
  potential energy increase, 342
  stability failure, 303
Two-phase filtration modeling, 55
  correlations, 53
  friction loss inside capillaries formed by
      corrugations, 334

hydrodynamics, 77
irreversibility minimum in, 55–63
Two-phase flow, 15, 147
    and stability of evaporation, 69
    dependence on heat input conditions, 43
    flow velocity and, 97
    homogeneous flow model, 48
    increased turbulence in, 93
    regimes at slits and capillaries, 116
    relative phase permeability model, 48
    stability through porous structure, 63
    velocity/subcooling regimes, 136
    with liquid slugs, 10
Two-phase flow friction loss, 335
Two-phase heat transfer zone, 74
Two-phase porous cooling, 64
    numerical results, stability and safety
        operation, 65
    physical model
        heat generating element (HGE), 65
        liquid flow and contact zones, 66
    thermal regimes, porous vapor generated
        surfaces, 68
Two-phase property parameter, and vapor
    quality, 49
Two-phase thermal regime, development of
    stable, 74
Two-scale porous coatings, 245
Two-side heat removal, heat pipes, 181

U

Unsteady processes, 45
    drying and vaporization, 44

V

Vapor blanket, 56, 84
Vapor bubble dynamics, 153
Vapor bubble movement, 144
Vapor bubbles, 93
    and heating surface changes, 132
    average growth velocity, 154
    bubbles flow boiling regime into horizontal
        annular silt, 137
    coalescence and effects on liquid microlayer
        thickness, 142
    departure size and gap size, 101
    diameter and slit gap, 107
    floating velocity into vertical channel, 146
    growth curves, boiling inside horizontal
        slits, 121
    growth into glass slit, 141
    growth regularities, water boiling into
        horizontal slits, 124
    growth velocity and two-phase flow velocity
        in slits, 116

liquid microfilm structure near phase
    border, 144
liquid microlayer thickness oscillograms
    under, 126
nucleation, growth, video movement, 145, 153
occupation of heating surface with increased
    bubble velocity, 128
on heating wall, 81
secondary, 132
sticking, 95
Vapor cavities, with deepened meniscus, 81
Vapor channel formation, inside porous
    structures, 54
Vapor film formation, 83
Vapor flow
    countercurrent/crosscurrent with liquid
        flow, 51
    separation from liquid flow, 85
Vapor flow filtration loss, 335
Vapor flow quality sensors, 122
    design by capacity method, 123
Vapor jet stability, loss of, and heat transfer
    crisis initiation, 299
Vapor phase, and temperature conditions, 54
Vapor phase border movement, 157
Vapor phase output, 156, 157
Vapor phase removal, 123
Vapor quality
    and two-phase property parameter, 49
    thermal parameters, 49
Vapor removal channels, 85
Vapor removal flow rate, 85
Vapor removal restriction model, 302
Vapor slug border displacement, 157
Vapor slug growth, 117, 125, 129
    accompanied by dry spots, 137
    volume deformation, 127
Vapor slug growth velocity, 129
Vapor slug movement pulsation, types of, 139
Vapor slug volume radius, 121
Vapor volume growth, 157
Vapor zones
    and heat flux values, 82
    at heat load approaching burnout value, 62
    growth inside MDL at heat load
        enlargement, 61
Vaporization
    and hydrodynamic phenomena in slit and
        capillary channels, 115–120
    and hysteresis phenomena, 213
    and nonlinear gas filtration regularities, 52
    and thermohydrodynamics in slit and
        capillary channels, 93
    calculation results, stabilized region, 64
    capillary input mechanism, 42
    circulation schematic, calculation base, 147
    combined mechanism, 42

conditions in capillaries/capillary-porous
structures, 41–45, 76
dependence between relative phase
permeabilities and phase
saturations at, 47
dependence of heat flux on temperature
drop, 77
dynamic effect on layer-wall contact, 91
experimental data at submerged disperse
layer, 73
general model on coated surfaces, 260–271
gravitational mechanism, 42
heat and mass losses, 63
heat carrier supply mechanisms, 42
heat input conditions inside porous
structures, 43
heat transfer crises at, 299
heat transfer inside dispersed multilayer
bed, 276
heat transfer inside dispersed submerged
layer, 272
heat transfer on fiber-metal surfaces, 183–184
heat transfer on surfaces covered by movable
capillary structures, 271–298
homogeneous flow model, 48
in porous structures with internal heat
generation, 63–71
inertial mechanism, 42
inside capillary-porous structures, 41, 45–54
inside motionless porous layer, 91
inside submerged porous structure, 261
instabilities of, 213
Leverett functions for capillary pressure, 48
liquid droplet throw flow rate at, 80
local heat transfer coefficient changes,
vertical slit channel, 96
mechanism on heat transfer surfaces, 44
models for surfaces with porous coatings,
226–260
on surfaces coated by screen/particle bed
coverings, 237
on surfaces with capillary-porous
coverings, 163
particle bed layer, 90
photograph at three-layer screen
structure, 82, 83
photograph at two-layer screen structure, 81
physical models on porous surfaces, 227
pulsation regimes in porous structure,
71, 74, 75
region margins and surrounding pressure, 66
regularities of, 78
steady thermal regimes, 83
surface of loop heat pipe evaporator, 89
theoretical models for heat transfer
horizontal and short slits and
capillaries, 153–162
vertical slits and capillaries, 147–153

video capture schematic layout, 79
viscosity phase interaction, 53
Vaporization heat transfer, on fiber-metal
surfaces, 182–190
Vaporization mechanism, experimental studies,
plain slits and annular channels,
121–146
Vaporization regimes, on surfaces coated by
porous coverings, 261
Velocity profile, inside tubes with step changes
in pressure gradient, 9
Vertical slit channels
generalization of experimental data, heat
transfer on boiling inside, 161
heat transfer or water and ethanol inside, 98
local heat transfer coefficients, 96
R-11 boiling heat transfer in, 111
theoretical models for heat transfer at
vaporization, 147–153
vapor bubbles floating velocity into, 146
water and ethanol boiling in, 98
Vertical tubes, boiling heat transfer in, 307, 309,
310, 311
Very small velocities/large subcooling regime, 136
Video capture, vaporization, 79
Viscosity, in porous media, 29
Viscosity coefficient, 4
Volumetric flow vapor quality, 118
and liquid microlayer thickness
measurement, 126
approach to critical value, 125
changes, boiling into horizontal plate slits, 128
measurement results, 123
pulsation frequency, 125
water boiling inside horizontal plate slit, 125
Volumetric heat generation, at vaporization, 43
Volumetric vapor flow quality sensors, 123

W

Wall overheat, 80
Wall superheat, 349
and inside layer crisis, 345
Wall temperature, 15
jump in porous media, 22
oscillations in boiling heat transfer, 94
Wall thermal conductivity, and laminar flow
heat transfer intensity, 8
Wall-structure contact, role in heat transfer
intensity for screen structures, 176
Wavy stratified two-phase flow regime, 140
Weber number, 42
Wetted porous structure, heat conduction
through, 36
Wicks
fiber-metal nickel and fiber-metal
copper, 165
water boiling on tube wrapped by, 166

Milton Keynes UK
Ingram Content Group UK Ltd.
UKHW021824071024
449327UK00021B/1421